Case studies in food safety and authenticity

Related titles:

Foodborne pathogens: hazards, risk analysis and control (Second edition)
(ISBN 978-1-84569-362-6)
Effective control of pathogens continues to be of great importance to the food industry. The first edition of *Foodborne pathogens* quickly established itself as an essential guide for all those involved in the management of microbiological hazards at any stage in the food production chain. This major new edition strengthens that reputation, with extensively revised and expanded coverage, including more than ten new chapters. Part I focuses on risk assessment and management in the food chain. Chapters in this section cover pathogen detection, microbial modelling, the risk assessment procedure, pathogen control in primary production, hygienic design and sanitation, among other topics. Parts II and III then review the management of key bacterial and non-bacterial foodborne pathogens.

Mycotoxins in food: detection and control
(ISBN 978-1-85573-733-4)
Mycotoxins are toxic compounds produced by fungi. They are produced in foods of plant origin and pose a significant contamination risk in cereal and other foods. With its distinguished editors and international team of contributors, *Mycotoxins in food* summarises the wealth of recent research on how to assess the risks from mycotoxins, detect particular mycotoxins and control them at differing stages in the supply chain.

Food chain integrity: a holistic approach to food traceability, safety, quality and authenticity
(ISBN 978-0-85709-068-3)
Issues related to food integrity are becoming increasingly important. Guaranteeing the authenticity, geographical origin, traceability and safety of foods requires new diagnostic tools and implementation of new information systems. This book outlines recent research and practice in this fast-moving field. It will increase the reader's ability to predict and implement actions to prevent a breach in the integrity of food production. Different sections of the book are devoted to the key areas of food tracing and tracking, food authenticity and origin, and food safety and quality.

Details of these books and a complete list of titles from Woodhead Publishing can be obtained by:

- visiting our web site at www.woodheadpublishing.com
- contacting Customer Services (e-mail: sales@woodheadpublishing.com; fax: +44 (0) 1223 832819; tel.: +44 (0) 1223 499140 ext. 130; address: Woodhead Publishing Limited, 80 High Street, Sawston, Cambridge CB22 3HJ, UK)
- in North America, contacting our US office (e-mail: usmarketing@woodheadpublishing.com; tel.: (215) 928 9112; address: Woodhead Publishing, 1518 Walnut Street, Suite 1100, Philadelphia, PA 19102-3406, USA)

If you would like e-versions of our content, please visit our online platform: www.woodheadpublishingonline.com. Please recommend it to your librarian so that everyone in your institution can benefit from the wealth of content on the site.

Woodhead Publishing Series in Food Science, Technology and Nutrition:
Number 240

Case studies in food safety and authenticity

Lessons from real-life situations

Edited by
J. Hoorfar

WOODHEAD PUBLISHING

Oxford Cambridge Philadelphia New Delhi

Published by Woodhead Publishing Limited,
80 High Street, Sawston, Cambridge CB22 3HJ, UK
www.woodheadpublishing.com
www.woodheadpublishingonline.com

Woodhead Publishing, 1518 Walnut Street, Suite 1100, Philadelphia, PA 19102-3406, USA

Woodhead Publishing India Private Limited, G-2, Vardaan House, 7/28 Ansari Road, Daryaganj,
New Delhi – 110002, India
www.woodheadpublishingindia.com

First published 2012, Woodhead Publishing Limited

British Library Cataloguing in Publication Data
A catalogue record for this book is available from the British Library.

Library of Congress Control Number: 2012936397

ISBN 978-0-85709-412-4 (print)
ISBN 978-0-85709-693-7 (online)
ISSN 2042-8049 Woodhead Publishing Series in Food Science, Technology and Nutrition (print)
ISSN 2042-8057 Woodhead Publishing Series in Food Science, Technology and Nutrition (online)

The publisher's policy is to use permanent paper from mills that operate a sustainable forestry policy, and which has been manufactured from pulp which is processed using acid-free and elemental chlorine-free practices. Furthermore, the publisher ensures that the text paper and cover board used have met acceptable environmental accreditation standards.

Typeset by RefineCatch Limited, Bungay, Suffolk
Printed by TJ International Ltd, Padstow, Cornwall, UK

Contents

Obituary for Niels Skovgaard

This book contains the last scientific publication of Professor Niels Skovgaard (Chapter 23), who died recently at the age of 87.

Niels Skovgaard had a remarkable career which spanned 60 years. Everybody engaged in food safety either knew him personally or knew of him. Niels and I used to travel to numerous meetings together and I always appreciated his sharp remarks and deep insight into the interaction of microbes with our food and animals.

After graduation as a veterinarian from the Royal Veterinary and Agricultural University (KVL) in Copenhagen in 1951 and a spell working as a veterinary practitioner, he joined the veterinary public health service in Denmark, dealing with meat inspection, food control and the administration of food safety. His teaching career started in 1965 as a lecturer in food microbiology and food hygiene at KVL. He was appointed professor in 1973, a position he held until his official retirement at the age of 70 in 1994.

During his time as professor and also after his retirement Niels Skovgaard contributed actively to several international organizations including the World Health Organization (WHO), Food and Agriculture Organization (FAO), EU, World Association of Veterinary Food Hygienists (WAVFH), International Organization for Standardization (ISO), International Union of Microbiological Societies (IUMS) and International Commission of Food Microbiology and Hygiene (ICFMH) in the roles of advisor and teacher. These organizations benefited from his profound knowledge. In addition, he frequently acted as a scientific consultant to the Danish Government and the Danish food industry.

Niels Skovgaard was also devoted to Nordic cooperation within the field of food microbiology and the development of analytical methods. For this cause he served terms as Secretary as well as President of the Nordic Committee on Food Analyses (NMKL). In the Nordic microbiology group he will be remembered for bringing forward proposals for new areas of study, such as emerging microbes of global interest. In this respect he acted as a true lecturer with deep insight in the subject. We will all miss him as a source of profound knowledge, and – it must not be forgotten – his ability to convey his messages and keep his audiences alert using his own unique brand of humor!

Niels Skovgaard was co-founder of the *International Journal of Food Microbiology* published by ICFMH, and was President of this organization from 1989 to 1996, followed by an appointment as Honorary President. He contributed more than 300 scientific articles in Danish, Nordic and international magazines, and also published several instructive text books for students. In particular during his retirement he spent considerable time reviewing scientific books and acting as referee for articles on food microbiology and hygiene. Niels Skovgaard was an honoury member of international organizations and universities, and in Denmark was made a Knight 1st Class of the Royal Order of Dannebrog.

However, his true legacy is neither his books nor his many publications; it is the many students that have taken over and are continuing his remarkable work in many different organizations, journals, panels, and committees.

Sven Qvist
Denmark

Contributor contact details

Editor and Chapters 4, 20, 24, 29 and 38

J. Hoorfar
National Food Institute
Technical University of Denmark
National Food Institute
Mørkhøj Bygade 28
2860 Søborg
Denmark

E-mail: jhoo@food.dtu.dk

Chapter 1

D. Schoder
Institute of Milk Hygiene and Food Science
Vienna, Austria

M. Wagner*
Bi-National Advisory Laboratory for Listeriosis
Austrian Agency for Health and Food Safety
Spargelfeldstrasse 171
1220
Vienna
Austria

E-mail: Martin.Wagner@vetmeduni.ac.at

F. Allerberger
Christian-Doppler Laboratory for Molecular Food Analytics
Vienna, Austria

Chapter 2

A. G. Stewart* and R. Vivancos
Consultants in Communicable Disease Control
Cheshire & Merseyside Health Protection Unit
5th Floor, Rail House
Lord Nelson Street
Liverpool L1 1JF
UK

E-mail: alex.stewart@hpa.org.uk

A. Keenan
Epidemiology & Surveillance Analyst
Cheshire & Merseyside Health Protection Unit
Liverpool
UK

C. Theodore
Environmental Health Officer
Liverpool City Council
Liverpool
UK

Chapter 3

A. G. Stewart* and S. Ghebrehewet
Consultants in Communicable Disease Control
Cheshire & Merseyside Health Protection Unit
5th Floor, Rail House
Lord Nelson Street
Liverpool L1 1JF
UK

E-mail: alex.stewart@hpa.org.uk

Chapter 4

M. A. Smith
Food and Drug Administration
Center for Food Safety and Allied Nutrition
Office of Food Safety (HFS-317)
5100 Paint Branch Parkway
College Park
MD 20740
USA

E-mail: michelle.smith@fda.hhs.gov

Chapter 5

L. D. Goodridge*
Center for Red Meat Safety and Quality
Department of Animal Sciences
Colorado State University
Fort Collins
CO 80523
USA

E-mail: Lawrence.Goodridge@Colostate.edu

J. T. LeJeune
Ohio Agricultural Research and Development Center
The Ohio State University
1680 Madison Ave
Wooster
OH 44691
USA

E-mail: Lejeune.3@osu.edu

L. Beutin
Nationales Referenzlabor für Escherichia coli (NRL-E.coli)
Bundesinstitut für Risikobewertung (BfR)
Diedersdorfer Weg 1
D-12277 Berlin
Germany

E-mail: Lothar.Beutin@bfr.bund.de

Chapter 6

Marja-Liisa Hänninen*
Department of Food Hygiene and Environmental Health
Helsinki University
P.O. Box 66
00014 Helsinki University
Finland

E-mail: marja-liisa.hanninen@helsinki.fi

Ari Hörman
The Finnish Defense Forces
P.O. Box 919
00131 Helsinki
Finland

Chapter 7

Prof. Dr Monika Ehling-Schulz*
Institute of Functional Microbiology
Department of Pathobiology
University of Veterinary Medicine
Veterinaerplatz 1
1210 Vienna
Austria

E-mail: monika.ehling-schulz@vetmeduni.ac.at

U. Messelhäusser
Bavarian Health and Food Safety Authority
Veterinaerstr. 2 85764 Oberschleissheim
Germany

E-mail: ute.messelhaeusser@lgl.bayern.de

Chapter 8

M. F. Dutton
Mycotoxin Research Group
University of KwaZulu Natal
Durban
South Africa

E-mail: mdutton@uj.ac.za

Chapter 9

Dr Brent Dixon*
Bureau of Microbial Hazards
Food Directorate
Health Canada
251 Sir Frederick Banting Driveway
P.L. 2204E
Ottawa
Ontario
Canada
K1A 0K9

E-mail: Brent.Dixon@hc-sc.gc.ca

L. Landry
Centre for Food-borne Environmental and Zoonotic Infectious Diseases
Public Health Agency of Canada
Guelph
Ontario
Canada

F. Milord
Direction de la Santé Publique de la Montérégie
Longueuil
Quebec
Canada

Chapter 10

Kieran N. Jordan*, Karen Hunt and Edward M. Fox
Teagasc Food Research Centre
Moorepark
Fermoy
Co. Cork
Ireland

E-mail: Kieran.Jordan@teagasc.ie

Chapter 11

V. Biguzzi
Responsible for Quality Control Unit in Food Industry (retired)

E-mail: valeriano.biguzzi@libero.it

G. Pezzotti
Istituto Zooprofilattico Sperimentale dell'Umbria e le Marche
Via Gaetano Salvemini, 1
Perugia
Italy

E-mail: g.pezzotti@izsum.it

M. V. Gianfranceschi and D. De Medici*
Department of Veterinary Public Health and Food Safety – Microbiological Foodborne Hazard Unit
Istituto Superiore di Sanità
Viale Regina Elena 299
00161 Rome
Italy

E-mail: dario.demedici@iss.it; monica.gianfranceschi@iss.it

Chapter 12

Erik Nordkvist
National Veterinary Institute
Uppsala
Sweden

E-mail: erik.nordkvist@sva.se

Chapter 13

N. C. Elviss* and F. Jørgensen
Health Protection Agency (HPA)
Food, Water and Environmental Microbiology Laboratories
Sutton Coldfield
Birmingham B75 7RR
UK

E-mail: Nicola.Elviss@hpa.org.uk

Chapter 14

Suresh D. Pillai
National Center for Electron Beam Research
Food Safety & Environmental Microbiology Program
Texas A&M University
College Station
Texas
USA

E-mail: s-pillai@tamu.edu

Chapter 15

Knut Rudi*
Hedmark University College
Hamar
Norway

and

The Norwegian University for Life Sciences
Ås
Norway

E-mail: knut.rudi@umb.no

Petr Ricanek
Department of Gastroenterology
Akershus University Hospital
Lørenskog
Norway

and

EpiGen Institute
Research Centre
Akershus University Hospital
Lørenskog
Norway

Tone Tannæs and Stephan Brackmann
University of Oslo
EpiGen-Institute
Akershus University Hospital
University of Oslo
Oslo
Norway

Gøri Perminow
Pediatric Department
Oslo University Hospital
Ullevål
Oslo
Norway

Morten H. Vatn
University of Oslo
EpiGen-Institute
Akershus University Hospital
University of Oslo
Oslo
Norway

and

Medical Clinic
Oslo University Hospital
Rikshospitalet
Norway

Chapter 16

A. Wingstrand*
National Food Institute
Technical University of Denmark
19 Mørkhøj Bygade
DK-2860 Søborg
Denmark

E-mail: awin@food.dtu.dk

T. Jensen
Danish Veterinary and Food Administration
19 Mørkhøj Bygade
DK-2860 Søborg
Denmark

E-mail: teje@fvst.dk

G. Sørensen
National Food Institute
Technical University of Denmark
19 Mørkhøj Bygade
DK-2860 Søborg
Denmark

E-mail: gits@food.dtu.dk

S. Gubbels
Statens Serum Institut
5 Artillerivej
DK-2300 Copenhagen S
Denmark

E-mail: gub@ssi.dk

and

European Programme for Intervention Epidemiology Training (EPIET)
European Centre for Disease Prevention and Control
SE-171 83 Stockholm
Sweden

L. J. Porsbo
National Food Institute
Technical University of Denmark
19 Mørkhøj Bygade
DK-2860 Søborg
Denmark

E-mail: ljpo@food.dtu.dk

S. Ethelberg
Statens Serum Institut
5 Artillerivej
DK-2300 Copenhagen S
Denmark

E-mail: set@ssi.dk

G. O. Bjerager
Danish Veterinary and Food Administration
19 Mørkhøj Bygade
DK-2860 Søborg
Denmark

E-mail: gorb@fvst.dk

M. K. Kjeldsen
Statens Serum Institut
5 Artillerivej
DK-2300 Copenhagen S
Denmark

E-mail: kjd@ssi.dk

M. Torpdahl
Statens Serum Institut
5 Artillerivej
DK-2300 Copenhagen S
Denmark

E-mail: mtd@ssi.dk

K. G. Kuhn
Statens Serum Institut
5 Artillerivej
DK-2300 Copenhagen S
Denmark

E-mail: kuh@ssi.dk

Chapter 17

R. Knutsson*, V. Båverud, M. Elvander and E. Olsson Engvall
National Veterinary Institute (SVA)
SE-751 89 Uppsala
Sweden

E-mail: rickard.knutsson@sva.se

K. Eliasson
National Board of Agriculture (SJV)
Jönköping
Sweden

S. Sternberg Lewerin
Swedish University of Agricultural Sciences (SLU)
Uppsala
Sweden

Chapter 18

D. Orlando, Dennis and F. M. C. Sigit Setyabudi*
Department of Food and Agricultural Product Technology
Faculty of Agricultural Technology
Universitas Gadjah Mada
Jalan Flora No.1, Bulaksumur, Yogyakarta
Indonesia – 55281

E-mail: francis.setyabudi@ugm.ac.id

M. Suci Hardianti
Division of Hematology Medical Oncology
Department of Internal Medicine
Faculty of Medicine
Universitas Gadjah Mada
Bulaksumur
Yogyakarta
Indonesia

Chapter 19

Anca Ioana Nicolau
Faculty of Food Science and Engineering
Dunarea de Jos University
Galati
Romania

E-mail: Anca.Nicolau@ugal.ro

Chapter 21

L. S. Christensen
Islands Organic Ltd
Horsekær
Sejerøvej 28
DK-4592 Sejerø
Denmark

E-mail: siig@horsekaer.dk

Chapter 22

Per Häggblom
National Veterinary Institute
Uppsala
Sweden

E-mail: per.haggblom@sva.se

Chapter 24

K. N. Jordan* and K. Hunt
Teagasc Food Research Centre
Moorepark
Fermoy
Co. Cork
Ireland

E-mail: Kieran.Jordan@teagasc.ie

Chapter 25

E. H. Drosinos* and P. N. Skandamis
Department of Food Science and Technology
Laboratory of Food Quality Control and Hygiene
Agricultural University of Athens
75, Iera Odos Street, Votanicos
GR-118 55 Athens
Greece

E-mail address: ehd@aua.gr; pskan@aua.gr

S. G. Charalambidou
Food Science Technologist – Agriculturist
14, Falirou Street, Ayios Pavlos
2362, Nicosia
Cyprus

E-mail: charalambidou.s@gmail.com

Chapter 26

D. W. Eddy* and I. N. Haynes
Dairy Food Safety Victoria
Level 1, 313 Burwood Road
Hawthorn
Victoria
Australia

Email: deddy@dairysafe.vic.gov.au

Chapter 27

Dr Peter Feng
Division of Microbiology
HFS-711, FDA
5100 Paint Branch Parkway
College Park
MD 20740-3835
USA

E-mail: peter.feng@fda.hhs.gov

Chapter 28

Dr Jeffrey L. Kornacki
Kornacki Microbiology Solutions, Inc.
9 Woodglen Court
Madison
WI 53716-2465
USA
www.kornackifoodsafety.com

E-mail: jlkorn731@gmail.com

Chapter 29

C. Löfström*, M. H. Josefsen and J. Hoorfar
National Food Institute
Technical University of Denmark
Mørkhøj Bygade 19
2860 Søborg
Denmark

E-mail: chalo@food.dtu.dk

F. Hansen
Danish Technological Institute
DMRI
Maglegårdsvej 2
4000 Roskilde
Denmark

Chapter 30

J. Sumner
Tasmanian Institute of Agriculture School of Agricultural Science University of Tasmania Hobart
Tasmania
Australia

E-mail: john_sumner2@bigpond.com

I. Jenson
Meat and Livestock Australia
North Sydney
Australia

E-mail: ijenson@mla.com.au

T. Ross*
Tasmanian Institute of Agriculture
School of Agricultural Science
University of Tasmania
Hobart
Australia

E-mail: tom.ross@utas.edu.au

Chapter 31

M. J. Walker
LGC Ltd
Queens Road
Teddington
UK

E-mail: michael.walker@lgcgroup.com

Chapter 32

J. Zentek* and F. Knorr
Institute of Animal Nutrition
Freie Universität Berlin
Department of Veterinary Medicine
Königin-Luise-Str. 49
14195 Berlin
Germany

E-mail: zentek.juergen@vetmed.fu-berlin.de

A. Mader
Institute of Animal Nutrition
Freie Universität Berlin
Department of Veterinary Medicine
Königin-Luise-Str. 49
14195 Berlin
Germany

and

Federal Institute for Risk Assessment
Diedersdorfer Weg 1
12277 Berlin
Germany

H. Schafft
Federal Institute for Risk Assessment
Diedersdorfer Weg 1
12277 Berlin
Germany

Chapter 33

Claudio F. Cardoso, Adriano G. Cruz* and José A. F. Faria
Universidade Estadual de Campinas (UNICAMP)
Faculdade de Engenharia de Alimentos (FEA)
Cidade Universitária 'Zeferino Vaz'
CEP 13083-862
Campinas

SP
Brazil

E-mail: food@globo.com

Uelinton M. Pinto
Universidade Federal de Ouro Preto
Departamento de Alimentos
Morro do Cruzeiro
Ouro Preto MG
CEP 35400-000
Brazil

Chapter 34

D. Schoder
Veterinarians Without Borders Austria

and

Institute of Milk Hygiene
University of Veterinary Medicine Vienna
Veterinärplatz 1
1210 Vienna
Austria

E-mail: dagmar.schoder@vetmeduni.ac.at

Chapter 35

N. Z. Ballin
Department of Food Chemistry
Regional Veterinary and Food Control Authority
Danish Veterinary and Food Administration
Søndervang 4
DK-4100 Ringsted
Denmark

E-mail: nixb@fvst.dk

Chapter 36

E. E. Nielsen*, J. Hemmer-Hansen and D. Bekkevold
National Institute of Aquatic Resources (DTU Aqua)
Technical University of Denmark
Vejlsøvej 39 8600
Silkeborg
Denmark

E-mail: een@aqua.dtu.dk

Chapter 37

A. Holst-Jensen*, A. R. Ali, L. Emanuelsen, T. M. Grønbeck, S. B. Harbo, C. Skjæret and B. Spilsberg
Norwegian Veterinary Institute
P.O. Box 750 Sentrum
0106 Oslo
Norway

E-mail: arne.holst-jensen@vetinst.no

H.-U. Waiblinger
State Institute of Chemical and Veterinary Analysis of Food Freiburg
Freiburg
Germany

Chapter 38

L. S. Christensen* and J. Hoorfar
National Food Institute
Technical University of Denmark
Mørkhøj Bygade 19
DK-2860 Søborg
Denmark

E-mail: siig@horsekaer.dk

J. Sørensen
Øhavet Aps
Korsvejen 2
Strynø
DK-5900 Rudkøbing
Denmark

E-mail: john@ohavet.dk

S. Bisgaard
Danish Veterinary and Food Administration
Mørkhøj Bygade 19
DK-2860 Søborg
Denmark

E-mail: sorb@fvst.dk

Woodhead Publishing Series in Food Science, Technology and Nutrition

1 **Chilled foods: a comprehensive guide** *Edited by C. Dennis and M. Stringer*
2 **Yoghurt: science and technology** *A. Y. Tamime and R. K. Robinson*
3 **Food processing technology: principles and practice** *P. J. Fellows*
4 **Bender's dictionary of nutrition and food technology Sixth edition** *D. A. Bender*
5 **Determination of veterinary residues in food** *Edited by N. T. Crosby*
6 **Food contaminants: sources and surveillance** *Edited by C. Creaser and R. Purchase*
7 **Nitrates and nitrites in food and water** *Edited by M. J. Hill*
8 **Pesticide chemistry and bioscience: the food-environment challenge** *Edited by G. T. Brooks and T. Roberts*
9 **Pesticides: developments, impacts and controls** *Edited by G. A. Best and A. D. Ruthven*
10 **Dietary fibre: chemical and biological aspects** *Edited by D. A. T. Southgate, K. W. Waldron, I. T. Johnson and G. R. Fenwick*
11 **Vitamins and minerals in health and nutrition** *M. Tolonen*
12 **Technology of biscuits, crackers and cookies Second edition** *D. Manley*
13 **Instrumentation and sensors for the food industry** *Edited by E. Kress-Rogers*
14 **Food and cancer prevention: chemical and biological aspects** *Edited by K. W. Waldron, I. T. Johnson and G. R. Fenwick*
15 **Food colloids: proteins, lipids and polysaccharides** *Edited by E. Dickinson and B. Bergenstahl*
16 **Food emulsions and foams** *Edited by E. Dickinson*
17 **Maillard reactions in chemistry, food and health** *Edited by T. P. Labuza, V. Monnier, J. Baynes and J. O'Brien*
18 **The Maillard reaction in foods and medicine** *Edited by J. O'Brien, H. E. Nursten, M. J. Crabbe and J. M. Ames*
19 **Encapsulation and controlled release** *Edited by D. R. Karsa and R. A. Stephenson*
20 **Flavours and fragrances** *Edited by A. D. Swift*
21 **Feta and related cheeses** *Edited by A. Y. Tamime and R. K. Robinson*

22 **Biochemistry of milk products** *Edited by A. T. Andrews and J. R. Varley*
23 **Physical properties of foods and food processing systems** *M. J. Lewis*
24 **Food irradiation: a reference guide** *V. M. Wilkinson and G. Gould*
25 **Kent's technology of cereals: an introduction for students of food science and agriculture Fourth edition** *N. L. Kent and A. D. Evers*
26 **Biosensors for food analysis** *Edited by A. O. Scott*
27 **Separation processes in the food and biotechnology industries: principles and applications** *Edited by A. S. Grandison and M. J. Lewis*
28 **Handbook of indices of food quality and authenticity** *R. S. Singhal, P. K. Kulkarni and D. V. Rege*
29 **Principles and practices for the safe processing of foods** *D. A. Shapton and N. F. Shapton*
30 **Biscuit, cookie and cracker manufacturing manuals Volume 1: ingredients** *D. Manley*
31 **Biscuit, cookie and cracker manufacturing manuals Volume 2: biscuit doughs** *D. Manley*
32 **Biscuit, cookie and cracker manufacturing manuals Volume 3: biscuit dough piece forming** *D. Manley*
33 **Biscuit, cookie and cracker manufacturing manuals Volume 4: baking and cooling of biscuits** *D. Manley*
34 **Biscuit, cookie and cracker manufacturing manuals Volume 5: secondary processing in biscuit manufacturing** *D. Manley*
35 **Biscuit, cookie and cracker manufacturing manuals Volume 6: biscuit packaging and storage** *D. Manley*
36 **Practical dehydration Second edition** *M. Greensmith*
37 **Lawrie's meat science Sixth edition** *R. A. Lawrie*
38 **Yoghurt: science and technology Second edition** *A. Y. Tamime and R. K. Robinson*
39 **New ingredients in food processing: biochemistry and agriculture** *G. Linden and D. Lorient*
40 **Benders' dictionary of nutrition and food technology Seventh edition** *D. A. Bender and A. E. Bender*
41 **Technology of biscuits, crackers and cookies Third edition** *D. Manley*
42 **Food processing technology: principles and practice Second edition** *P. J. Fellows*
43 **Managing frozen foods** *Edited by C. J. Kennedy*
44 **Handbook of hydrocolloids** *Edited by G. O. Phillips and P. A. Williams*
45 **Food labelling** *Edited by J. R. Blanchfield*
46 **Cereal biotechnology** *Edited by P. C. Morris and J. H. Bryce*
47 **Food intolerance and the food industry** *Edited by T. Dean*
48 **The stability and shelf-life of food** *Edited by D. Kilcast and P. Subramaniam*
49 **Functional foods: concept to product** *Edited by G. R. Gibson and C. M. Williams*
50 **Chilled foods: a comprehensive guide Second edition** *Edited by M. Stringer and C. Dennis*
51 **HACCP in the meat industry** *Edited by M. Brown*
52 **Biscuit, cracker and cookie recipes for the food industry** *D. Manley*
53 **Cereals processing technology** *Edited by G. Owens*
54 **Baking problems solved** *S. P. Cauvain and L. S. Young*
55 **Thermal technologies in food processing** *Edited by P. Richardson*
56 **Frying: improving quality** *Edited by J. B. Rossell*
57 **Food chemical safety Volume 1: contaminants** *Edited by D. Watson*

58 **Making the most of HACCP: learning from others' experience** *Edited by T. Mayes and S. Mortimore*
59 **Food process modelling** *Edited by L. M. M. Tijskens, M. L. A. T. M. Hertog and B. M. Nicolaï*
60 **EU food law: a practical guide** *Edited by K. Goodburn*
61 **Extrusion cooking: technologies and applications** *Edited by R. Guy*
62 **Auditing in the food industry: from safety and quality to environmental and other audits** *Edited by M. Dillon and C. Griffith*
63 **Handbook of herbs and spices Volume 1** *Edited by K. V. Peter*
64 **Food product development: maximising success** *M. Earle, R. Earle and A. Anderson*
65 **Instrumentation and sensors for the food industry Second edition** *Edited by E. Kress-Rogers and C. J. B. Brimelow*
66 **Food chemical safety Volume 2: additives** *Edited by D. Watson*
67 **Fruit and vegetable biotechnology** *Edited by V. Valpuesta*
68 **Foodborne pathogens: hazards, risk analysis and control** *Edited by C. de W. Blackburn and P. J. McClure*
69 **Meat refrigeration** *S. J. James and C. James*
70 **Lockhart and Wiseman's crop husbandry Eighth edition** *H. J. S. Finch, A. M. Samuel and G. P. F. Lane*
71 **Safety and quality issues in fish processing** *Edited by H. A. Bremner*
72 **Minimal processing technologies in the food industries** *Edited by T. Ohlsson and N. Bengtsson*
73 **Fruit and vegetable processing: improving quality** *Edited by W. Jongen*
74 **The nutrition handbook for food processors** *Edited by C. J. K. Henry and C. Chapman*
75 **Colour in food: improving quality** *Edited by D MacDougall*
76 **Meat processing: improving quality** *Edited by J. P. Kerry, J. F. Kerry and D. A. Ledward*
77 **Microbiological risk assessment in food processing** *Edited by M. Brown and M. Stringer*
78 **Performance functional foods** *Edited by D. Watson*
79 **Functional dairy products Volume 1** *Edited by T. Mattila-Sandholm and M. Saarela*
80 **Taints and off-flavours in foods** *Edited by B. Baigrie*
81 **Yeasts in food** *Edited by T. Boekhout and V. Robert*
82 **Phytochemical functional foods** *Edited by I. T. Johnson and G. Williamson*
83 **Novel food packaging techniques** *Edited by R. Ahvenainen*
84 **Detecting pathogens in food** *Edited by T. A. McMeekin*
85 **Natural antimicrobials for the minimal processing of foods** *Edited by S. Roller*
86 **Texture in food Volume 1: semi-solid foods** *Edited by B. M. McKenna*
87 **Dairy processing: improving quality** *Edited by G. Smit*
88 **Hygiene in food processing: principles and practice** *Edited by H. L. M. Lelieveld, M. A. Mostert, B. White and J. Holah*
89 **Rapid and on-line instrumentation for food quality assurance** *Edited by I. Tothill*
90 **Sausage manufacture: principles and practice** *E. Essien*
91 **Environmentally-friendly food processing** *Edited by B. Mattsson and U. Sonesson*
92 **Bread making: improving quality** *Edited by S. P. Cauvain*
93 **Food preservation techniques** *Edited by P. Zeuthen and L. Bøgh-Sørensen*
94 **Food authenticity and traceability** *Edited by M. Lees*

95 **Analytical methods for food additives** *R. Wood, L. Foster, A. Damant and P. Key*
96 **Handbook of herbs and spices Volume 2** *Edited by K. V. Peter*
97 **Texture in food Volume 2: solid foods** *Edited by D. Kilcast*
98 **Proteins in food processing** *Edited by R. Yada*
99 **Detecting foreign bodies in food** *Edited by M. Edwards*
100 **Understanding and measuring the shelf-life of food** *Edited by R. Steele*
101 **Poultry meat processing and quality** *Edited by G. Mead*
102 **Functional foods, ageing and degenerative disease** *Edited by C. Remacle and B. Reusens*
103 **Mycotoxins in food: detection and control** *Edited by N. Magan and M. Olsen*
104 **Improving the thermal processing of foods** *Edited by P. Richardson*
105 **Pesticide, veterinary and other residues in food** *Edited by D. Watson*
106 **Starch in food: structure, functions and applications** *Edited by A.-C. Eliasson*
107 **Functional foods, cardiovascular disease and diabetes** *Edited by A. Arnoldi*
108 **Brewing: science and practice** *D. E. Briggs, P. A. Brookes, R. Stevens and C. A. Boulton*
109 **Using cereal science and technology for the benefit of consumers: proceedings of the 12th International ICC Cereal and Bread Congress, 24 – 26th May, 2004, Harrogate, UK** *Edited by S. P. Cauvain, L. S. Young and S. Salmon*
110 **Improving the safety of fresh meat** *Edited by J. Sofos*
111 **Understanding pathogen behaviour: virulence, stress response and resistance** *Edited by M. Griffiths*
112 **The microwave processing of foods** *Edited by H. Schubert and M. Regier*
113 **Food safety control in the poultry industry** *Edited by G. Mead*
114 **Improving the safety of fresh fruit and vegetables** *Edited by W. Jongen*
115 **Food, diet and obesity** *Edited by D. Mela*
116 **Handbook of hygiene control in the food industry** *Edited by H. L. M. Lelieveld, M. A. Mostert and J. Holah*
117 **Detecting allergens in food** *Edited by S. Koppelman and S. Hefle*
118 **Improving the fat content of foods** *Edited by C. Williams and J. Buttriss*
119 **Improving traceability in food processing and distribution** *Edited by I. Smith and A. Furness*
120 **Flavour in food** *Edited by A. Voilley and P. Etievant*
121 **The Chorleywood bread process** *S. P. Cauvain and L. S. Young*
122 **Food spoilage microorganisms** *Edited by C. de W. Blackburn*
123 **Emerging foodborne pathogens** *Edited by Y. Motarjemi and M. Adams*
124 **Benders' dictionary of nutrition and food technology Eighth edition** *D. A. Bender*
125 **Optimising sweet taste in foods** *Edited by W. J. Spillane*
126 **Brewing: new technologies** *Edited by C. Bamforth*
127 **Handbook of herbs and spices Volume 3** *Edited by K. V. Peter*
128 **Lawrie's meat science Seventh edition** *R. A. Lawrie in collaboration with D. A. Ledward*
129 **Modifying lipids for use in food** *Edited by F. Gunstone*
130 **Meat products handbook: practical science and technology** *G. Feiner*
131 **Food consumption and disease risk: consumer-pathogen interactions** *Edited by M. Potter*
132 **Acrylamide and other hazardous compounds in heat-treated foods** *Edited by K. Skog and J. Alexander*

133 **Managing allergens in food** *Edited by C. Mills, H. Wichers and K. Hoffman-Sommergruber*
134 **Microbiological analysis of red meat, poultry and eggs** *Edited by G. Mead*
135 **Maximising the value of marine by-products** *Edited by F. Shahidi*
136 **Chemical migration and food contact materials** *Edited by K. Barnes, R. Sinclair and D. Watson*
137 **Understanding consumers of food products** *Edited by L. Frewer and H. van Trijp*
138 **Reducing salt in foods: practical strategies** *Edited by D. Kilcast and F. Angus*
139 **Modelling microorganisms in food** *Edited by S. Brul, S. Van Gerwen and M. Zwietering*
140 **Tamime and Robinson's Yoghurt: science and technology Third edition** *A. Y. Tamime and R. K. Robinson*
141 **Handbook of waste management and co-product recovery in food processing Volume 1** *Edited by K. W. Waldron*
142 **Improving the flavour of cheese** *Edited by B. Weimer*
143 **Novel food ingredients for weight control** *Edited by C. J. K. Henry*
144 **Consumer-led food product development** *Edited by H. MacFie*
145 **Functional dairy products Volume 2** *Edited by M. Saarela*
146 **Modifying flavour in food** *Edited by A. J. Taylor and J. Hort*
147 **Cheese problems solved** *Edited by P. L. H. McSweeney*
148 **Handbook of organic food safety and quality** *Edited by J. Cooper, C. Leifert and U. Niggli*
149 **Understanding and controlling the microstructure of complex foods** *Edited by D. J. McClements*
150 **Novel enzyme technology for food applications** *Edited by R. Rastall*
151 **Food preservation by pulsed electric fields: from research to application** *Edited by H. L. M. Lelieveld and S. W. H. de Haan*
152 **Technology of functional cereal products** *Edited by B. R. Hamaker*
153 **Case studies in food product development** *Edited by M. Earle and R. Earle*
154 **Delivery and controlled release of bioactives in foods and nutraceuticals** *Edited by N. Garti*
155 **Fruit and vegetable flavour: recent advances and future prospects** *Edited by B. Brückner and S. G. Wyllie*
156 **Food fortification and supplementation: technological, safety and regulatory aspects** *Edited by P. Berry Ottaway*
157 **Improving the health-promoting properties of fruit and vegetable products** *Edited by F. A. Tomás-Barberán and M. I. Gil*
158 **Improving seafood products for the consumer** *Edited by T. Børresen*
159 **In-pack processed foods: improving quality** *Edited by P. Richardson*
160 **Handbook of water and energy management in food processing** *Edited by J. Klemeš, R. Smith and J.-K. Kim*
161 **Environmentally compatible food packaging** *Edited by E. Chiellini*
162 **Improving farmed fish quality and safety** *Edited by Ø. Lie*
163 **Carbohydrate-active enzymes** *Edited by K.-H. Park*
164 **Chilled foods: a comprehensive guide Third edition** *Edited by M. Brown*
165 **Food for the ageing population** *Edited by M. M. Raats, C. P. G. M. de Groot and W. A Van Staveren*
166 **Improving the sensory and nutritional quality of fresh meat** *Edited by J. P. Kerry and D. A. Ledward*

167 **Shellfish safety and quality** *Edited by S. E. Shumway and G. E. Rodrick*
168 **Functional and speciality beverage technology** *Edited by P. Paquin*
169 **Functional foods: principles and technology** *M. Guo*
170 **Endocrine-disrupting chemicals in food** *Edited by I. Shaw*
171 **Meals in science and practice: interdisciplinary research and business applications** *Edited by H. L. Meiselman*
172 **Food constituents and oral health: current status and future prospects** *Edited by M. Wilson*
173 **Handbook of hydrocolloids Second edition** *Edited by G. O. Phillips and P. A. Williams*
174 **Food processing technology: principles and practice Third edition** *P. J. Fellows*
175 **Science and technology of enrobed and filled chocolate, confectionery and bakery products** *Edited by G. Talbot*
176 **Foodborne pathogens: hazards, risk analysis and control Second edition** *Edited by C. de W. Blackburn and P. J. McClure*
177 **Designing functional foods: measuring and controlling food structure breakdown and absorption** *Edited by D. J. McClements and E. A. Decker*
178 **New technologies in aquaculture: improving production efficiency, quality and environmental management** *Edited by G. Burnell and G. Allan*
179 **More baking problems solved** *S. P. Cauvain and L. S. Young*
180 **Soft drink and fruit juice problems solved** *P. Ashurst and R. Hargitt*
181 **Biofilms in the food and beverage industries** *Edited by P. M. Fratamico, B. A. Annous and N. W. Gunther*
182 **Dairy-derived ingredients: food and neutraceutical uses** *Edited by M. Corredig*
183 **Handbook of waste management and co-product recovery in food processing Volume 2** *Edited by K. W. Waldron*
184 **Innovations in food labelling** *Edited by J. Albert*
185 **Delivering performance in food supply chains** *Edited by C. Mena and G. Stevens*
186 **Chemical deterioration and physical instability of food and beverages** *Edited by L. H. Skibsted, J. Risbo and M. L. Andersen*
187 **Managing wine quality Volume 1: viticulture and wine quality** *Edited by A. G. Reynolds*
188 **Improving the safety and quality of milk Volume 1: milk production and processing** *Edited by M. Griffiths*
189 **Improving the safety and quality of milk Volume 2: improving quality in milk products** *Edited by M. Griffiths*
190 **Cereal grains: assessing and managing quality** *Edited by C. Wrigley and I. Batey*
191 **Sensory analysis for food and beverage quality control: a practical guide** *Edited by D. Kilcast*
192 **Managing wine quality Volume 2: oenology and wine quality** *Edited by A. G. Reynolds*
193 **Winemaking problems solved** *Edited by C. E. Butzke*
194 **Environmental assessment and management in the food industry** *Edited by U. Sonesson, J. Berlin and F. Ziegler*
195 **Consumer-driven innovation in food and personal care products** *Edited by S. R. Jaeger and H. MacFie*
196 **Tracing pathogens in the food chain** *Edited by S. Brul, P. M. Fratamico and T. A. McMeekin*

197 **Case studies in novel food processing technologies: innovations in processing, packaging, and predictive modelling** *Edited by C. J. Doona, K. Kustin and F. E. Feeherry*
198 **Freeze-drying of pharmaceutical and food products** *T.-C. Hua, B.-L. Liu and H. Zhang*
199 **Oxidation in foods and beverages and antioxidant applications Volume 1: understanding mechanisms of oxidation and antioxidant activity** *Edited by E. A. Decker, R. J. Elias and D. J. McClements*
200 **Oxidation in foods and beverages and antioxidant applications Volume 2: management in different industry sectors** *Edited by E. A. Decker, R. J. Elias and D. J. McClements*
201 **Protective cultures, antimicrobial metabolites and bacteriophages for food and beverage biopreservation** *Edited by C. Lacroix*
202 **Separation, extraction and concentration processes in the food, beverage and nutraceutical industries** *Edited by S. S. H. Rizvi*
203 **Determining mycotoxins and mycotoxigenic fungi in food and feed** *Edited by S. De Saeger*
204 **Developing children's food products** *Edited by D. Kilcast and F. Angus*
205 **Functional foods: concept to product Second edition** *Edited by M. Saarela*
206 **Postharvest biology and technology of tropical and subtropical fruits Volume 1: Fundamental issues** *Edited by E. M. Yahia*
207 **Postharvest biology and technology of tropical and subtropical fruits Volume 2: Açai to citrus** *Edited by E. M. Yahia*
208 **Postharvest biology and technology of tropical and subtropical fruits Volume 3: Cocona to mango** *Edited by E. M. Yahia*
209 **Postharvest biology and technology of tropical and subtropical fruits Volume 4: Mangosteen to white sapote** *Edited by E. M. Yahia*
210 **Food and beverage stability and shelf life** *Edited by D. Kilcast and P. Subramaniam*
211 **Processed Meats: improving safety, nutrition and quality** *Edited by J. P. Kerry and J. F. Kerry*
212 **Food chain integrity: a holistic approach to food traceability, safety, quality and authenticity** *Edited by J. Hoorfar, K. Jordan, F. Butler and R. Prugger*
213 **Improving the safety and quality of eggs and egg products Volume 1** *Edited by Y. Nys, M. Bain and F. Van Immerseel*
214 **Improving the safety and quality of eggs and egg products Volume 2** *Edited by F. Van Immerseel, Y. Nys and M. Bain*
215 **Animal feed contamination: effects on livestock and food safety** *Edited by J. Fink-Gremmels*
216 **Hygienic design of food factories** *Edited by J. Holah and H. L. M. Lelieveld*
217 **Manley's technology of biscuits, crackers and cookies Fourth edition** *Edited by D. Manley*
218 **Nanotechnology in the food, beverage and nutraceutical industries** *Edited by Q. Huang*
219 **Rice quality: A guide to rice properties and analysis** *K. R. Bhattacharya*
220 **Advances in meat, poultry and seafood packaging** *Edited by J. P. Kerry*
221 **Reducing saturated fats in foods** *Edited by G. Talbot*
222 **Handbook of food proteins** *Edited by G. O. Phillips and P. A. Williams*
223 **Lifetime nutritional influences on cognition, behaviour and psychiatric illness** *Edited by D. Benton*

224 **Food machinery for the production of cereal foods, snack foods and confectionery** *L.-M. Cheng*
225 **Alcoholic beverages: sensory evaluation and consumer research** *Edited by J. Piggott*
226 **Extrusion problems solved: food, pet food and feed** *M. N. Riaz and G. J. Rokey*
227 **Handbook of herbs and spices Volume 1 Second edition** *Edited by K. V. Peter*
228 **Handbook of herbs and spices Volume 2 Second edition** *Edited by K. V. Peter*
229 **Bread making: improving quality Second edition** *Edited by S. P. Cauvain*
230 **Emerging food packaging technologies: principles and practice** *Edited by K. L. Yam and D. S. Lee*
231 **Infectious disease in aquaculture: prevention and control** *Edited by B. Austin*
232 **Diet, immunity and inflammation** *Edited by P. C. Calder and P. Yaqoob*
233 **Natural food additives, ingredients and flavourings** *Edited by D. Baines and R. Seal*
234 **Microbial decontamination in the food industry: novel methods and applications** *Edited by A. Demirci and M. O. Ngadi*
235 **Chemical contaminants and residues in foods** *Edited by D. Schrenk*
236 **Robotics and automation in the food industry: current and future technologies** *Edited by D. G. Caldwell*
237 **Fibre-rich and wholegrain foods: improving quality** *Edited by J. A. Delcour and K. Poutanen*
238 **Computer vision technology in the food and beverage industries** *Edited by D.-W. Sun*
239 **Encapsulation technologies and delivery systems for food ingredients and nutraceuticals** *Edited by N. Garti and D. J. McClements*
240 **Case studies in food safety and authenticity** *Edited by J. Hoorfar*

Preface

Why is this book different?

Because this is not a simple collection of scientific reviews written by experts for experts, but short stories of typical situations that might happen to any quality manager, food safety advisor, or policy maker.

With today's busy lifestyle, reading (technical) books is not the first priority. On the contrary, fewer and fewer read books, but most update their technical knowledge by going to the internet; quick and searchable access to the most recent information at their fingertips!

In so many instances around the world we hear about incidents of foodborne illness, adulteration suspects or quality problems but never get to understand how they originated and how they were then managed. This book will hopefully address that deficiency.

Food is like air, consumed by everyone every day, and is therefore of major public interest and one of the best-studied subjects on the planet. Nevertheless, the Byzantine complexity of global food trade that has given rise to so many outbreaks still keeps food scientists up at night, and continues to provide fodder for experimentation.

Despite working for more than 20 years in this field, I only learned about bits and pieces here and there, sometimes missing the big picture. What triggers food problems in the first place, how do food detectives work, how is trouble-shooting planned?

These, and many other questions, led me to the next question: how can we share our hard-earned experience with our colleagues, students, fellow citizens, and the food industry, which is struggling every day to produce safe food but still make enough money to survive?

That was when I decided to discuss the idea of this book with some colleagues, who immediately saw the point and the need for more real-life (and less publication-style) knowledge sharing.

The chapters needed to be written by the 'Sherlock Holmeses' of the food sector, who struggle daily with new food outbreaks, quality issues, or fraud products. At the same time, it was important to include as many diverse stories as possible to show different kinds of situations in different parts of the world.

That is why the chapters are kept rather short, leaving most scientific details to conventional publications. Another notable change (as compared with the classic review books) is the inclusion of information on the regulatory, economics and cultural settings where the food problems were found.

More than 100 experts from 20 countries around the world have shared their valuable experience with the rest of us. These are the everyday heroes of food safety.

J. Hoorfar
Denmark

Foreword

The book titled *Case studies in food safety and authenticity: lessons from real-life situations*, edited by J. Hoorfar, is not only different, it is actually unique, which makes it not only impressive but necessary for all interested in food safety. Two major characteristics make it different, unique and impressive. One is the approach it uses to present important knowledge on food safety. Instead of the traditional style and format of disseminating food safety knowledge by describing either foodborne hazards and their presence and control in various foods, or types of foods and associated hazards and their control, this book follows a case-study description and analysis approach. Such an approach not only includes information that is emphasized in traditional format books, but does so in a way that makes reading more interesting. Disseminating knowledge through the description of real foodborne illness outbreaks and episodes is more attractive to readers as it deals with exploration of real-life events. It is as though the reader participated in the analysis of a real problem. I find this approach very effective and practical, as well as attractive to the reader, who receives knowledge needed in understanding and dealing with food safety hazards by reading how real-life food safety and authenticity problems developed, were handled, and eventually were resolved.

While following this unique approach, this book does not lack in breadth of content, which is its second impressive characteristic. The contents are divided into six parts, covering outbreak investigations, source tracing, crisis management, farm-level interventions, safe food production, and food adulteration and authenticity. The total of 38 chapters within these parts of the book provide a wealth of knowledge that is easy to understand, as it is linked with descriptions of real-life events.

The unique approach of this book does not reduce its ability to cover a wide range of topics and issues. The book manages to address numerous hazards, including *Bacillus cereus*, *Bacillus anthracis*, *Campylobacter*, *Listeria monocytogenes*, *Salmonella*, Shiga toxin-producing *Escherichia coli*, parasites, viruses, bovine spongiform encephalopathy, mycotoxins, dioxins, and genetically modified organisms. Products examined in the case studies presented include bakery products, fresh produce, sprouted seeds, drinking water, chicken products, cattle, milk and raw milk, fresh herbs, dog chews, meat, street foods, yoghurt, dairy foods, dry foods, low pH foods, animal feed, milk powder, cod, rice, and so on. Additionally, other topics covered include outbreak investigation and tracing, ethnic community food safety issues, catering facilities, small food-manufacturing plants, persistent contamination, inflammatory bowel disease, animal vaccination, free-range animal production, pathogen control in the food chain, food safety certification, regulatory issues, laboratory methods, predictive microbiology, food adulteration, forensic investigation of sabotage, melamine contamination, food taste disturbance, molecular tools, and so on. It is also noteworthy that, in addition to the above wide breadth of topics and hazards, the book covers food safety cases originating from numerous countries throughout the world, including Africa, Australia, Austria, Brazil, Canada, Cyprus, Denmark, Finland, France, Germany, Indonesia, Ireland, Italy, Norway, Romania, South Africa, Sweden, Thailand, the United Kingdom, and the United States.

This is a truly unique and impressive book of wide breadth and interest. Its approach makes it easy to use, and of interest for students of food safety as well as food safety and quality managers, epidemiologists, public health officials, food safety regulators, and food safety consultants.

John N. Sofos
Colorado State University

How to use this book

Each chapter is organized as an independent story that has evolved over a certain period of time. Although each is related to a specific problem, the authors have given us the broader view by explaining the regulatory, financial and consumer relevance of the story.

The trouble-shooting part describes the painstaking process of fact-finding, not only by following the guidelines but mostly by making mistakes; because we primarily do learn from our mistakes.

Each chapter ends with a personal reflection of the author's own experience with the case, and lines up a number of questions to stimulate the learning process. The questions can certainly be used for classroom exams when the book is used as a textbook.

Some chapters are supported by illustrations, graphs, photos or maps, but, with the almost unlimited access to online information today, the reader is encouraged to supplement the chapters with videos, animations and other illustrations.

Acknowledgements

With today's demanding work culture, it is not easy to find time for extracurricular work such as contributing to a book; in particular, writing case stories in an atypical format can be arduous for scientists, who are used to more traditional research article formats. That is why I express my heartfelt gratitude to the large number of my colleagues who have contributed chapters. Their enthusiasm to share their stories, expertise and knowledge has made this book possible.

Coordinating the European project BIOTRACER (FOOD-2006-CT-036272), one of the world's largest food microbiology research projects, provided me with the opportunity to collaborate closely with many excellent scientists from different disciplines of food safety and quality. One of the outcomes of this collaboration is this book. The European project BIOTRACER supported a number of scientists who have contributed to this book.

The advice and dedication of senior commissioning editor Sarah Whitworth and project editor Sarah Lynch were significant to the quality of this book.

I thank my family for putting up with me sitting many nights in front of a computer to do this work.

J. Hoorfar
Denmark

BIOtracer™

Part I

Outbreak investigations

1

Tracing an outbreak of listeriosis in Austria: what an investigative microbiologist needs to consider

D. Schoder, Institute of Milk Hygiene and Food Science, Austria, M. Wagner, Austrian Agency for Health and Food Safety, Austria and F. Allerberger, Christian-Doppler Laboratory for Molecular Food Analytics, Austria

Abstract: An unusual cluster of three cases of human listeriosis occurred in Austria in the summer of 2009. As the months went by, the number of cases went up but no common source was initially identified. The outbreak seemed to be over, but then a second peak occurred. Since most of the cases occurred in elderly males, food typically consumed by males rather than females was suspected to have been involved. Then a bright idea came into play: could assessment of purchase behaviour help to identify the contaminated food commodity?

Key words: outbreak, *Listeria monocytogenes*, epidemiology, tracking.

1.1 Introduction to the case

On 14 August 2009 the Austrian Reference Centre for Listeria in Vienna first noticed the occurrence of a new pulsed-field gel electrophoresis (PFGE) pattern (clone 1) that was shared in human isolates of *Listeria monocytogenes* serotype 1/2a. Most cases occurred in elderly males. In the first outbreak phase, 14 clinical cases (12 Austrians and two Germans) infected with this new clone 1, including five with fatal outcome (two of them German), were identified, with onset of the disease ranging from June 2009 to January 2010 (Fretz *et al.*, 2010a).

1.2 Significance of the case

Listeriosis is a rare but serious infection caused by *Listeria (L.) monocytogenes*. This organism can be found throughout the environment: in soil, vegetation, and animals. The main route of transmission is known to be through consumption of contaminated food. However, infection can also (in very rare cases) be transmitted directly from infected animals to humans. In neonatal infections, *L. monocytogenes* can be transmitted from mother to child *in utero* or during passage through the infected birth canal. The bacterium is particularly successful in causing foodborne disease, because it survives food processing technologies that rely on acidic or salty conditions, and, unlike many other pathogens, can continue to multiply slowly at low temperatures, allowing growth even in properly refrigerated foods (Allerberger and Wagner, 2010).

1.2.1 Regulatory aspects

Food contamination with *L. monocytogenes* is intensively regulated by international and national authorities. *L. monocytogenes* is prominent in Annex A of the EU 'Zoonoses' Directive 99/2003 and thus must be monitored in all member states. To harmonize microbiological food safety criteria, *L. monocytogenes* was promptly listed in the EU Directive 2073/2005. At a glance, the regulation states that *L. monocytogenes* must be absent from foodstuffs that facilitate the growth of the organisms, but it can be present up to a limit of 100 CFU/g if the physico-chemical profile of a food commodity prevents growth (the bacteria must not exceed 100 CFU/g food at the end of shelf-life; Anonymous, 2006). Since listeriosis is usually caused by a higher number of *L. monocytogenes* ($>10^5$ CFU/g), the search began for a food item that fosters growth and that is predominantly eaten by elderly persons.

1.2.2 Economic and market aspects

L. monocytogenes infections have the third highest mortality rate of all foodborne infections in the USA (Scallan *et al.*, 2011). The pathogen is a contaminant of multiple food chains, including meat, dairy, fish, seafood and plant foods. However, from our own investigations we knew that three food categories contribute to the major burden of cases of contamination in Austria: smoked fish products, soft cheeses and cooked sausages (Wagner *et al.*, 2007).

1.3 The case history continues

1.3.1 What made outbreak investigations difficult?

Since the incubation period of listeriosis in humans can be up to 70 days, the outbreak pattern is quite scattered, and thus a source of infection often remains undetected. Furthermore, due to the prolonged incubation period, foodstuffs have usually been eaten or already thrown away before an outbreak investigation starts. Due to the complexity of most outbreak scenarios, no pre-specified formulae dictate the path that an outbreak investigation is supposed to take.

1.3.2 What fostered the search for the source?

We collected grocery receipts of purchases made by seven patients in December 2009, after their discharge from hospital, and screened them for a common food item. The survey of the purchase behaviour allowed us to generate a hypothesis that was then tested by a case–control study using case–case comparisons. For this study, a case was defined as a person in Austria from whom the *L. monocytogenes* outbreak clone 1 was isolated (case group). Controls were patients from Austria with *L. monocytogenes* infections in 2009, whose isolates showed typing profiles other than the outbreak clone 1 (control group). Patients were asked about consumption of 12 cheese products in the six-month period prior to disease onset. Persons in the control group were requested to provide information on consumption of the same products and a parameter called the 'odds ratio' was computed (Fretz *et al.*, 2010b).

1.4 Resolution and outcomes

1.4.1 Clinical dynamics

In total, the outbreak involved 34 cases of invasive listeriosis: 25 outbreak cases originated from seven of nine Austrian provinces. Another eight patients were from four German federal states, and one patient was from the Czech Republic. Eight of the 34 cases in this outbreak had a fatal outcome. The median age of the cases was 72 years (range: 57–89 years), and 26 patients were male (Fretz *et al.*, 2010b). There were no materno-neonatal cases. Prevalence of underlying disease in the case group was not different from the prevalence of underlying disease in the control group.

1.4.2 Epidemiology did the job

In the case and control groups, 10 and 24 out of 12 and 33 possible cases (response rates: 83.3% and 72.2%), respectively, provided answers to the questionnaire. Consumption of Quargel cheese was identified as the only significant risk factor highly associated with the illness in question. Nine of the ten patients with clone 1 had consumed the product; the tenth provided no answer concerning this food item. Of 22 controls all but two denied having eaten this specific cheese; the remaining two provided no answer concerning this food item. The computed odds ratio was 76.6 (95% confidence interval (CI): 9.3–infinity; p-value <0.001).

Interestingly, molecular typing revealed that two clones of the same serovar 1/2a of *L. monocytogenes* had contributed to the outbreak. Whereas the first clone

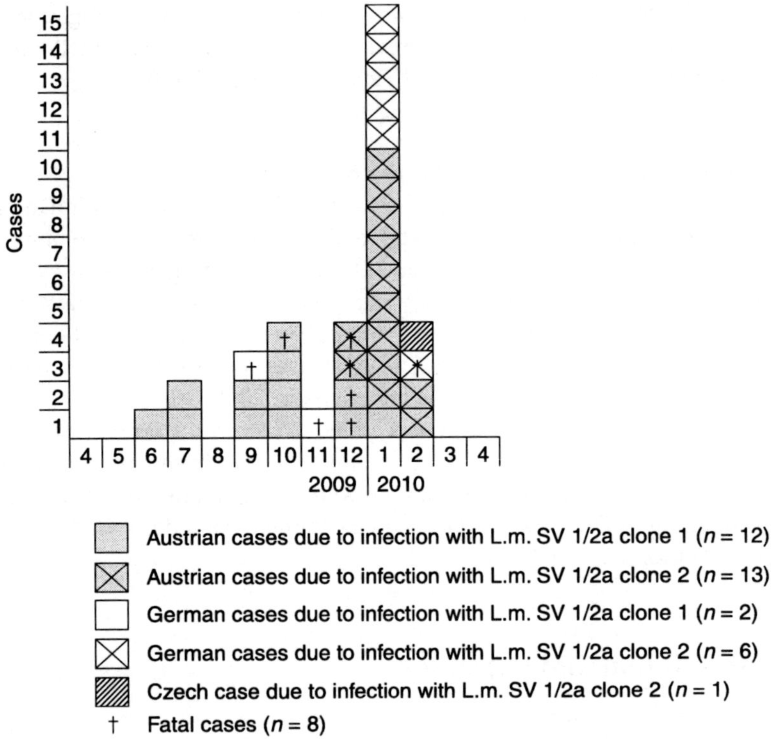

Fig. 1.1 Outbreak cases of listeriosis by onset of illness and final outcome, Austria, Germany and the Czech Republic, 2009–2010. $n = 34$; Austria = 25, Germany = 8, Czech Republic = 1.

caused 14 cases in the first phase of the outbreak (Summer–Autumn 2009), the second outbreak clone accounted for 20 cases (13 Austrian cases, six German cases, one Czech case) that were reported from December 2009 until the end of February 2010 (Fig. 1.1).

1.4.3 Short description of Quargel processing technology

This cheese is made of curdled milk which ripens, after addition of starter cultures, for one day at 28 °C, and for another two days at 14 °C by spraying the surface with *Brevibacterium linens*. The shelf life after packing is up to 50 days. With regard to the incriminated plant, 16 tons of Quargel cheese were produced per week. Almost 50% of the cheese was exported to the German market and small amounts to the Czech Republic and Slovakia.

1.4.4 Contamination rates of samples and recalled lots of Quargel

Only one leftover specimen, stored in the patient's refrigerator and sampled on March 3rd, could be tested by the public health authorities and yielded

2 100 000 CFU/g of *L. monocytogenes*. Despite the high profile of the topic in the mass media, the patient and his wife were not aware of the product recall.

Two of 64 isolates available for testing (44 isolates cultured from cheese produced in 2010 and provided by the manufacturer, 20 isolates cultured from samples officially obtained during outbreak investigation) showed the PFGE pattern of clone 1 associated with the outbreak. Sixty-two remaining food isolates showed a different PFGE pattern (clone 2) that proved indistinguishable from the pattern of a human isolate from a listeriosis patient hospitalized at the time.

The company launched a spectacular recall action in January 2010. Eighteen lots of cheese were transported to the Institute for Milk Hygiene, Milk Technology and Food Science and a total of 260 acid curd cheese samples were investigated. Examination of the samples (according to ISO 11290-1 and -2) resulted in 16 *L. monocytogenes* positive and two negative lots. These results were confirmed by a combined enrichment/real-time polymerase chain reaction (PCR) method that was developed by the Christian-Doppler Laboratory for Molecularbiological Food Analytics (Rossmanith *et al.*, 2010). The values of contamination obtained (by quantitative ISO 11290-2) varied from $\leq$100 to 1.2×10^8 CFU/g. However, contamination rates varied throughout production dates. There were no tendencies for increased or decreased CFU levels when samples were taken on continuous days (Schoder *et al.*, 2010).

Three out of 16 *L. monocytogenes* positive lots showed a contamination rate of $\leq$100 CFU/g at the beginning of the shelf-life when stored at 4 °C. Nevertheless, by increasing the storage time or the storage temperature (15 °C, 22 °C) the contamination rate could be raised to 10^5–10^6 CFU/g.

In general, storage at 4 °C resulted in an average of 5.6 log units, a significantly minor contamination rate compared with storage at 15 °C or 22 °C with mean log values of 7.2 and 6.8, respectively. Lower storage temperatures are supposed to slow down bacterial growth and may restrict amounts to non-harmful doses. This was not the case in this outbreak. Although there was an overall higher variation in the samples stored at 4 °C than in samples kept at 15 °C or 22 °C, and even though the mean values differed significantly, maximum levels were similar. This finding indicates a high potential for growth even when Quargel is stored at 4 °C (Schoder *et al.*, 2010; Schoder *et al.*, pers. comm.).

1.4.5 At-site contamination scenarios

An environmental *L. monocytogenes* 1/2a isolate from the suspect production plant, from a gully (drainage pit) sample collected in December 2009, became available on 19 January 2010 and proved indistinguishable from the outbreak strain by genotyping.

1.5 Commentary and personal reflections

Outbreak investigations are complex operations involving many institutions and people with very different expertise. They are influenced greatly by many parameters, and the majority of outbreak investigations do not result in the source being found. In terms of *L. monocytogenes*, large-scale food production combined with complex transnational retailing systems, long incubation periods, the low proportion of a population usually affected, and the fact that the food has usually already been eaten, hinder outbreak investigations that follow traditional concepts.

However, the considerable potential of molecular subtyping of human *L. monocytogenes* isolates increases the probability of recognizing outbreak clusters and emphasizes the immense importance of cross-border cooperation for elucidating chains of infections in multinational outbreaks (Pichler *et al.*, 2011). Further improvements in pathogen typing technology, interlaboratory exchange of typing data, and thorough collection of isolates may increase the chances of foodborne illnesses being linked to specific food products and facilities more efficiently.

The outbreak dynamics show clearly that a decline in case numbers does not necessarily imply that the underlying problem has disappeared. The occurrence of a different outbreak clone in December 2009/January 2010 impeded the epidemiological investigation. The fact that the outbreak scenario was eventually completely unravelled rested on the unusually high abundance of the outbreak clones. The shift from one clone to the other could have been caused by the change to a new commercial yeast-ripening culture used in the cheese factory in late November 2009 due to the short supply of the original culture.

With regard to *L. monocytogenes*, experts agree that food environment testing is essential to ensure that incidental contamination of a processing facility is not causing an unacceptably high concentration in the end product. Albeit 'looking for the proverbial needle in the haystack' in many cases, testing for environmental samples is mandatory, as laid down in Regulation 2073/2005. In the outbreak described here, a *L. monocytogenes* isolate from an environmental sample taken by the quality manager in the facility in May 2009 was the first evidence of a contamination problem. A consecutive source tracking approach was not able to eliminate the strain from the facility and repeated testing of the food lots revealed no evidence that the food was contaminated. From post-outbreak epidemiological examinations, we hypothesize that *L. monocytogenes* was introduced into the plant during major construction work (Schoder and Wagner, 2010). From 23 February 2009 until 27 May 2009, a ripening room in direct proximity to the central production facility was remodelled, and this coincides with the time of the first documented detection of *L. monocytogenes* in a smear sample, allegedly after years of no microbiological proof of *L. monocytogenes* in the facility.

This finding emphasizes the unequivocal need for implementing a self-control system capable of detecting contamination scenarios with highly sensitive methods and as soon as possible after transmission of the pathogen into the critical production area. This strong recommendation becomes a 'must' whenever the physico-chemical profile of a food item allows multiplication of *L. monocytogenes*, or even if the physico-chemical profile is unknown (Asperger *et al.*, 2001).

To our knowledge, this was the first outbreak in which grocery bills were collected prospectively – after the patients' discharge from hospital. The thorough epidemiological investigation successfully revealed the food consumption patterns of the patients and finally indicated Quargel sour milk curd cheese as the source of this outbreak. Taking into account that thousands of consumers were exposed to a food that was contaminated at very high levels, the number of infections (34) is surprisingly low. This again poses the question of the virulence heterogeneity of isolates of the species *L. monocytogenes* and demonstrates that, fortunately, *L. monocytogenes* is an organism with rather low pathogenicity that, unfortunately, is perfectly adapted to modern food production systems.

1.6 Critical questions for discussion

1. What makes *L. monocytogenes* outbreak investigations special in comparison to outbreak investigations of other foodborne pathogens?
2. What information should a perfect self-control system deliver? How should it be constructed?
3. How difficult is the interpretation of food legislation with regard to *L. monocytogenes* contamination?
4. Where and how do *Listeria* survive outside the foodstuff?
5. What are the factors intrinsic and extrinsic to food that allow growth of *Listeria*?

1.7 References

ALLERBERGER F., and WAGNER M. 2010. Listeriosis: a resurgent foodborne infection. *Clin. Microbiol. Infect.* **16**: 16–23.

ANONYMOUS (2006) *Microbiological criteria for food.* Dir. EU 2073/2005.

ASPERGER H., WAGNER M., and BRANDL E. 2001. An approach towards public health and foodborne listeriosis: The Austrian *Listeria* monitoring. *Berl. Muench. Tierärztl. Wschr.* **114**: 446–52.

FRETZ R., SAGEL U., RUPPITSCH W., PIETZKA A.T., STÖGER A. *et al.* 2010a. Listeriosis outbreak caused by acid curd cheese 'Quargel', Austria and Germany 2009. *Euro. Surveill.* **15**(5): pii=19477. Available at: http://www.eurosurveillance.org/ViewArticle.aspx?ArticleId=19477. Accessed 19 February 2011.

FRETZ R., PICHLER J., SAGEL U., MUCH P., RUPPITSCH W. *et al.* 2010b. Update: Multinational listeriosis outbreak due to 'Quargel', a sour milk curd cheese, caused by two different *L. monocytogenes* serotype 1/2a strains, 2009–2010. *Euro. Surveill.* **15**(16): pii=19543. Available at: http://www.eurosurveillance.org/ViewArticle.aspx?ArticleId=19543. Accessed 19 February 2011.

PICHLER J., APPL G., PIETZKA A., and ALLERBERGER F. 2011. Lessons to be learned from an outbreak of food-borne listeriosis, Austria 2009–2010. *Food Prot. Trends* **31**: 268–73.

ROSSMANITH P., MESTER P., WAGNER M., and SCHODER D. 2010. Demonstration of the effective performance of a combined enrichment/real-time PCR method targeting the *prf*A gene of *Listeria monocytogenes* by testing fresh naturally contaminated acid curd cheese. *Lett. App. Microbiol.* **51**: 480–4.

SCALLAN E., HOEKSTRA R., ANGULA F.J., TAUXE R.V., WIDDOWSON M.A. *et al.* 2011. Foodborne illness acquired in the United States-major pathogens. *Emerg. Inf. Dis.* **17**: 16–22.

SCHODER D., STESSL B., ALLERBERGER F., NEPF R., and WAGNER M. 2010. Growth of and genetic diversity of *L. monocytogens* in acid cured milk cheese recalled during the Austrian episode of listeriosis 2009/2010. Abstract S11-2, pp. 58–9. *Abstract-Proceedings of the 32nd Meeting of the Austrian Society for Hygiene, Microbiology and Preventive Medicine*, 17–20 May 2010, Vienna, Austria.

SCHODER D., and WAGNER M. 2010. The role of insects in the transmission of *L. monocytogenes* in an outbreak of listeriosis caused by acid curd cheese. Abstract S11-5, pp. 61. *Abstract-Proceedings of the 32nd Meeting of the Austrian Society for Hygiene, Microbiology and Preventive Medicine*, 17–20 May 2010, Vienna, Austria.

WAGNER M., AUER B., TRITTREMMEL C., HEIN I., and SCHODER D. 2007. Survey on the *Listeria* contamination of ready-to-eat food products and household environments in Vienna, Austria. *Zoon. Publ. Health* **54**: 16–22.

2

A point-source outbreak of *Salmonella* in an ethnic community in the UK

A. G. Stewart, R. Vivancos and A. Keenan, Cheshire & Merseyside Health Protection Unit, UK and C. Theodore, Liverpool City Council, UK

Abstract: It is important to investigate the source of outbreaks of food poisoning in order to understand what has happened so that suitable lessons may be learned and measures put in place to reduce the likelihood of a similar outbreak arising again. This case study follows an outbreak in a closed community and highlights important issues in outbreak investigation and management. It should be read alongside Chapter 3 'Investigating the source of a food poisoning incident in Cheshire, UK' by Stewart and Ghebrehewet.

Key words: case–control, epidemiology, pathogen, *Salmonella enterica* subsp. *enterica* serovar Enteritidis, small retail outlets, outside catering, multi-disciplinary responses, culture.

2.1 Introduction

One Saturday morning (18 July), approximately 1000–1200 people gathered in an urban worship centre for a festival. They were all members of a local immigrant community that is united by ethnicity and religion. Many of them were health care workers. The feast was a family occasion and all ages attended. Most family groups came from the surrounding conurbation and nearby towns, but a few had travelled from other parts of England.

At 1 pm, at the end of the ceremonies, while the Very Important Persons were fed in the centre, boxes of precooked food were distributed to each of the 1000-plus attendees as they left for home. Some ate the food in their cars, some waited until they got home; some ate the food cold, some reheated it.

On Monday 20 July, two cases from the same family reported symptoms of diarrhoea and vomiting to Cheshire and Merseyside Health Protection Unit. At this stage, it was not clear that there was a link between the cases, or a possible link to the festival.

On Tuesday 21 July, two cases from another town were reported to the Health Protection Unit. The Unit was told that these cases had attended the festival and knew of others who had also attended and developed symptoms of diarrhoea and vomiting. A further three cases were reported to Environmental Health Officers of the same two towns over the next two days.

It became clear that there was an outbreak of gastroenteritis linked to the festival and which was spread across at least two local authorities. The festival food had been supplied by a local restaurant.

Environmental Health Officers visited the restaurant on 23 July, during which a preliminary inspection of the premises was made. Suppliers, food preparation, handling and cooking, methods of food storage and distribution were all investigated. Refrigerator and freezer temperatures were checked. Samples were taken as follows from the restaurant kitchen for microbiological analysis:

- Samples of the cooked chicken curry and rice prepared for the event and subsequently frozen in the restaurant freezer,
- Samples of raw shell eggs,
- Swabs from two chopping boards, freezer and refrigerator door handles and the washcloth.

Full menu details for the festival were also collected from the restaurant, to inform further investigations of the clinical cases. Samples of the distributed food were also taken from the freezer of a reported case in a nearby town on 23 July. The Health Protection Agency's Food, Water and Environmental Microbiology Laboratory in Preston undertook microbiological analysis of the food samples and items from the kitchen.

On Friday 24 July, an Outbreak Control Meeting was convened in the Health Protection Unit, to ascertain the situation and decide on the appropriate actions to investigate and manage the outbreak.

It was confirmed that each attendee at the festival was offered one box of chicken curry and rice, in separated compartments within the box, and one box containing four starters: spring rolls, chicken lollipops, vegetable samosas and beef cutlets. A hot meal was served at the centre for about 30 Very Important Guests. Lamb curry and fish curry were served to these guests in addition to the above main course and starters; all the Very Important Guest meals were reheated at the centre.

2.1.1 Challenges

The challenges that the outbreak control team faced at the beginning included:

- Identifying the attendees at the festival.
- The best access to an ethnic community with a different culture from the main community.

- Organising and administering a timely and potentially large epidemiological investigation related to a one-off, take-away food distribution at a festival.
- Linking environmental data with patient data to ensure the best evidence for the regulating local authority.
- The outbreak occurred during the early days of pandemic influenza, resulting in a number of the cases initially being misdiagnosed as influenza, not food poisoning. This also meant that faecal samples were not taken at the initial visit to the patients' General Practitioners.

2.2 Significance of the situation

Since the initial evidence pointed to the food distributed at the festival (hence the term 'a point-source outbreak'), the question of how much investigation should be pursued needed to be considered. By its nature, a point source is of limited duration and will not, therefore, continue to generate cases. Nevertheless, there are several reasons for investigating an outbreak that might very well be over before the incident control team meets.

Every outbreak is different. It is important to examine each one and learn from a review of the outbreak, investigation and control measures; without such review and learning, there will be no progress in food safety and outbreak management.

There is a need to understand the size and extent of any outbreak in terms of how this has affected the health of the population. Without investigation, there is unlikely to be certainty as to why the outbreak started, with resulting uncertainty as to the efficacy of control measures in place. There may be questions asked by regulatory authorities about any food handlers or establishments that were involved, or appeared to be involved, in the outbreak. There may be questions about the breakdown or efficacy of established measures to control disease that need to be answered, even if just to check that such measures continue to work. Without investigation, new strains of microbiological agents and unusual transmission routes may remain unexplored.

This outbreak also raises the issue of culture and how that affects disease control; this point will be discussed in more detail below. The outbreak also raises issues of how to help a defined community, which will want to continue celebrating its festivals, to do so in the safest possible manner.

Outbreaks should be investigated and reported in peer reviewed journals in order to provide evidence when further outbreaks come to light at some point in the future (O'Brien, 2006).

2.2.1 Regulatory aspects

The catering establishment that supplied 1000–1200 meals to the festival goers had not given the Environmental Health Department any concerns up to this time. However, it had not undertaken outside catering before, nor was it used to such a large-scale operation. The restaurant owners had not consulted with the local

Environmental Health Department about the implications of extending its catering operation to include outside catering, so had not been able to fully consider the requirements for providing over 1000 meals compared with the more modestly sized parties that could be accommodated within the restaurant.

The restaurant, which had been open for 18 months at the time of the outbreak, offered full cooperation to the Environmental Health Officers during and after the investigation. At a previous inspection carried out by an Environmental Health Officer the restaurant had been rated Category C (satisfactory).

The enforcement of food safety and hygiene is the responsibility of Local Authority Environmental Health Officers in England and other parts of the UK. European legislation, which has applied in the UK since 2006, requires food businesses to meet minimum standards to ensure that the food they produce is safe and to have food safety management systems based on the principles of Hazard Analysis and Critical Control Points (HACCP).

2.2.2 Economic and market aspects

It was clear early on in the investigation that the likely cause of the outbreak was related to the cooking and storage of the food before distribution at the festival. However, there is a difference between suspicion and proof.

The investigation of the food premises needed to balance ensuring that forensic evidence of any possible link between the restaurant and the outbreak was robust, while at the same time maintaining good relationships with the restaurant, being mindful that any adverse publicity or decision was likely to have major effects on the business with resulting consequences for staff employment. Investigations should not put a business at risk without good reason.

2.3 The case history continues

Although there were five different community groups (congregations) that attended the festival, it was not possible to gain a full attendance list from the organisers. As a result, the total number of attendees remained unknown, between 1000 and 1200. Access to the community was enhanced by excellent support from a senior community leader from the early days of the investigation.

Salmonella enterica subsp. *enterica* serovar Enteritidis (commonly known as *S.* Enteritidis) Phage Type 5c was identified in faecal samples from the early clinical cases. The local Primary Care Trusts wrote to General Practitioners asking them to submit faecal samples from anyone presenting with appropriate symptoms.

Cases of food poisoning related to the festival continued to be identified through routine notifications of food poisoning to the Health Protection Unit from General Practitioners, from laboratory reports of faecal samples positive for *Salmonella* (untyped), and through the surveillance questionnaires administered to each identified case. All untyped *Salmonella* specimens were sent from the local laboratories to the national reference laboratory for sero, geno and phage typing.

Positive results for *S.* Enteritidis Phage Type 5c were also obtained from the food stored in a family freezer and the restaurant dish cloth. No restaurant staff or food packers (five were brought in especially for the event) had any history of diarrhoea and vomiting; neither was *Salmonella* detected in any of their faecal samples. *S.* Enteritidis Phage Type 5c was detected on a sample taken from the restaurant kitchen washcloth, used to wipe work surfaces. It was not detected in food samples or swabs taken from kitchen surfaces in contact with food and hands, or on the shell eggs.

All persons notified as having food poisoning possibly related to the festival were interviewed by Environmental Health Officers using a routine questionnaire that had been augmented by specific questions about attendance at the festival and foods eaten from the distributed boxes. Other questions asked about any knowledge of other possible cases.

Such possible cases that were identified through the questionnaires were then interviewed by the Environmental Health Officers or by staff from the Health Protection Unit and asked, if they fitted the case definition, to attend their General Practitioner and give a faecal specimen. Not everyone so requested submitted a faecal sample.

The questionnaire was also actively distributed through the community with the help of community leaders in a snowball manner (cases identifying other cases known to them), in order to obtain information about unreported cases. Members of the investigating team also visited worship in the largest congregation and gave questionnaires to worshippers to obtain further cases as well as asymptomatic controls. In all, about 600 questionnaires were distributed.

An unmatched, analytical case–control study was conducted, following discussions with epidemiological colleagues. The hypothesis tested was that 'eating chicken curry (and/or rice?) from the food distributed at the event on 18 July resulted in illness'.

A primary case was identified as someone who ate food that was distributed on 18 July and, within 72 hours of eating, had developed symptoms of diarrhoea and fever, plus possibly abdominal pain, headache and nausea. Taking into account the incubation period of *Salmonella*, secondary cases could occur within families after 21 July; they were later removed from subsequent statistical analysis. Controls were defined as those who had eaten food distributed at the festival but had not developed any symptoms.

In total, 148 cases were reported to the Health Protection Unit (clinical cases: 140 primary cases and eight secondary; 67 positive for *S.* Enteritidis Phage Type 5c). Of these, two cases were hospitalised briefly. We obtained information on 52 controls. Two members of one family that attended the event from another part of the country and who became sick also tested positive for *S.* Enteritidis Phage Type 5c.

We estimated the odds ratio for individual food items using single variable (Table 2.1) and multivariate logistic regression. Since people from the same

Table 2.1 Single variable analysis – logistic regression results (independent variables)

Variable	Odds ratio (95% confidence intervals)	*N*	P-value
Age	1.02 (0.99–1.04)	188	0.051
Sex	1.21 (0.61–2.38)	174	0.586
Spring rolls*	5.33 (2.68–10.60)	192	<0.01
Chicken lollipop*	6.65 (3.31–13.35)	192	<0.01
Beef*	6.78 (3.32–13.86)	192	<0.01
Samosa*	4.19 (2.14–8.21)	192	<0.01
Chicken curry	16.97 (7.60–37.86)	192	<0.01
Rice	16.39 (7.04–38.14)	192	<0.01

*Starters; other foods were main courses.

household are likely to have similar eating habits, we also added households (i.e. people with the same address) to the multivariate analysis. Households were identified manually by postcode and name.

The table shows that all the foods eaten were positively associated with becoming ill (P is less than 0.01, or, better, the confidence intervals for the odds ratios do not include 1), although the odds of becoming ill having eaten the chicken curry and rice were higher than becoming ill after eating the starters. Age and sex made no difference to the chances of becoming ill (P is greater than 0.05, or, better, the confidence intervals for the odds ratios include 1).

In the second stage of the analysis all the statistically significant variables identified were analysed together by multiple logistic regression to identify which variables were independently associated with being ill. The use of this technique identifies variables in the analysis (here, foods) that are not themselves directly linked to the question (here, being ill), but are linked through another variable (here, another food); these are known as confounding variables (Last, 2007; Porta, 2008).

To take another example, lower socio-economic status is often associated with respiratory disease but the link is poor housing, and it is the effects of living in poor housing (e.g. mould, damp) that are the direct link with respiratory disease, not the socio-economic status, which is a confounding variable.

Only five people did not belong to a family cluster; four patients and one control. There were 62 clusters identified, ranging from ten with one member to eight with five and one with six (mode four members). The age structure of those reporting illness showed a double peak, in children and young adults (parents). The epidemic curve (Fig. 2.1) clearly showed a point source of infection.

People who ate food distributed at the festival were 25 times more likely to be sick than those who did not (odds ratio = 24.9; 95% confidence interval 8.0–77.7). Including the 62 family clusters in the analysis clarified the multivariate analysis result, so that chicken lollipops (odds ratio 5.13 (95% confidence interval

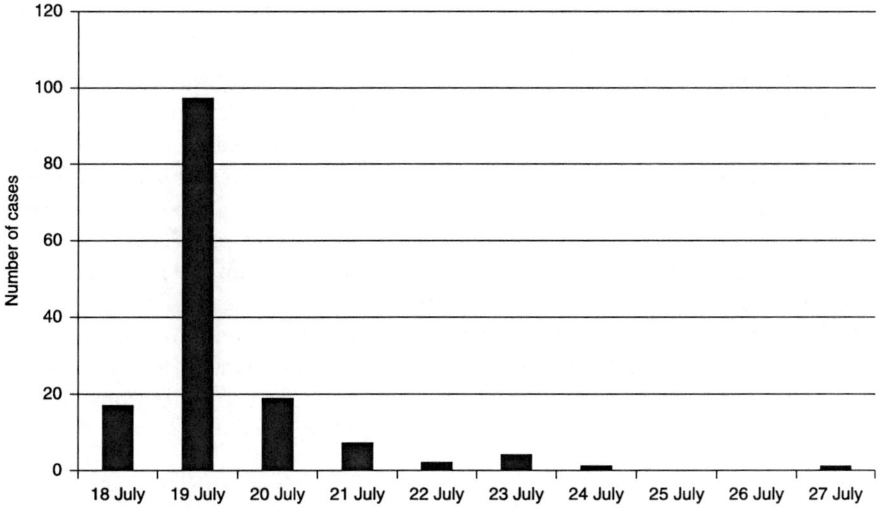

Fig. 2.1 Epidemiological curve, illustrating a point-source outbreak (HPA data).

1.09–24.20)) and rice (3.28 (1.35–7.92)), the two main courses, stood out as the main foods giving rise to the risk of disease.

Using the rates from the reporting pyramid described in the large study of infectious intestinal disease in England (Wheeler *et al*, 1999, table 3) we estimate that, in this outbreak, there were up to 440 people ill from *Salmonella*. This was confirmed by the fact that three or four members of a family were often sick, but only one submitted a stool sample through their General Practitioner or returned a questionnaire.

Since estimates of the number attending the conference vary, only an approximate crude attack rate of 12–44% could be estimated. These rates are in keeping with those reported in other outbreaks.

2.4 Resolution and outcomes

Thirty Very Important Guests were served a hot meal consisting of the same food as was distributed to the congregations, plus two other dishes, but reheated on site. None of these guests were ill following the meal. Two cases were hospitalised for short periods of time before the diagnosis became apparent. There were no deaths.

The food for the festival had been cooked in several batches during Friday evening (17 July) and the early hours of Saturday morning. Some batches had their temperatures checked, but these temperatures were not recorded. After the food was cooked, it was cooled down with the assistance of fans and packed into individually portioned plastic cartons by food packers and then stored at ambient temperatures until transportation to the venue. No blast chilling or other appropriate methods were used.

The investigation showed that people who ate the chicken lollipop or rice at the event were highly likely to become ill with *Salmonella*. Laboratory tests confirmed that all the salmonellas in this outbreak were Phage Type 5c, as was the *Salmonella* isolated on the washcloth in the restaurant. As this is a rare phage type, it means that we can say with a high degree of certainty that all these cases are linked to the same source. It seems highly likely that the source was contaminated chicken, either directly contaminated or from the eggs with which the lollipops were prepared, and that the rice was contaminated by the storage practices.

The restaurant cooperated fully with the investigating team and responded well to control measures. Control measures included a deep clean of the restaurant kitchen as a precaution. The restaurant management agreed not to undertake any outside catering without seeking advice from Environmental Health Officers. Subsequently, it has been confirmed that they have not carried out any further outside catering operations in the two years since the incident.

2.5 Commentary

Given the high number of cases arising in those who attended the event, it was fortunate that there were no fatalities or more serious illness recorded. Several cases were initially diagnosed as a flu-like illness, which slowed the recognition of the true diagnosis and the fact of an outbreak.

The epidemic curve is typical of a point-source outbreak, with a quick rise and fall in the number of cases. This fits well with the incubation period of *Salmonella* food poisoning (6–72 hours, but usually 12–36 hours). Later cases with an onset longer than 72 hours after the initial distribution of food at the festival are likely to be secondary cases and were treated as such. Such cases can occur where there is person-to-person spread, often within a household. There were only eight cases identified as secondary, indicating that either limited spread occurred or such cases were not reported.

Finding the source of a point-source outbreak is important for learning about the dynamics of why and how outbreaks occur; by the time an investigation is initiated many point-source outbreaks are resolving naturally, and control measures may offer little in the situation under investigation but can be vital in preventing a repeat outbreak. Compare this curve with that in Stewart and Ghebrehewet (2012).

The fact that this outbreak was confined to the members of a clearly defined ethnic–religious group was both positive and negative. On the positive side, initial identification of cases was simplified and communication across the community benefited from one point of access. One of the Outbreak Control Team had experience of a similar community, which also helped with communication and understanding issues, particularly when speaking, and distributing questionnaires, at the worship session.

On the negative side, identification of controls for the analytical study proved more difficult than expected, possibly because the community focus was centred on illness and recovery, and the need for non-ill controls was not clearly understood. It is possible that other, unseen, cultural issues further hampered the attempts by the Outbreak Control Team to convey the message about the need for controls. Similarly, there could have been an unknown element of bias in the returned questionnaires, since usually only one questionnaire was returned for each affected family, whereas best practice is to obtain a completed questionnaire for each ill person. However, this was probably largely resolved by the household cluster factor in the statistical analysis.

In outbreaks, secondary cases are usually far more common than were found here. This might be due to the culture and training of the community.

Culture is a system of shared concepts, beliefs and values where individuals interpret and make sense of life and the world around them. Hiebert (1983) suggests that there is more to learning culture than learning how to conform to external patterns of behaviour; culture is more than only learning a particular way to dress or talk. While these external patterns of behaviour are a large part of what makes up culture, they do not comprise the entirety. Instead, Hiebert proposes that culture is a complex network of shared ideas, beliefs and values; the way that people interpret and make sense of the world around them is important. He argues for an active learning of culture, rather than an unconscious absorption, from the surrounding environment. The learned culture is then passed on by each generation to the next, the process beginning at birth. We would argue that culture is a combination of both active and passive learning; for example, eating habits are learnt (by being taught) but also unconsciously absorbed from those around us. These concepts, beliefs and values find expression in actions by individuals that mark them as members of a cultural group.

Although we did not ask about toilet usage at the festival, the culture of the community includes frequent washing. It is possible that this higher than normal level of cleanliness in the home reduced the number of secondary cases.

The large number of family clusters reflects the make-up of the community: many small, nuclear families, as observed by members of the Outbreak Control Team at the worship gathering they attended. This would help explain the double-peaked age distribution of the cases. Normally, *Salmonella* is most commonly reported in the under-five-year-olds.

Phage Type 5c is an uncommon *Salmonella* phage type: of the 32 phage types reported in the North West Region of England that year, 14B was the commonest and was reported in 26 weeks of the year, while 24 phage types, including Phage Type 5c, were reported in fewer than 10 weeks of the year. All reported cases of Phage Type 5c were associated with this outbreak at the festival.

The microbiological finding of *Salmonella* in the food sample taken from the household and in the sample taken from the washcloth in the restaurant kitchen is consistent with what is known about *Salmonella*, that it can be transmitted to humans through the consumption of contaminated poultry or eggs. However, undercooked meat, particularly mince, and any fresh product contaminated with animal waste, have also been implicated in other situations, although the most common source is usually eggs. The chicken lollipops were chicken wings marinated with eggs and flour, seasoned and deep-fried.

In the 1980s and 1990s *S.* Enteritidis was associated with many foodborne outbreaks, with shell eggs being the major source of contamination. Much work has been done with poultry farms and egg distributors as well as food outlets and restaurants to reduce this burden of disease. Since the late 1990s reported infections with *S.* Enteritidis have declined sharply in the UK but still occur, particularly in the summer months.

The restaurant staff were very supportive and provided samples without hesitation. This highlights the advantage of the benefit of a forceful but fair relationship when dealing with outbreaks related to food premises.

It is clear that poor practice in mass catering was the underlying reason for the outbreak. A key issue was how the food was cooked, stored and reheated. One possibility is that the chicken was not properly cooked at the outset. Although the restaurant reported probing some of the cooked food, they did not keep a written record of the temperatures found.

A second risk factor identified was the storage of the cooked food outside a refrigerator, at ambient temperatures, which were reasonably high as this outbreak occurred in July. Guidelines state that cooked foods should be cooled down and refrigerated as soon as practicable. The longer that the food was kept at ambient temperature, the more likely it is that the *Salmonella* bacteria multiplied so that, by the time the food was served, it would have been well established in the food.

A third risk factor was how the food was reheated. None of the Very Important Guests became sick. Their food was reheated at the venue and served immediately. If the food was properly reheated, any *Salmonella* therein would have been destroyed. The other guests took the food home. Some reported eating it in the car on the way home, others reported reheating it at home. It is possible that, in many cases, the food was not adequately reheated. As the investigation progressed, the hypothesis that people became ill due to not reheating the food adequately was put forward. Unfortunately, the questionnaire did not ask any questions around reheating, so this theory could not be tested in the epidemiological analysis.

Combinations of factors often combine to create the conditions for outbreaks: in this case, poor practice in the restaurant kitchen with no recording of the cooked food temperature, long storage of the cooked food at ambient temperature and, possibly, how guests reheated the food. To prevent a similar outbreak arising in these circumstances, the caterers were asked to inform Environmental Health Officers if outside catering is part of their undertaking so that advice on proper practices can be given; they should also provide information to consumers on how foods should be reheated.

2.6 Acknowledgements

Outbreak control team colleagues; North West Centre for Food, Water and Environmental Microbiology Services, Preston; microbiology laboratory staff in University Hospital, Aintree; staff in the Centre for Infections, Colindale, London; regional and national epidemiological colleagues; Martin Stewart.

2.7 Critical questions for discussion

1. Compare this point-source outbreak with the propagated-source outbreak from a bakery (Stewart and Gebrehewet 2012) – what do the epidemic curves indicate about the two situations, what similarities are there, what differences?
2. What is culture? How does it affect health? How important is culture in the presentation, investigation and response to disease in general and in food safety issues? If possible, find other examples.
3. When should food premises be prosecuted for failures? Why?
4. Microbiological, epidemiological and environmental investigations take time and effort by highly trained professionals. Should every outbreak be investigated? To what extent should any outbreak be investigated? What outcomes matter in an outbreak investigation and response?
5. What is the difference between a primary case and a secondary case?
6. What is the 'odds ratio' in case–control studies?

2.8 Further reading

ANDERSON J., 2010. *Understanding Cultural Geography: Places and Traces*. Routledge, Abingdon, UK.

BOWLING A., 2002. *Research Methods in Health, Investigating Health and Health Services*, 2nd ed. Open University Press, Buckingham, UK.

DONALDSON L.J., and SCALLY G. (eds), 2009. *Essential Public Health*, 3rd ed. Radcliffe, Oxford, UK.

HOFSTEDE G., HOFSTEDE G.H., and HOFSTEDE G.J., 2005. *Cultures and Organizations: Software of the Mind*. McGraw-Hill, Columbus, USA. Also see web-site: Geert Hofstede. *Geert Hofstede™ Cultural Dimensions*, 1967–2009. http://www.geert-hofstede.com/ (accessed 7 July 2011).

PEREIRA-MAXWELL F., 1998. *A–Z of Medical Statistics. A Companion for Critical Appraisal.* Hodder Arnold, London, UK.

2.9 References

HIEBERT P.G., 1983. *Cultural Anthropology*, 2nd ed. Baker Book House, Grand Rapids, USA.

LAST J.M., 2007. *A Dictionary of Public Health*. Oxford University Press, New York, USA.

O'BRIEN S., 2006. Preventing epidemics of communicable disease. In D. Pencheon, C. Guest, D. Melzer, and J. A. Muir Gray (eds), *Oxford Handbook of Public Health Practice*, 2nd ed., pp. 206–16. Oxford University Press, Oxford, UK. (A very helpful chapter on outbreak investigation and response.)

PORTA M. (ed.), 2008. *A Dictionary of Epidemiology*, 5th ed. Oxford University Press, New York, USA. (Probably the single most useful book – dip in and read – it goes well beyond simple definitions; previously edited by John M. Last, whose 4th edition (2000) is available electronically at http://www.ebook3000.com/A-Dictionary-of-Epidemiology_109179.html.)

STEWART A.G., and GHEBREHEWET S., 2012. Investigating the source of a food poisoning incident in Cheshire, UK. In J. Hoorfar (ed.), *Case Studies in Food Safety and Authenticity*, pp. 23–32. Woodhead Publishing, Cambridge, UK.

WHEELER J.G., SETHI D., COWDEN J.M., WALL P.G., TOMPKINS D.S. *et al.*, 1999. Study of infectious intestinal disease in England: rates in the community, presenting to general practice, and reported to national surveillance. *BMJ* **318**: 1046–50.

3

Investigating the source of a food poisoning incident in Cheshire, UK

A. G. Stewart and S. Ghebrehewet, Cheshire & Merseyside Health Protection Unit, UK

Abstract: It is important to investigate the source of outbreaks of food poisoning in order to plan and instigate control measures that are suitable, sensible and proportionate. This case study follows an outbreak that was somewhat atypical, but which, by that very strangeness, highlights important issues in outbreak investigation and management. It illustrates the challenges associated with investigating the contamination of a food product through a common ingredient and with implementing national or international control measures that are outside the influence of the local control team. It should be read alongside Chapter 2 'A point-source outbreak of *Salmonella* in an ethnic community in the UK' by Stewart, Vivancos, Keenan and Theodore.

Key words: case–control, epidemiology, pathogen, small retail outlets, multi-disciplinary responses, international aspects, contamination, biosecurity, *Salmonella enterica* subsp. *enterica* serovar Enteritidis.

3.1 Introduction

The story began on Wednesday 2 October, when a hospital microbiologist in Cheshire noticed an increase in the number of stool samples positive for *Salmonella*. She had, she said, 'a bench full of samples'. Nine cases of *Salmonella enterica* were identified that day (eight being serogroup D), with a further three having been identified the day before; normally, she would only expect to see two or three cases a week.

An outbreak control team was convened in the local authority Environmental Health Department closest to the hospital, co-ordinated by the local Health Protection Unit. The team consisted initially of local Public Health doctors (one particularly responsible for health protection issues (Consultant in Communicable

Disease Control), the other responsible through the local health board (Primary Care Trust) for the health of the community), Environmental Health Officers from the two local authorities with affected residents and the hospital microbiologist. The team set out to identify the source of the outbreak and determine what control measures would be necessary and what communications should be undertaken.

The Environmental Health Officers questioned the initial cases using a standard questionnaire; there was no obvious link with food. The national reference laboratory was contacted by the outbreak control team, who were informed of an outbreak of *Salmonella enterica* subsp. *enterica* serovar Enteritidis (commonly known as *S.* Enteritidis) Phage Type 14b associated with a bakery in London the previous month.

On Thursday, the Health Protection Unit sought advice from the head of the Gastrointestinal Disease Unit in the national reference laboratory in Colindale, London. The unit was informed that the bakery outbreak was ongoing in and around London.

By Friday, the first three local (Cheshire) cases had been typed as *S.* Enteritidis Phage Type 14b. Questioning by the Environmental Health Officers identified that four of the first nine cases had eaten at a particular bakery chain, partaking of the same sort of food (cakes and pastries) as the London cases. Cases continued to present to the health services over the weekend.

Food and environmental samples were taken from the bakery and faecal samples from the staff. The use of raw shell egg was identified and Spanish eggs were found on site. There was a potential for cross-contamination of food within the bakery before the products reached the counter. However, such identification did not establish the source of the outbreak beyond reasonable doubt. This needed further investigations.

3.1.1 Challenges

The challenges that the outbreak control team faced at the beginning included:

- The identification of the source. *Salmonella* is not usually associated with a bakery and it took some time to be sure that the bakery chain was the source.
- The lack of appropriate food to sample: eggs and poultry are the usual vehicles of *Salmonella* transmission but were not sold as such in the bakery. Nor was the use of raw eggs expected to be a major issue in a bakery, since bakeries in the UK use pasteurised eggs and outbreaks of *Salmonella* related to eggs in the UK are more usually associated with restaurants and other catering establishments.
- Although the bakery used eggs in its products, the batch that could have been the source of the infections was all used and unavailable for sampling.
- The number of eggs needed for adequate sampling is of the order of 250 per sample, since *Salmonella* is not distributed on every egg. The rate of infection within eggs (indicating infection of the hen's oviduct) is much lower than on the shell, which is more easily contaminated by infected faeces. Of 1744 boxes

of six or more non-UK eggs tested for *Salmonella* in the North West of England and London in 2006, 157 were positive for surface contamination on the shell (weighted prevalence 3.3.% all *Salmonella* species or one in 30 boxes of six eggs infected; *S.* Enteritidis 136 boxes, 2.6%, one in 40 boxes) and 10 were also contents positive (Little *et al.*, 2006).
- Eggs were sourced both within the UK and from European suppliers.
- It was difficult to obtain evidence that linked the bakery and eggs to the patients and that was legally 'beyond reasonable doubt'. This complicated the investigation and made appropriate control measures more difficult to implement.

3.2 Significance of the situation

Salmonella infection is a common cause of gastroenteritis, often resulting in large outbreaks that can affect all ages. *Salmonella* is a bacterium that is transmitted to humans through the consumption of contaminated food, mainly poultry, eggs, meat, and unpasteurised milk. The most likely source of infection through a bakery chain is contaminated eggs. Raw eggs can be infected inside or on the outside of the shell (Pegues and Miller, 2010).

There is a consistent seasonal pattern in *Salmonella* infections, with a peak in late summer (Hawker *et al.*, 2005), which is thought to be related to more rapid multiplication at higher ambient temperatures and, perhaps, seasonal variation in the consumption of raw food. Around 500 cases of *Salmonella* infection are reported every year in the two million people served by the Cheshire and Merseyside Health Protection Unit (Cheshire and Merseyside Health Protection Unit, surveillance data on *Salmonella* reported infections, 2004–2008).

Around 2500 different strains of *Salmonella* have been identified to date. *S.* Enteriditis Phage Type 14b was isolated in this outbreak, being one of a number of group D *Salmonella*. In 2001 in England, out of approximately 11 000 reported cases of *S.* Enteritidis (the commonest type of *Salmonella*) only 393 cases of phage type 14b were identified, making it a rare form of *Salmonella*. Some local laboratories can identify the serogroup, while genotyping needs specialist laboratory work, undertaken in England at the reference laboratory in Colindale, London.

3.2.1 Regulatory aspects

The regulatory implications of the outbreak centred around the advice from the Food Standards Agency concerning the use of raw eggs.

The Food Standards Agency is an independent Government department set up by an Act of the UK Parliament in 2000 to protect the public's health and consumer interests in relation to food. It provides advice and information to the public and Government on food safety from farm to fork. It also protects consumers through effective food enforcement and monitoring.

In this incident, there was a question on the appropriate use of eggs without the Lion mark, which is a mark stamped on eggs laid in farms that adhere to a strict code of practice regulated by the industry. The code includes the vaccination of flocks against *Salmonella*. The Foods Standards Agency does not endorse the Lion mark as it cannot be seen to be promoting or preferring one food company over another.

This raised difficulties, therefore, about the best way to advise on the use of raw and shell eggs. The Foods Standards Agency had already issued its advice on the use of raw eggs and this was freely available (http://www.food.gov.uk/). Within the agencies dealing with the outbreak there were differing outlooks on repeatedly issuing the advice.

In 1999 there were 166 000 officially reported cases of *Salmonella* across the countries of the European Union. In 2000 the number of reported cases had dropped to 150 000. The recommended targets for December 2006 were:

- Layer flocks – reduction in flocks infected with *S. enteritidis* and *S. typhimurium* to 5%,
- Broilers – reduction in all *Salmonella* serotypes to 10%.

European Union directives are enforced to minimize infection in layers and breeders. In addition to general European Union legislation, many countries have enforced their own regulations to reduce *Salmonella* infections in poultry. Regulations are often supported by voluntary codes of practice. See http://www.safe-poultry.com/controllegislation.asp for details of the legislation around *Salmonella* and http://europa.eu/rapid/pressReleasesAction.do?reference=IP/06/1082&format=HTML&aged=0&language=EN&guiLanguage=en for a useful summary.

Regarding European strategies and directives for the control of *Salmonella*, different countries have adopted different control strategies for *Salmonella* with varying degrees of success. The Zoonoses Directive contains:

1. Rules for the monitoring of flocks for *Salmonella* infection,
2. Measures to be taken if breeding flocks test positive,
3. Rules for the monitoring of *Salmonella* at processing and slaughter,
4. Rules for the monitoring of *Salmonella* contamination of feed,
5. Targets for the reduction of zoonoses.

The relevance of *Salmonella* control is also reflected in the EUREP code of practice for poultry (EUREP is the European organization of retailers, started in 1997). The EUREP Good Agricultural Practice code was developed by EUREP as a benchmark for regional and national Integrated Farm Assurance programs. See http://www2.globalgap.org/about.html for further information.

To comply with this program:

1. Broiler flocks must be tested between 21 and 28 days for *Salmonella*.
2. Test all other flocks 10–14 days prior to slaughter.
3. Positive flocks must be slaughtered.

4. Repopulation of the site is forbidden until confirmed as *Salmonella* negative.
5. Hatcheries cannot receive eggs from flocks infected with *Salmonella enteritidis* or *Salmonella typhimurium*. Egg supplies from these flocks must cease immediately and all unhatched eggs must be recovered, destroyed and disposed of appropriately.

Despite all the regulations and codes of practice, international relationships can suffer when one country has an outbreak of an infectious disease that is attributed, rightly or wrongly, to a source within another country. Communication issues and control measures can become very sensitive, particularly when there is pressure for a quick result and preliminary findings are released that may not be confirmed by further testing and analysis. Early action under international media scrutiny and strong political pressure for results, when correct, may lead to a reduction in disease but, when wrong, may lead to bad science and the unnecessary prolongation of an outbreak. Markets and politics affect international trade and relationships, jobs and economies; outbreaks of foodborne disease, such as *Salmonella*, are both susceptible to and a result of these larger forces, and can affect the practice of local health and public health practitioners in unpredictable ways.

3.2.2 Economic and market aspects

A large proportion of the eggs used in the UK are imported, often from Spain or Eastern Europe. Implicating eggs in the two bakery outbreaks raised international questions about European relationships, particularly when the evidence implicating imported eggs was seen as being weak by the regulatory bodies.

In addition to the discomfort in international relationships, the approach of the authorities in the country of origin of the eggs to ensuring bio-security in its poultry is different from that accepted as standard in the UK. The poultry flocks in the source country are not vaccinated: they rely on stringent hygiene precautions to protect their flocks from *Salmonella*. In such a situation, it is easy to think in derogatory terms about the way others regulate their own system, further disrupting relationships. Negotiations need much diplomacy in these circumstances.

3.3 The case history continues

The Principal Environmental Health Officer for the local authority covering the main bakery of the chain talked the bakers through their usage of raw eggs and identified the recent introduction of egg glaze on pastries and buns. Fondant icing was made with raw egg white and used on several items, including iced buns.

The local egg supplier was investigated; they provided non-UK eggs, mainly from Europe; overall, 20% of the eggs supplied came from unvaccinated flocks. As well as the bakery chain under investigation, this egg supplier also supplied eggs to a sandwich shop in a nearby city, where there was an outbreak of *S.* enteriditis Phage Type 1.

Of 80 relevant *Salmonella* notifications in Cheshire, 69 were confirmed as *S.* Enteriditis Phage Type 14b and 64 (41% male; 33% under age 5 years) fitted the case definition. The clinical case definition was 'an individual who experienced an episode of gastrointestinal disease (i.e. diarrhoea (three or more loose stools in a 24 hour period), and/or vomiting, and/or abdominal pain) with a faecal and/or blood isolate of an aerogenic strain of *S.* Enteriditis Phage Type 14b confirmed by the Laboratory of Enteric Pathogens (Colindale, London) since 22nd September'.

All patients fitting the case definition reported diarrhoea, 90% reported abdominal pain, 84% fever, 40% bloody diarrhoea, 59% nausea and 41% vomiting. Most reported cases (96%) had consulted their own doctor and 14% (n = 9) had been admitted to hospital. Food histories showed a wide variety of shops where food had been bought, but most of these were by single patients, except for the bakery chain, which had been visited by 33 (52%) people.

Nationally, there were 224 cases of *S.* Enteriditis with the same genetic fingerprint and phage type (PT14b), with two main foci, one in Cheshire and one in London. A national case–control study was undertaken, with 64 cases and 113 controls being interviewed; power calculations had shown that 60 cases and 120 controls would be sufficient to detect an odds ratio of 3 (50% controls exposed) to 4 (10% controls exposed).

The case nominated controls were interviewed. Controls were excluded if they had a gastrointestinal illness in their household; one case was excluded as a possible secondary case. Further cases and controls were effectively removed from the study as they had no nominated control or case, leaving 47 cases and 99 controls in the national dataset, of which 27 cases were from the Cheshire cluster.

Initial interviews with cases had suggested several hypotheses to be explored in the case–control study: exposure to poultry, eggs, milk, cheese, salads/vegetables, and cakes/desserts. None of these hypotheses were specific to a particular brand or product type, but they were able to indicate a general problem with a particular food type of retail outlet. A single variable analysis was undertaken and any positive (any odds ratio >1 and $p<0.03$) association (e.g. specific foods, food types, retail outlets associated with illness) entered into a multivariate analysis.

All possible two-way interactions between significant exposures were tested and also included ($p<0.01$). All excluded (not significant) exposures were tested against the final model and any strongly significant association ($p<0.05$) was included into the model again. All exposures relating to specific hypotheses were also included to check none had a strong positive association with illness. A regional sub-analysis was undertaken for the Cheshire situation and a further analysis of the national data without the Cheshire cluster.

When using all 47 cases and their nominated controls, a total of 19 exposures showed evidence of an association with the illness in the single variable analysis (eating at a sandwich shop, eating at a baker's shop, eating at a pub, dried milk, custard slice, custard doughnut, cream slice, eating buns, shopping at co-operative, eating cakes, eating custard filled cakes, eating cakes with fondant icing, eating cream-filled dessert, eating out at a pub or hotel, shopping at a mini-market). After removing the non-significant exposures, seven were left in the multivariate analysis (eating at a sandwich shop, eating at a pub, drank dried milk, custard doughnut, shopping at one named multi-national supermarket, shopping at a mini-market, eating buns).

The sub-analysis for Cheshire showed an odds ratios of 11.5 (95% confidence interval 2.6–50.9) for having shopped at the bakery chain, with an odds ratio of 7.2 (2.0–25.9) for having eaten products with fondant icing. Country-wide, significant odds ratios were found for eating at a sandwich shop (4.6; 1.2–11.0), shopping at a local butcher (3.6; 1.2–10.6) and shopping at a local baker (4.9; 1.7–14.2). Two people died during this outbreak, one directly from *Salmonella* infection, the other from unrelated causes.

3.4 Resolution and outcomes

Along with the results of the microbiological analysis, visits to the bakery chain shops in Cheshire and a knowledge of the food chain and the usual provenance of *S.* Enteriditis Phage Type 14b, it was concluded that the Cheshire outbreak was caused by contaminated eggs from Spain that had been distributed through major egg distributors to the bakery that had initiated a promotion through its chain of shops for buns with fondant icing on them. Such use of raw eggs was stopped in the bakery following discussions with the Environmental Health Officers. No new cases were found in Cheshire following that change in the bakery's practice.

3.5 Commentary

Such a complex outbreak as this had several strands in the investigation, locally in Cheshire, in London and in tracing the source of the eggs outside the UK.

The epidemic curve (Fig. 3.1) is typical of a propagated outbreak with a continuing source of infection. This should be compared with the point source curve in Chapter 2 by Stewart *et al.* (2012). Our experience is that the distinction is helpful for a greater understanding of the dynamics of outbreaks and the role of control measures. In propagated outbreaks (whether from an ongoing source or from person to person spread), the priority for the outbreak control team, as in the situation discussed in this chapter, is limiting the spread of infection through appropriate intervention measures.

Drawing an epidemic curve should be almost the first stage of an investigation (once it has been decided that there actually is an outbreak) and, although it appears simple, can give a helpful pointer as to the source of the outbreak (O'Brien, 2006). It may also aid in reviewing the continuation or end of an outbreak but it does not necessarily show the impact of control measures. This is particularly true when the contamination is limited (even if ongoing), such as a few eggs in a batch (the limitation) that are used, as here, to baste a large batch of buns (resulting in an ongoing outbreak since the buns were sold over a number of days).

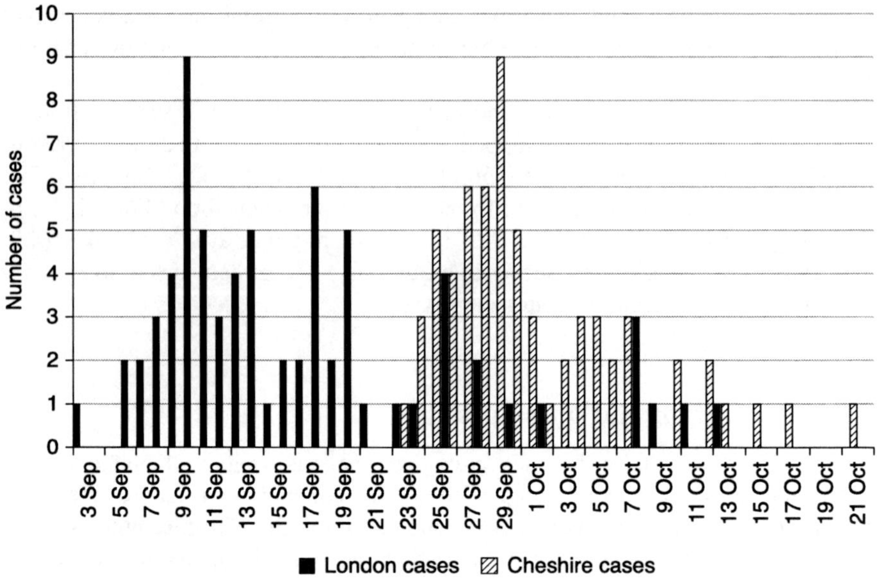

Fig. 3.1 Epidemiological curve, illustrating an ongoing or propagating outbreak (HPA data).

Even when a source farm has an infected flock, not all eggs will be affected, so the resulting outbreak can be patchy in its distribution. However, an understanding of the outbreak and communicating back to the egg producer ensures that appropriate control measures are instigated to reduce the possibility or effect of any future outbreak.

Descriptive epidemiology (age and sex of clinical cases and a review of places relevant (e.g. restaurants and food markets)), preferably with a case–control study or other statistical measures such as a cohort study if all the people exposed are known, further describe the outbreak and point to sources of transmission (Last, 2007, Porta, 2008).

Even with good epidemiology, it can be difficult to find strong evidence linking source to case because of the small number of cases and the range of food items that need to be ruled out. Nevertheless, statistical epidemiology should be performed more often than it is.

The testing of eggs is a difficult practical issue, given the low rate of infection in (or on) eggs and the large number of eggs needing to be tested to obtain a positive result. Samples of 250 eggs are usually needed and, while obtaining this size of sample from wholesalers is possible, end users do not often hold that number of eggs.

3.5.1 Research and development

There is a need to consider one system across the European Union:

- Can biosecurity measures produce the equivalent *Salmonella*-free flocks as does vaccination?
- How can suitable biosecurity measures reduce contamination beyond the farm in a safe and sustainable manner?

3.6 Acknowledgements

Dr Bernard Schlecht (deceased) and other members of the Cheshire and Merseyside Health Protection Team including Professor John Reid and Dr Chris Whiteside, outbreak team in Colindale led by Dr Sarah O'Brien, North West Regional Epidemiologists, Environmental Health Officers of the Cheshire local authorities (particularly Annette Potts and Congleton and Macclesfield District Council officers), Food Standards Agency.

3.7 Critical questions for discussion

1. Compare this propagated outbreak with the point source outbreak in an ethnic community (Chapter 2, Stewart *et al.*, 2012) – what do the epidemic curves indicate about the two situations, what similarities are there, what differences?
2. How important is a multi-disciplinary approach to outbreak management? Who should participate? Who should lead?
3. Should the bakery owners have been part of the outbreak management team? What are the advantages and the disadvantages of including them?
4. How should international relationships be taken into account at a local level in an outbreak situation?
5. In foodborne outbreaks, more than one organism may be present (here *S.* Enteritidis Phage Type 14b, *S.* Enteriditis Phage Type 1). How might this affect an investigation?

3.8 Further reading

BOWLING A., 2002. *Research Methods in Health, Investigating Health and Health Services*, 2nd ed. Open University Press, Buckingham, UK.

DONALDSON L.J. and SCALLY G. (eds), 2009. *Essential Public Health*, 3rd ed. Radcliffe, Oxford, UK.

PEREIRA-MAXWELL F., 1998. *A–Z of Medical Statistics. A Companion for Critical Appraisal.* Hodder Arnold, London, UK.

3.9 References

HAWKER J., BEGG N., BLAIR I., REINTJES R. and WEINBERG J., 2005. *Communicable Disease Control Handbook*, 2nd ed. Blackwell, Oxford, UK.

LAST J.M., 2007. *A Dictionary of Public Health*. Oxford University Press, New York, USA.

LITTLE C.L., WALSH S., HUCKLESBY L., SURMAN-LEE S., PATHAK K. *et al.*, 2006. *Survey of* Salmonella *Contamination of Non-UK Produced Shell Eggs on Retail Sale in the North West of England and London. Final Report – Project B18012*. Food Standards Agency and Health Protection Agency, London, UK. Available at http://www.food.gov.uk/news/newsarchive/2006/nov/eggs.

O'BRIEN S., 2006. Preventing epidemics of communicable disease. In D. Pencheon, C. Guest, D. Melzer and J. A. Muir Gray (eds), *Oxford Handbook of Public Health Practice*, 2nd ed., pp. 206–16. Oxford University Press, Oxford, UK. (A very helpful chapter on outbreak investigation and response.)

PEGUES D.A. and MILLER S.I., 2010. *Salmonella* species, including *Salmonella Typhi*. In G. L. Mandell, J. E. Bennett and R. Dolin (eds), *Principles and Practice of Infectious Diseases*, 7th ed., pp. 2887–903. Churchill Livingstone, Philadelphia, USA.

PORTA M. (ed.), 2008. *A Dictionary of Epidemiology*, 5th ed. Oxford University Press, New York, USA. (Probably the single most useful book – dip in and read – it goes well beyond simple definitions; previously edited by John M. Last, whose 4th edition (2000) is available electronically at http://www.ebook3000.com/A-Dictionary-of-Epidemiology_109179.html.)

STEWART A.G., VIVANCOS R., KEENAN A. and THEODORE C., 2012. A point-source outbreak of *Salmonella* in an ethnic community in the UK. In J. Hoorfar (ed.), *Case Studies in Food Safety and Authenticity*, Ch 2, pp. 11–22. Woodhead Publishing, Cambridge, UK.

4

Investigating a large outbreak of salmonellosis in the US involving fresh produce

M. A. Smith, Food and Drug Administration (FDA), USA and J. Hoorfar, Technical University of Denmark, Denmark

Abstract: Fresh produce is an important part of a healthy diet. However, produce is susceptible to contamination with pathogens. Identifying the food vehicle responsible for an outbreak is critical to minimizing the public health and economic impacts of an outbreak. Identifying the food vehicle and its source may also help prevent future outbreaks. The case study describes an especially challenging investigation of an outbreak of salmonellosis that occurred in the US in 2008.

Key words: fresh produce, tracing, recall, pathogen, outbreak.

4.1 Introduction to the case

The number of illnesses associated with fresh produce is a continuing concern for the US Food and Drug Administration (FDA). Fresh produce presents special challenges. For example, consumption of produce in its fresh form, particularly 'ready-to-eat' products, has increased dramatically during the past decade. This is a positive development from a nutrition perspective, but also a new dynamic that challenges our food safety efforts. Because most produce is grown in an outdoor environment, it is vulnerable to contamination from pathogens that may be present in the soil, in agricultural irrigation or process water and in manure used as fertilizer, and due to the presence of animals in or near fields or packing areas. Produce also may be vulnerable to contamination due to inadequate worker health and hygiene protection, environmental conditions, inadequate production safeguards, and inadequate sanitation of tools, equipment and buildings.

Fresh produce is produced on tens of thousands of farms, and contamination at one step in the growing, packing, and processing chain can be amplified throughout subsequent steps in the chain. The fact that produce is often consumed raw or with only minimal processing, without any type of intervention that would eliminate pathogens (if they are present) prior to consumption, contributes to its potential as a source of foodborne illness. Consequently, addressing the way fresh produce is grown, harvested, packed and moved from field to fork is crucial to minimizing the risk of microbial contamination.

In 1998, FDA issued guidance to industry entitled 'Guide to Minimize Microbial Food Safety Hazards for Fresh Fruits and Vegetables' (FDA 1998). This guide recommends good agricultural practices (GAPs) that growers, packers, and shippers can undertake to address common risk factors associated with the growing, harvesting and packing of fresh produce consumed in the US. The GAPs Guide was based on available, but limited, information. To be most effective, FDA suggested that firms use the general recommendations in the GAPs Guide to tailor practices to their individual operations.

Because the GAPs Guide is guidance, not a regulation, FDA primarily promoted adoption and implementation of the recommendations in the Guide through education and outreach, including providing technical assistance to the US National GAPs Program in developing educational materials for domestic growers and packers, and collaborating with the Joint Institute of Food Safety and Nutrition (JIFSAN) in developing GAPs train-the-trainer programs for international audiences.

Additional activities to address produce safety concerns include: working with industry, academia, and other Federal, State and foreign agencies to develop guidance on ways to prevent or minimize potential contamination, outreach to consumers on safe food handling practices, sampling and analyzing both domestic and imported produce for pathogens, and working with industry and foreign countries to promote the use of good growing, harvesting, packing, transportation, and processing practices.

4.1.1 How did it all begin?

On 22 May 2008, The New Mexico Department of Health in the US notified the Centers for Disease Control and Prevention (CDC) about 19 human cases of *Salmonella* infections. Seven isolates from the patients were identified as *Salmonella enterica* serotype Saintpaul, mostly with the same pulsed field gel electrophoresis (PFGE) pattern (see Chapter 10, Jordan *et al.* 2012 for PFGE typing).

In the same month CDC identified three additional *Salmonella* isolates with the same PFGE pattern from Colorado and Texas. By September 2008 a total of about 1500 human cases of serotype Saintpaul in 43 states, the District of Columbia and Canada were confirmed by laboratory testing. Among the 1500 case subjects, 21% were hospitalized, and infection may have contributed to two deaths.

On 26 May 2008, CDC informed FDA of the hypothesis of a possible association between ill persons and the consumption of raw tomatoes. On 31 May, CDC formally notified FDA of a significant statistical association between consumption of certain types of tomatoes and a multi-state outbreak of *Salmonella* Saintpaul infections. In response, FDA issued numerous public advisories to protect consumers from this rapidly expanding outbreak. FDA initiated investigations attempting to trace the tomatoes reported to have been eaten by ill persons back to their sources.

4.2 Significance of the case

The CDC estimates that 48 million Americans become ill as a result of foodborne pathogens each year, 128 000 are hospitalized, and 3000 die. *Salmonella* is the leading bacterial cause of gastroenteritis in the US, and the incidence has reduced little in the past decade. This outbreak was one of the largest *Salmonella* outbreaks ever identified in the US.

Between 1973 and 1997, outbreaks of foodborne illness in the US associated with fresh produce increased both in absolute numbers and as a proportion of all reported foodborne outbreaks. These outbreaks were caused by a wide variety of pathogens, including bacteria (e.g. 20 different serotypes of *Salmonella*, *E. coli* O157:H7, non-0157:H7 STEC's O145, O111, O104:H4), viruses (e.g., Hepatitis A and Norovirus), and parasites (e.g. *Cyclospora cayetanensis*).

4.2.1 Regulatory aspects

FDA is the Federal agency that regulates 80% of the US food supply: almost everything Americans eat except for meat, poultry, and processed egg products, which are regulated by FDA partners at the United States Department of Agriculture (USDA). FDA and USDA are committed to ensuring that America's food supply continues to be among the safest in the world.

4.2.2 Market and economic issues

In the US, produce is grown on approximately 180 000 farms, representing over 7 million acres. Produce production occurs throughout the US, with the largest acreage in California, North Dakota (primarily potatoes and pulse crops), Idaho, Michigan, Minnesota, Washington, and Wisconsin. California, Florida, Georgia, Arizona, and New York harvest the largest acreage for the fresh market.

While not specific to produce farms, nearly all farms (97% in 2006) are family farms. Most farms have only one operator. About 40% of farms have more than one operator, but three-quarters of these are operated by a husband–wife team. Approximately 90% of fresh vegetable farms cultivate approximately 11% of the total fresh vegetable acreage. Relatively few farms account for the majority of commercial produce production.

Organic production, value-added (fresh-cut) products, and specialty crop sectors are experiencing rapid growth. In addition, consumers are increasingly looking to buy locally grown produce, with the number of operating farmers markets growing at an average of 15.4% per year. Consumer demand for year-round access to a wide variety of produce items, international trade agreements, and improved technologies have led to substantial growth in the volume and variety of US fresh fruit and vegetable imports.

Imported foods make up approximately of 10% of retail food sales, with fruits, juices, nuts, and vegetables comprising approximately 30% of imported food products. Fresh vegetable imports are largely from the North American Free Trade Agreement (NAFTA) region (Mexico and Canada). US fruit trade has more diverse foreign trade partners compared with vegetable trade. Mexico has been the dominant supplier for US fresh vegetable imports for several decades and, since the establishment of NAFTA, imports from that country have further increased.

Issues such as those described in the current case can have a profound economic impact, not just for the producer or packer of the implicated product, but for the entire commodity sector. A large foodborne illness outbreak associated with consumption of fresh produce can have a long-term effect on consumer confidence and a negative impact on the entire produce industry. Estimates of the economic cost of the current case to the tomato industry in Florida alone have been more than $100 million and in Georgia close to $14 million.

4.3 Background information

4.3.1 Conducting a foodborne illness outbreak investigation

A brief description of the typical stages of the outbreak investigation process in the US follows. These stages may overlap: product tracing may start while case–control studies are still ongoing; information from either may spur follow-up interviews with cases or revisions to food history questionnaires. At any point in the investigation, FDA may determine that there is insufficient information to go further and the identification of the food vehicle, or its source, may be deemed 'inconclusive'.

Recognizing an outbreak

CDC, State and local health departments, and other agencies collaborate to compile and track illness data. Data include case findings, such as laboratory isolates from patients reporting illness and PFGE fingerprinting of the isolate. CDC traditionally defines a foodborne outbreak as two or more cases of a similar illness associated with a common food.

Identifying the food vehicle

Local health department(s) interview patients or cases, using a standard questionnaire to collect consumption and product information; food histories are reviewed for common features. All parties may work collaboratively to refine questionnaires or conduct case studies. A food vehicle is confirmed if there is a significant epidemiological association linking the pathogen with a food or the pathogen is recovered from a food. In the absence of a confirmed food vehicle, a food may be suspect based on a case series where the food was consumed by a majority of cases or the food has a history of association with the pathogen. If one or more foods are suspect, but not confirmed, tracing suspect food(s) back through the supply chain may provide information that could confirm (or rule out) the food as a vehicle in the outbreak.

Once CDC, through its epidemiological investigation with State and local governments, identifies the possible food(s) associated with a foodborne illness outbreak, CDC notifies FDA. At that point, FDA starts a product-tracing investigation, working with industry and with local, State, and Federal officials and, when needed, with foreign governments, to identify the source of the food.

FDA does this by tracing the suspected food(s) back through the supply chain from point of purchase or consumption, and inspecting or investigating points throughout the supply chain to determine where the contamination most likely occurred. Tracing food through a supply chain requires FDA to find and examine documents such as bills of lading, invoices, and other records maintained by the firm. FDA also obtains information on the practices and conditions under which the product was stored and handled at each point to better determine shipments of interest and whether contamination may have occurred at each point. If FDA or State partners determine that contamination most likely occurred during preparation in a restaurant, for example, there is no need to trace product further back in the supply chain.

Product-tracing investigations involving fresh produce are especially difficult because the food is perishable and is usually no longer available for testing by the time consumers become ill and an outbreak is identified. In addition, fresh fruits and vegetables may be sold loose or otherwise without packaging that could provide information about its source. Further, practices such as commingling, packing, and repacking produce from multiple sources add complexity to product-tracing process investigations. Each product-tracing investigation is different, presenting different degrees of difficulty.

In contrast to the current case, in 2006, CDC informed FDA of a multi-state outbreak of illnesses associated with the consumption of fresh spinach contaminated with *E. coli* O157:H7. Although this outbreak involved a perishable food, the food was sold in a package. Several patients still had packages of spinach in their refrigerators, making it easier to obtain information about the identity and source of the product. In addition, packaged spinach was sampled and found to match the outbreak strain. Investigators were able to identify the processor who produced the implicated product and to narrow the focus of the investigation to the specific production lot of product that had been responsible for the outbreak. By looking at the processor's records, the investigators were able to identify the farms that supplied product associated with the identified production lot of bagged spinach. This is an example of product tracing of medium complexity that was aided by the information on the package.

Identifying the source of contamination
The seasonal production of fresh produce, compared with continuous food processing, for example, and challenges associated with product tracing during outbreaks, including the passage of time between the contamination event and when FDA conducts an investigation, limit what investigators are able to see with respect to potential causes of contamination and the conclusions FDA can draw. By the time FDA traces product to a packing facility, it may no longer be in operation; if FDA traces product to one or a few fields, growing operations may be finished for the season and the fields ploughed under.

Between 1998 and 2006, FDA conducted 40 produce farm investigations (20 domestic, 20 foreign) in response to outbreaks and positive sample findings. During these investigations, FDA looked for contributing factors to the outbreaks and positive sample findings. Key areas of potentially unsafe farm practices that FDA observed during these inspections include: worker health and hygiene issues, animal management issues, poor quality irrigation water and/or water used for pesticide/herbicide applications, sewage disposal issues and problems related to cleanability and cleanliness of harvest tools and equipment.

4.3.2 What are the challenges involved in outbreak investigations?

Foodborne illness outbreak investigations, especially for outbreaks associated with fresh produce, reflect a number of challenges that affect the conclusions FDA can draw from these investigations. These challenges include:

- Most cases of foodborne illness are unreported and some outbreaks are undetected.
- Some outbreaks are recognized by State or local health departments but never reported to CDC.
- Even when an outbreak is recognized, the food vehicle remains undetected in approximately half of reported outbreaks.
- As discussed above, the short shelf-life of fresh produce, lack of available packaging, labeling and complex distribution systems can hinder efforts to trace an outbreak to its source. In many outbreaks, it has not been possible to trace product back to a single source.

The *Salmonella* Saintpaul outbreak investigation, which initially focused on certain types of raw tomatoes, provides an example of one of the most challenging investigations of foodborne illness associated with fresh produce FDA has ever faced.

4.4 The case history continues

As the outbreak and the investigation continued, data were not supporting tomatoes as the vehicle. FDA conducted a total of 12 tomato tracebacks based on both sporadic cases and clusters of patients who ate at the same restaurants. Those tracebacks did not converge on any single geographic area, grower or distributor. In addition, extensive sampling of domestic and imported tomatoes, and environmental swabs from tomato operations in Florida and Mexico (over 1700 samples), did not detect *Salmonella*.

Due to early implication of raw tomatoes in the outbreak, FDA initially advised consumers not to eat Roma and red, round tomatoes. In an attempt to limit the impact on the tomato industry, FDA included lists of sources of tomatoes that were not implicated, including tomatoes from regions where production had not yet started when the outbreak began and that, therefore, could not have been the source of illness. Despite this, as mentioned in Section 4.2.2, the economic cost to the tomato industry has been estimated to have been high. Following the advisories, many restaurants stopped serving raw tomatoes, and others switched from using raw to canned tomatoes in dishes such as salsa. However, the outbreak continued.

4.5 Resolution and outcomes

After further investigation, an association between illnesses and eating Mexican-style foods began to emerge. All indications, taken together, indicated that an unidentified food vehicle might be involved in this outbreak. Subsequent epidemiologic and microbial evidence implicated jalapeño and Serrano peppers. Ultimately, the outbreak strain was found in a jalapeño pepper at a Texas importer facility, in a water source used for drip irrigation and crop spray, and in Serrano peppers at Farm B in Mexico. The Colorado Department of Health also recovered the outbreak strain in a jalapeño pepper from the home of a case subject. The jalapeño pepper was traced back to a distributor but FDA was unable to identify a farm.

4.5.1 Addressing the challenges of epidemiological investigations for FDA, CDC and State public health agencies

Speed is essential for identifying the food vehicle in an outbreak and pinpointing its source to minimize public health impact. Improvements in product-tracing systems and the ability of the systems to work together are needed for more rapid tracing of implicated products through the supply chain.

Epidemiologists from CDC said that better traceability, better understanding of how foods become contaminated on farms, and continued exploration of alternative hypotheses during an outbreak are necessary to reduce the impact of such occurrences in the future (Behravesh *et al.*, 2011).

This outbreak investigation highlights the recurring challenges of epidemiologic identification of ingredients in foods that are commonly consumed, rapid identification and investigation of local clusters, and produce tracing in the supply chain.

Following this outbreak, high-level FDA and CDC officials met to examine internal improvements, especially to address the challenges of epidemiological investigations and product tracing for multi-ingredient foods. FDA and CDC also met with academic and industry experts on product tracing, including an exercise to conduct a commercial-scale mock tomato traceback. FDA held two public meetings in 2008 to seek information on procedures to facilitate produce tracing and a joint public meeting in 2009 with USDA on tracing all food products. In February 2011 FDA, CDC and State agencies met to discuss improving procedures for complicated, multi-state outbreak investigations.

This outbreak significantly informed key provisions of the Food Safety Modernization Act (FSMA), signed into law on 4 January 2011. FSMA directs FDA to initiate rulemaking on product tracing and preventive controls for manufactured/processed foods, and establish standards for the growing and on-farm packing of produce. FSMA also contains provisions for Federal, State and local capacity building and collaboration.

4.6 Commentary and personal reflections

This outbreak was complicated by two significant factors: the presence of multiple ingredients in foods, like salsa, that were connected with illnesses, and the complexity of the tomato distribution chain, which involves repacking of tomatoes at repack and distribution centers throughout the country where they were or might have been commingled with tomatoes from many different sources. This commingling has the potential to multiply the quantity of food that is associated with contaminated product. It also increases the difficulty in determining which tomatoes were the sources of the illnesses.

A further complicating factor was poor record keeping and unconnected records in the distribution chain. For example, one party might describe the tomatoes as 'hothouse' or 'greenhouse' tomatoes while the next party in the chain might describe the same tomatoes simply as 'tomato bulk'. Another party might use a descriptor such as 'green six-by-six'. This inconsistent nomenclature made it more difficult and more time-consuming to connect the links in the chain and to identify the source of the tomatoes. It also made it difficult to determine whether lack of convergence was due to a misidentified food vehicle or simply poor records.

Similar cases around the world, VTEC 104-H4 outbreak in Europe being the latest (see Chapter 5, Goodridge *et al.*, 2012), point towards imported produce and berries (see Chapter 13, Elviss and Jørgensen, 2012) as an emerging source of human gastroenteritis. Since consumption of minimally processed or raw foods, especially those produced in low-cost countries, is increasing, stronger national regulation and local inspection are required.

The European Union has conducted several tracing studies of food products. These were developed as part of a 5-year project that began in 2007, in which tools and models to improve the tracing of accidental and deliberate microbial contamination of food and feed products were constructed (Hoorfar *et al.*, 2011). The goal of the project was to put in place systems enabling the total tracing of a food or feed product along the whole chain from production to consumption, should an outbreak occur. As part of this effort, the EU is developing, testing, and evaluating two full pilot product-tracing systems, including one for the tomato food chain.

Research is also a critical element of FDA efforts to improve the safety of fresh produce. FDA's current research agenda includes improving the identification and detection of disease-causing bacteria and toxins in a variety of foods. More rapid and precise testing methods to identify contaminants are important for detecting contamination if it is present and minimizing the spread of foodborne disease once it occurs. In addition, FDA is working with academia, industry, other Federal agencies, and State governments to develop both risk-based microbiological research programs and technology transfer programs to ensure that the latest food technology reaches the appropriate end users along the supply chain.

4.7 Critical questions for discussion

1. What are the main sources of contamination of raw vegetables?
2. Which pathogens other than *Salmonella* can contaminate raw fruits and vegetables?
3. What is the source of irrigation water for produce production in your country, and what does the law say about the quality control of the irrigation water?
4. What is the main reason for using case–control studies for identifying foods suspected to be vehicles in an outbreak?
5. What is the difference between tracing suspect food and identifying the cause of contamination?

4.8 Further reading

P.S. MEAD, L. SLUTSKER, C. DIETZ, L.F. MCCAIG, J.S. BRESEE *et al.* (2000). Food-related illness and death in the United States. *Journal of Environmental Health*, **62**(March), 9–18.

R.J. MODY, S.A. GREENE, L. GAUL, A. SEVER, S. PICHETTE *et al.* (2011). National outbreak of *Salmonella* Serotype Saintpaul infections: Importance of Texas restaurant investigations in implicating jalapeño peppers. *PLoS One*, **6**(2), e16579.

S. SIVAPALASINGAM, C.R. FRIEDMAN, L. COHEN, and R.V. TAUXE (2004). Fresh produce: A growing cause of outbreaks of foodborne illness in the United States, 1973 through 1997. *Journal of Food Protection*, **67**(10), 2342–53.

Produce Safety Alliance at Cornell University (2011). Providing fundamental, science-based, on-farm food safety knowledge to fresh fruit and vegetable farmers, packers, regulatory personnel and others interested in the safety of fresh produce. Available at http://producesafetyalliance.cornell.edu/psa.html.

4.9 References

C.B. BEHRAVESH, R.K. MODY, J. JUNGK, L. GAUL, J.T. REDD *et al.* for the *Salmonella* Saintpaul Outbreak Investigation Team (2011). 2008 Outbreak of *Salmonella* Saintpaul infections associated with raw produce. *N Engl J Med*, **10**(364), 918–27.

N.C. ELVISS and F. JØRGENSEN (2012). Investigations prompted by the discovery of high levels of *Salmonella* in fresh herbs in the UK. In J. Hoorfar (ed.), *Case Studies in Food Safety and Authenticity*, Ch 13, pp. 116–122. Woodhead Publishing, Cambridge, UK.

L.D. GOODRIDGE, J.T. LEJEUNE, and L. BEUTIN (2012). What was the source of the 2011 outbreak of *Escherichia coli* in Germany and France? In J. Hoorfar (ed.), *Case Studies in Food Safety and Authenticity*, Ch 5, pp. 43–53. Woodhead Publishing, Cambridge, UK.

J. HOORFAR, M. WAGNER, K. JORDAN, S.L. BOUQUIN, and J. SKIBY (2011). Towards biotracing in food chains. *International Journal of Food Microbiology*, **145**: S1–S4.

JIFSAN (2002). *Improving the Safety and Quality of Fresh Fruits and Vegetables: A Training Manual for Trainers*. Available at http://www.jifsan.umd.edu/training/gaps_manual.php.

K.N. JORDAN, K. HUNT., and E.M. FOX (2012). Troubleshooting the environmental source of contamination with *Listeria monocytogenes* in a typical small food manufacturing plant in Ireland. In J. Hoorfar (ed.), *Case Studies in Food Safety and Authenticity*, Ch 10, pp. 95–101. Woodhead Publishing, Cambridge, UK.

US FDA (1998). *Guide to Minimize Microbial Food Safety Hazards for Fresh Fruits and Vegetables*. Available at http://www.cfsan.fda.gov/~acrobat/prodguid.pdf. Accessed 31 October 2011.

5

What was the source of the 2011 outbreak of *Escherichia coli* in Germany and France?

L. D. Goodridge, Colorado State University, USA, J. T. LeJeune, The Ohio State University, USA and L. Beutin, Bundesinstitut für Risikobewertung (BfR), Germany

Abstract: Shiga toxin-producing *Escherichia coli* (STEC) O104:H4 is a very rare cause of foodborne illness. In the European Union, sporadic cases have been previously linked to travel to North Africa, the Middle East and Central Asia. In Europe there have been cases of STEC O104:H21, which have led to the identification of cattle as an animal reservoir of this pathogen. In North America, an outbreak of STEC O104:H21 was linked to consumption of milk contaminated post-pasteurization. This case study summarizes the epidemiologic and laboratory investigations of the largest outbreak of *E. coli* O104:H4 that occurred in Germany and France during the summer of 2011 and discusses its implications. This outbreak became the deadliest outbreak of STEC ever recorded.

Key words: outbreak, fresh produce, emerging pathogen, tracing, Shiga toxin-producing *Escherichia coli* (STEC), enteroaggregative *E. coli* (EAggEC), diagnostics.

5.1 Introduction to the case

Shiga toxin-producing *Escherichia coli* (STEC) are important agents of foodborne disease worldwide. While STEC serotype O157:H7 is recognized as causing many outbreaks of foodborne disease, more than 400 other serotypes of STEC, called non-O157 STEC, have been isolated from human patients (Scheutz and Strockbine 2005).

STEC O104 is a rare serogroup and, while several O104 serotypes, such as O104:H2, O104:H7 and O104:H21, have been isolated from humans, outbreaks caused by these bacteria are rare. Several sporadic cases of STEC O104:H4 have been reported, including two cases of hemolytic uremic syndrome (HUS) in Germany in 2001 (Bielaszewska *et al.* 2011), one isolate from France in 2004, and one from a case of HUS in South Korea in 2005 (Kim *et al.* 2011). In 2009, STEC

O104:H4 was isolated from two patients in the Republic of Georgia, who were part of a cluster of diarrheal illness. No food was identified as a source of the infections (CDC 2011).

Sporadic cases with STEC O104:H4 in the EU have been previously linked to travel to North Africa, the Middle East and Central Asia. In the US, an outbreak of acute gastroenteritis caused by STEC O104:H21 occurred in Helena, Montana, in 1994. Eleven confirmed and seven suspected case patients were identified as part of this outbreak, which was linked to consumption of milk contaminated post-pasteurization (CDC 1995).

Although most outbreaks of STEC are sporadic in nature or limited to a few or several hundred people, depending upon the vehicle, several extremely large outbreaks have occurred. For example, in July 1996, the largest outbreak of STEC infection ever reported occurred in Sakai City, Japan. The outbreak caused over 7000 illnesses and was traced to *E. coli* O157:H7-contaminated white radish sprouts served through a centralized lunch program (Mermin and Griffin 1999). In May 2000, the largest waterborne outbreak of STEC illness was recorded. An outbreak of *E. coli* O157:H7 in the town of Walkerton, Ontario, Canada led to 2300 illnesses and seven deaths. The outbreak was traced to the town's drinking water supply, which was not properly chlorinated, and had become contaminated with agricultural runoff during an extreme rain event (O'Connor 2002).

While the historical data associated with foodborne outbreaks related to STEC O104 indicate that these outbreaks are sporadic with a small number of cases, during the summer of 2011 a related serotype, O104:H4, was associated with an outbreak of STEC illness with the highest fatality rate ever recorded. The Robert Koch Institute (RKI) in Germany first reported an ongoing outbreak of STEC serotype O104:H4 on 23 May 2011. From a case–control study, the outbreak was initially associated with the consumption of fresh salad vegetables.

This outbreak is noteworthy due to several novel aspects of the bacterial isolate implicated in the outbreak, and represents what may be viewed as a seminal moment in the study of foodborne disease outbreaks.

5.2 Significance of the case

This outbreak led to a total of 3816 confirmed cases, including 845 cases of HUS and 54 deaths (Frank *et al.* 2011). The number of deaths that occurred as a result of the outbreak makes it the deadliest known outbreak of STEC illness on record. The significance of the outbreak in terms of the emergence of foodborne pathogens with combined virulence profiles is discussed in Section 5.5.

5.3 The case continues

Initial indications suggested that the outbreak was linked to the consumption of fruits or vegetables, including cucumbers, tomatoes and leafy green salads, and it was recommended that these foods were not consumed. Subsequent investigations showed that the risk of infection was specifically associated with the consumption of fresh sprouted seeds. Further epidemiological analysis indicated that most illnesses could be attributed to consumption of sprouted seeds produced by a single producer in Germany. While investigation of the production site showed no evidence of environmental contamination, several employees at the sprout farm were found to be infected. Specifically, three employees had diarrhea, and one tested positive for STEC O104:H4. There were also two asymptomatic infected employees, and both tested positive for STEC O104:H4. As it was unknown how long the asymptomatic infected employees had carried the strain, it is theoretically possible that they contaminated the production without knowing and without becoming sick (i.e. showing symptoms). Still, because they had not become ill prior to the outbreak, it was concluded that the employees were not the source of the contamination, and that contaminated seeds used for the sprout production were the most likely source. Subsequently, another cluster of patients with bloody diarrhea was reported, after having participated in a community event in France on 8 June 2011. The seeds in France were from the same lot of fenugreek seeds (15 tons) that was incriminated as the possible source. However, the French seeds were obtained via a UK trader whereas the German seeds were from a German importer. The French outbreak was based on a single batch of 50 g of seeds that were taken from a private sprouting at a single home. Consumption of sprouted seeds was also associated with occurrence of the disease in this cluster. Furthermore, the STEC isolates responsible for the outbreaks in France and Germany were found to be indistinguishable. It was therefore concluded that there was a common source for both outbreaks. A specific lot of fenugreek seeds imported from Egypt was the most likely link between the outbreaks, although it could not be excluded that other lots from the same exporter and importer were also implicated.

5.3.1 Clinical and diagnostic characteristics of the outbreak strain

Polymerase chain reaction (PCR) testing of the outbreak strain showed that it contained the *aggR* gene, which is a typical characteristic of enteroaggregative *E. coli* (EAggEC). Continued analysis indicated that the strain was positive for *aatA*, *aap*, *aggR*, *aagC*, and *aggA*. These genes were present on the EAggEC virulence plasmid. Both *aagC* and *aggA* are genes responsible for fimbrial adhesion, with *aggC* responsible for control of fimbrial gene expression, while *aggA* encodes an AAF/I fimbrial subunit (Anonymous 2011, Scheutz *et al.* 2011).

Furthermore, the strain was shown to contain a variant of Shiga toxin (stx) 2, stx2a (Scheutz *et al.* 2011). Sequence analysis of the stx2a sequence indicated that it had 100% predicted homology at the amino acid level to stx2a from *E. coli* O157:H7 strain EDL933. At the nucleotide level, the sequence from the outbreak strain was found to differ by only one nucleotide at position 867 (C instead of T), making it identical to *E. coli* O157 strains isolated in Germany in 2002 and 2005, *E. coli* O121:H19 isolates from Canada and the US (Idaho), and *E. coli* O111:H8 (US, Idaho) (Scheutz *et al.* 2011).

Table 5.1 Virulence factors associated with the STEC O104:H4 outbreak strain

Virulence factor	Positive/negative	Function
Shigatoxin 1*	Negative	Shigatoxin 1 gene
Shigatoxin 2a†	Positive	Shigatoxin 2 gene
Intimin (*eae*)*	Negative	Attaching and effacing phenotype
Enterohemolysin*	Negative	Hemolysin protein gene
Saa†	Negative	STEC associated adhesion gene
subAB†	Negative	Cytotoxin subtilase gene
*aatA**	Positive	ABC transporter protein gene
*aggR**	Positive	Master regulator gene of Vir-plasmid genes
*aap**	Positive	Secreted protein dispersin gene
*aagA**	Positive	AAF/I-fimbrial subunit gene
*aagC**	Positive	AAF/I-fimbrial operon gene
iutA†	Positive	Aerobactin receptor gene of extraintestinal pathogenic *E. coli*

* Anonymous 2011, †Scheutz *et al.* 2011.

The virulence factors of the *E. coli* O104:H4 outbreak strain are summarized in Table 5.1.

Collectively, these data demonstrate the EAggEC/STEC hybrid nature of the outbreak strain, which appears to be a member of the EAggEC that presumably acquired the stx2a gene via a bacteriophage. These characteristics also explain the pathogenicity of the outbreak strain. Increased adherence of the outbreak strain, via the AAF/I fimbriae, could have facilitated systemic adsorption of stx2a and could explain the high numbers of patients with HUS.

In addition, the *E. coli* O14:H4 outbreak strain is resistant to 14 antibiotics, including ampicillin, amoxicillin/clavulanic acid, piperacillin/sulbactam, piperacillin/tazobactam, cefuroxim, cefuroxime-axetil, cefoxitin, cefotaxim, cetfazidim, cefpodoxim, streptomycin, nalidixinsaure, tetracycline, and trimethoprim/sulfamethoxazol (Anonymous 2011). The strain was also positive for CTX-M-15, which is an extended spectrum beta-lactamase (ESBL), and contained the TEM-1 beta-lactamase (Anonymous 2011).

An agglutination assay can be used for quick detection of *E. coli* O104 isolates, using K9 antiserum, because the O104 O antigen is identical to the K9 capsular antigen (Orskov and Orskov 1984). The serotype can then be confirmed as O104:H4 using conventional O and H antisera (Scheutz *et al.* 2011). Biochemical analysis showed the outbreak strain to be a typical *E. coli* isolate, as it was lactose positive, sorbitol fermenting, and beta-galactosidase positive (Scheutz *et al.* 2011). The *E. coli* O104:H4 strain was also demonstrated to be capable of producing biofilms when grown in Dulbecco's minimum essential medium (DMEM) modified by addition of 0.45% glucose, which is a defining feature of EAggEC strains (Scheutz *et al.* 2011).

The bacterium was detectable using the standard EAggEC clumping test, indicating the presence of the AAF/I fimbriae (Beutin 2011). The strain also grew on Chromagar STEC and O104 agars, where it produced mauve colonies (Beutin

2011). The outbreak strain also produces light red colonies on cefixime tellurite sorbitol MacConkey (CT-SMAC) agar (Scheutz *et al.* 2011). On Brilliance ESBL agar, colonies of the outbreak strain appear black (Beutin 2011) and plating clinical samples on other ESBL media including tryptone bile X-glucuronide (TBX) medium allows growth of the outbreak strain while inhibiting the majority of other *E. coli* (Scheutz *et al.* 2011).

5.3.2 Identifying the vehicle of contamination

One of the unique aspects of this outbreak was the controversy that arose regarding the food source of the bacterial contamination. While sprouts have been identified as the most likely source of the outbreak, they were not initially suspected. As of the beginning of May 2011, a significant increase in the number of enterohemorrhagic *E. coli* (EHEC) cases was reported to the Robert Koch Institute (RKI). On 20 May, the RKI began an outbreak investigation and determined that affected patients had consumed raw tomatoes, cucumbers and lettuce more often than healthy study participants. On 2 June, 2842 samples of cucumber, tomatoes, and leafy salads were examined. All of the samples were negative for STEC O104, although two cucumber samples yielded isolates of STEC O8:H19, which was unrelated to the outbreak isolate.

A variety of foods of animal origin and fresh produce, including sprouts, had previously been suspected during the initial investigation conducted by the RKI on 20 and 21 May, but at that time only three of 12 patients responded that they had consumed sprouts. Therefore, sprouts were initially rejected as the vehicle of contamination. On 5 June, further details emerged leading to the suspicion that sprouts originating from a farm in Lower Saxony were the vehicle of contamination. Subsequently, on 10 June, consumers were issued a warning not to consume sprouts, while tomatoes, salad and cucumbers were no longer considered as suspected vehicles.

Almost simultaneously, on 24 June, a small outbreak of STEC O104:H4 near Bordeaux, France was identified. Fenugreek sprouts were suspected as the vehicle in this outbreak, and the isolates from the French and German outbreaks were determined to be identical. Subsequently, on 29 June, the European Food Safety Authority (EFSA) confirmed the initial warning to the public regarding sprouts and seeds as the vehicles of contamination, and concluded that the outbreaks in both France and Germany were caused by fenugreek seeds produced in Egypt.

5.3.3 The search for the source of the outbreak strain

While the official source of the outbreak has been determined to be fenugreek seeds produced in Egypt, it is important to note that none of the seed samples tested positive for STEC O104:H4, leading to questions regarding whether the seeds were the true source of the outbreak.

One interesting aspect of the STEC O104:H4 outbreak is the fact that no animal reservoirs of STEC O104:H4 strains have been identified to date. This raises additional questions about how the sprouts became contaminated. For example, Cassar *et al.* (2004) determined the prevalence of EAggEC in the UK carried by healthy animals at slaughter. Twelve hundred and twenty-seven *E. coli* isolates (401 from cattle, 406 from sheep and 400 from pigs), selected at random, were probed with an EAggEC-specific probe. The results indicated that none of the isolates tested was positive via the probe-based method, showing that none of the *E. coli* isolated in this study were EAggEC.

Uber *et al.* (2006) determined the phenotypic and genotypic relationships among 86 EAggEC strains of human and animal (calves, piglets and horses) origin. Strains harboring *aggR* (typical EAggEC) of human origin were isolated, while the strains of animal origin were classified as atypical EAggEC (strains lacking *aggR*). The authors concluded that the sampled animals may not represent a reservoir of human pathogenic typical EAggEC.

Recently, Martin and Beutin (2011) did not find STEC O104:H4 strains among 593 serotype STEC strains isolated from food of animal origin. Finally, the investigation of more than 300 STEC strains isolated from food samples of animal origin (maintained in the culture collection at the National Reference Laboratory for *Escherichia coli* of the Bundesinstitut für Risikobewertung (BfR)) for the presence of the aggR gene (indicative of EGggEC strains) showed negative results (L. Beutin and Y. Burgos, 2011 unpublished data).

Since *E. coli* O104:H4 is considered to be an EAggEC strain that has acquired stx2a, these studies show that pathogenic EAggEC strains (including O104:H4) to date have only been associated with human sources.

The lack of an animal reservoir for STEC O104:H4 may indicate a human source of infection. In the current outbreak, three employees who worked at the incriminated sprout farm in Lower Saxony tested positive for STEC O104:H4. One of the employees subsequently developed HUS, while the other two employees were asymptomatic shedders of STEC O104:H4. Therefore, while no microbiological data are available for the fenugreek seeds, such data suggest that the sprout farm may have been the source of the contamination. It is possible that, during the humid and warm environment of sprout production, the infected employees may have inadvertently contaminated the sprouts, with the intrinsic and extrinsic properties of the sprouts facilitating pathogen growth. Figure 5.1 details how both sources (fenugreek seeds and the sprout farm) may have contributed to the outbreak.

5.4 Resolution and outcomes

In July 2011, the European Commission (EC) announced an emergency ban on the import of fenugreek and certain seeds, sprouted seeds and beans imported from Egypt until 31 October 2011, following the two outbreaks of *E. coli* O104:H4 in Germany and France. As of 11 October, fresh and chilled peas and beans were removed from the list. The EC reassessed the risk from these products following an audit by officials of production sites in Egypt.

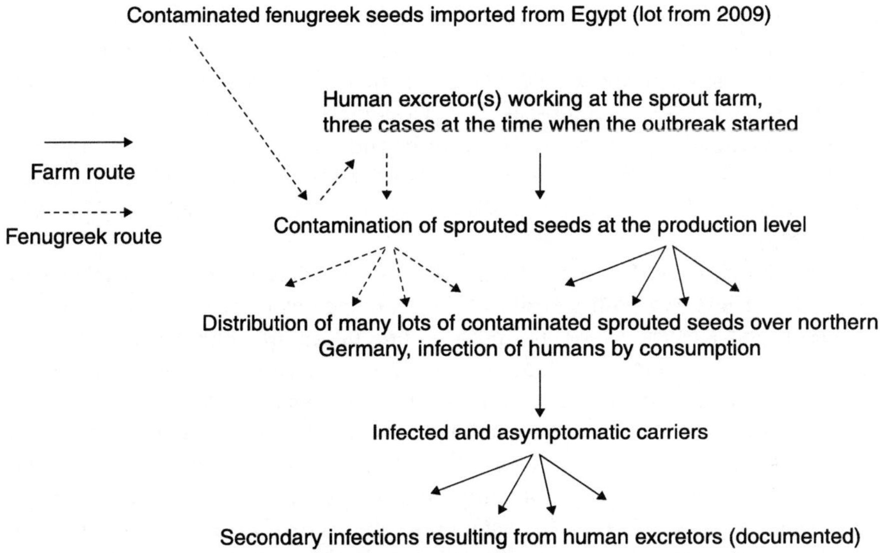

Fig. 5.1 Potential sources and routes of STEC O104:H4 contamination.

At the time of writing, following a request from the EC, EFSA's Biological Hazards Panel is carrying out a risk assessment on the EU production chain for sprouts and sprouting seeds. In addition, the EC has informed EFSA that the EU Member States have now completed tracing activities across the food chain with respect to foods suspected as the vehicle of the outbreak. With the removal from the market in all Member countries of the most likely source of the contaminated food, coupled with ongoing importation restrictions, EFSA revised its consumer advice on the consumption of sprouts and sprouting seeds. Specifically, EFSA is no longer advising consumers to cease growing sprouts for their own consumption and is also no longer advising against consumption of sprouts or sprouted seeds unless they have been cooked thoroughly.

As of 11 October 2011, EFSA recommends that consumers refer to national food safety agencies for any specific advice regarding sprout consumption. Consumers are reminded to follow good food safety practices when preparing and consuming fresh vegetables, including washing hands before food preparation, washing food properly under running potable water and separating raw foods from ready-to-eat or cooked foods to avoid cross-contamination.

5.5 Commentary and personal reflections

This outbreak demonstrates that hybrid virulence profiles, in which the outbreak strain contains characteristics of two or more pathogenic classes, can have dangerous consequences when introduced into susceptible populations. This is not the first instance of non-EHEC bacteria that have acquired the Shiga toxins; at least two examples of non-*E. coli* bacteria acquiring the Shiga toxins have been recognized. In 1994, an outbreak of gastroenteritis at a nursery school and kindergarten in Germany caused 20 illnesses and one death. Nine of the infected children developed HUS. The outbreak was traced to fresh parsley that had been contaminated with a strain of *Citrobacter freundii* that expressed stx2 (Tschape *et al.* 1995).

In Australia in 1995, a patient who developed severe diarrhea and HUS was shown to be infected with a strain of *Enterobacter cloacae* that produced stx2 (Acheson and Keusch 1996). In addition, production of Shiga toxins has been reported in other bacteria, including *Aeromonas hydrophila*, *Aeromonas caviae*, *Shigella sonnei* and *Shigella dysenteriae* type 4 (Gupta *et al.* 2007).

These occurrences have highlighted the role of horizontal gene transfer (HGT) in the enhancement of bacterial virulence. Acquisition of virulence factors and antibiotic resistance by many clinically important bacteria can be traced to HGT between related or evolutionarily distant microflora (Brody *et al.* 2008). Virulence factors of pathogenic bacteria, including adhesins, toxins, invasins, protein secretion systems, and iron uptake systems, may be encoded by pathogenicity islands: large genomic regions that are present on the genomes of pathogenic strains but not nonpathogenic members of the same or related species (Hacker and Kaper 2000). The observation that pathogenicity islands are associated with transfer RNA genes, the presence of integrase determinants and other mobility loci, indicates their ability to be transferred by HGT (Hacker and Kaper 2000).

Several specific examples of the spread of pathogenicity islands by HGT include the common locus of enterocyte effacement (LEE) in typical and atypical enteropathogenic *E. coli* (Gomes *et al.* 2004), and type IV secretion systems (Juhas *et al.* 2008). More recently, Brody *et al.* (2008) found that the hospital epidemic methicillin-resistant *Staphylococcus aureus* MRSA252 strain shares multiple putative HGT DNA sequences with different causative agents of bovine mastitis that are not found in the other human *S. aureus* isolates.

For example, MRSA252 shares more than 14 different DNA sequence blocks with the bovine mastitis ET3 *S. aureus* strain RF122 and, of importance to the topic discussed here, many of the HGT DNAs encode virulence factors. Further analysis of the MRSA252 chromosome uncovered virulence-factor encoding HGT events with the genome of *Listeria monocytogenes* and a *Staphylococcus saprophyticus* plasmid. Both of these bacteria are also implicated as agents of contagious bovine mastitis (Brody *et al.* 2008).

The emergence of hybrid bacterial pathogens should lead to a discussion regarding new and effective methods of detection. Many current molecular methods rely on either detection of virulence genes that are linked to known bacterial serotypes or detection of virulence genes alone, without linking them to any particular bacteria. This approach is problematic, since, in the first scenario, current molecular assays for EHEC would have led to a false negative test result when used on the O104:H4 outbreak strain, a strain that resulted in the deadliest known outbreak of *E. coli* illness ever observed. In the second scenario, detection of virulence genes alone (for example, the Shiga toxins) would not have delivered any information about the causative infectious agent. In the future, then, diagnostic assays should include a combination of phenotypic (biochemical tests, clumping assays, Shiga toxin enzyme-linked immunosorbent assays (ELISA)) and genotypic (PCR detection of known virulence traits) methods to effect reliable detection of emerging pathogens with hybrid virulence profiles.

5.6 Acknowledgments

This work was supported by a grant from the United States Department of Agriculture Specialty Crops Research Initiative 2009-01208.

5.7 Critical questions for discussion

1. What are some environmental situations and production practices that can lead to microbial contamination of fresh produce?
2. What is the global impact of emerging pathogens with hybrid virulence profiles?
3. What improvements in diagnostic assays need to be achieved to detect these emerging pathogens?
4. What is horizontal gene transfer, and how does this mechanism contribute to enhanced virulence of pathogens?
5. What are the genetic mechanisms by which virulence genes and pathogenicity islands are transferred to bacteria?
6. Is organic fresh produce production at more risk for foodborne bacterial contamination than conventional fresh produce production? Give the reasons for your answer.
7. What are Good Agricultural Practices (GAPS) and how can they be used to reduce the risk of contamination of fresh produce?
8. Which GAPS could have been put in place to prevent this outbreak from occurring?
9. Is locally grown fresh produce safer than produce that is transported over long distances?
10. What are microbial genotyping methods and what are the main advantages and disadvantages of genotyping methods in comparison to phenotyping methods for detection of emerging pathogens?

5.8 References

ACHESON D.W.K. and KEUSCH G.T. 1996. Which Shiga toxin-producing types of *E. coli* are important? *ASM News* **62**: 302–6.

ANONYMOUS. 2011. EHEC O104:H4. Robert Koch Institut. Available at http://www.iss.it/binary/seur/cont/EHEC_O104_H4_car_RKI.pdf. Accessed 26 October 2011.

BEUTIN L. 2011. Outbreak with aggregative EHEC O104:H4 in Germany: Specific characteristics and search for possible sources. International Association for Food Protection Annual Meeting, Milwaukee, Wisconsin, 31 July–3 August 2011. Available at http://www.foodprotection.org/files/annual_meeting/2011-breakout-session-lothar-beutin.pdf. Accessed 26 October 2011.

BIELASZEWSKA M., MELLMANN A., ZHANG W., KÖCK R., FRUTH A. *et al.* 2011. Characterisation of the *Escherichia coli* strain associated with an outbreak of haemolytic uraemic syndrome in Germany, 2011: A microbiological study. *Lancet*, **3099**: 70165–7.

BRODY T., YAVATKAR A.S., LIN Y., ROSS J., KUZIN A. *et al.* 2008. Horizontal gene transfers link a human MRSA pathogen to contagious bovine mastitis bacteria. *PLoS One*, **3(8)**: e3074.

CASSAR C.A., OTTAWAY M., PAIBA G.A., FUTTER R., NEWBOULD S. *et al.* 2004. Absence of enteroaggregative *Escherichia coli* in farmed animals in Great Britain. *Vet Rec*, **154**: 237–9.

CDC 1995. Outbreak of acute gastroenteritis attributable to *Escherichia coli* serotype O104:H21 – Helena, Montana, 1994. *MMWR*, **44**: 501–3.

CDC 2011. Investigation update: Outbreak of Shiga toxin-producing *E. coli* O104 (STEC O104:H4) infections associated with travel to Germany. Available at http://www.cdc.gov/ecoli/2011/ecolio104/. Accessed 25 October 2011.

FRANK C., WERBER D., CRAMER J.P., ASKAR M., FABER M. *et al.* 2011. Epidemic profile of Shiga-toxin-producing *Escherichia coli* O104:H4 outbreak in Germany. *NEJM*, 10.1056/nejmoa1106483. Available at http://www.nejm.org/doi/pdf/10.1056/nejmoa1106483. Accessed 26 October 2011.

GOMES T.A.T., IRINO K., GIRÃO D.M., GIRÃO V.B.C., GUTH B.E.C. *et al.* 2004. Emerging enteropathogenic *Escherichia coli* strains? *Emerg Infect Dis*, **10**: 1851–5.

GUPTA S.K., STROCKBINE N., OMONDI M., HISE K., FAIR M.A. *et al.* 2007. Emergence of Shiga toxin 1 genes within *Shigella dysenteriae* type 4 isolates from travelers returning from the Island of Hispañola. *Am J Trop Med Hyg*, **76**: 1163–5.

HACKER J. and KAPER J.B. 2000. Pathogenicity islands and the evolution of microbes. *Annu Rev Microbiol*, **54**: 641–79.

JUHAS M., CROOK D.W. and HOOD D.W. 2008. Type IV secretion systems: Tools of bacterial horizontal gene transfer and virulence. *Cell Microbiol*, **10**: 2377–86.

KIM J., OH K., JEON S., CHO S., LEE D. *et al.* 2011. *Escherichia coli* O104:H4 from 2011 European outbreak and strain from South Korea [letter]. *Emerg Infect Dis*, September 2011. Available at http://dx.doi.org/10.3201/eid1709.110879. Accessed 21 October 2011.

MARTIN A. and BEUTIN L. 2011. Characteristics of Shiga toxin-producing *Escherichia coli* from meat and milk products of different origins and association with food producing animals as main contamination sources. *Int J Food Microbiol*, **146**: 99–104.

MERMIN J.H. and GRIFFIN P.M. 1999. Invited commentary: Public health in crisis: Outbreaks of *Escherichia coli* O157:H7 infections in Japan. *Am J Epidemiol*, **150**: 797–803.

O'CONNOR D.R. 2002. *Report of the Walkerton Inquiry*. Available at http://www.attorneygeneral.jus.gov.on.ca/english/about/pubs/walkerton/part1/. Accessed 14 January 2012.

ORSKOV F. and ORSKOV I. 1984. Serotyping of *Escherichia coli. Meth Microbiol*, **14**: 43–112.

SCHEUTZ F., MØLLER NIELSEN E., FRIMODT-MØLLER J., BOISEN N., MORABITO S. *et al.* 2011. Characteristics of the enteroaggregative Shiga toxin/verotoxin-producing *Escherichia coli* O104:H4 strain causing the outbreak of haemolytic uraemic syndrome in Germany, May to June 2011. *Euro Surveill*, **16**(24): pii=19889. Available at http://www.eurosurveillance.org/ViewArticle.aspx?ArticleId=19889. Accessed 26 October 2011.

SCHEUTZ F. and STROCKBINE N.A. 2005. *Escherichia.* In G. M. Garrity, D. J. Brenner, N. R. Krieg, and J. T. Staley (eds), *Bergey's Manual of Systematic Bacteriology*, pp. 607–24. Springer, New York.

TSCHAPE H., PRAGER R., STRECKEL W., FRUTH A., TIETZE E. *et al.* 1995. Verotoxinogenic *Citrobacter freundii* associated with severe gastroenteritis and cases of haemolytic uraemic syndrome in a nursery school: green butter as the infection source. *Epidemiol Infect*, **114**: 441–50.

UBER A.P., TRABULSI L.R., IRINO K., BEUTIN L., GHILARDI A.C. *et al.* 2006. Enteroaggregative *Escherichia coli* from humans and animals differ in major phenotypical traits and virulence genes. *FEMS Microbiol Lett*, **256**: 251–7.

6

Tracing the source of a community-wide gastroenteric outbreak caused by *Campylobacter* in Finland

M.-L. Hänninen, Helsinki University, Finland and A. Hörman, The Finnish Defence Forces, Finland

Abstract: Community-associated foodborne or waterborne outbreaks are rather common. Waterborne outbreaks often occur in communities where groundwater is used as a source and is distributed nondisinfected. If it is microbiologically contaminated, the barriers preventing distribution of contaminated tap water are minor. Local public health authorities require strategies and plans to clarify how to manage future waterborne outbreaks and prevent such incidents. The case deals with a waterborne outbreak in a small community in Finland, requisition of patient and epidemiological data, and how the outbreak was identified and controlled.

Key words: waterborne campylobacteriosis, drinking water, sampling, groundwater well, epidemiological analysis.

6.1 Introduction to the case

In a community of 18 000 residents, an estimated 400 people with various symptoms and severity of gastroenteritis contacted local medical health care authorities in October–November 2001. A potential common source for the outbreak, either food or drinking water, was suspected. The first faecal samples for analysis for the most common enteric pathogens were taken between 9 and 12 October. The first results, made available on 22 October, showed that the patients were positive for *Campylobacter jejuni*. Additional analyses identified a total of 56 patients positive for *C. jejuni*. The last laboratory-confirmed *C. jejuni* case associated with the outbreak was diagnosed on 6 November. In the preliminary investigation, no common food items or catering sources were identified.

Water samples were collected extensively from the water supply system, starting on 23 October.

6.2 Significance of the case

Campylobacter jejuni is the most commonly reported enteric bacterial pathogen causing illness in humans in the industrialized countries, including those of the EU. In 2009, approximately 200 000 cases of campylobacteriosis were reported from 17 of the 27 EU member states. The vast majority of cases are sporadic and most of them are linked to foodborne or food-associated transmission of the pathogen. However, according to the European Food Safety Authority (EFSA), large foodborne outbreaks caused by *Campylobacter* are uncommon (EFSA 2011a). On the other hand, waterborne outbreaks are rather common in many developed countries, even those with advanced water treatment facilities and systems (WHO 2004).

For example, between 2000 and 2007 a total of 354 drinking water-associated outbreaks, in which a total of 47 617 people fell ill, were reported in 14 European countries. The reporting countries included Spain, UK (England and Wales), Norway, Finland, Sweden and the Czech Republic among others (WHO 2009). *Campylobacter* was the common causative agent in the outbreaks. Together with *Aeromonas* spp. and *Shigella sonnei*, it was responsible for 44.9% of all registered waterborne outbreaks.

Contaminated drinking water has been associated with several outbreaks in Finland, where a total of 45 outbreaks were reported between 2000 and 2007. Characteristically, Finland has a large number (approximately 1500) of small groundwater plants that distribute drinking water without disinfection. These plants usually serve relatively small communities with approximately 50–10 000 residents. Despite the fact that these types of plants distribute only a minor proportion of the total drinking water supplied in Finland, most of the waterborne outbreaks have been associated with these small plants (http://www.ktl.fi/portal/english/research__people___programs/environmental_health/).

6.2.1 Regulatory aspects

The quality and safety of drinking water are based on the assessment of raw water quality and treatment processes as well as regular monitoring and control of the processes and the water distributed. The European Drinking Water Directive (DWD, Council Directive 98/83/EC, Article 4) specifies that drinking water must be free from any microorganisms and parasites and from any substances that, in numbers or concentrations, constitute a potential danger to human health. The safety and other quality parameters are based on the recommendations of the World Health Organization (WHO 2004). The Water Safety Plan (WSP), analogous to the Hazard Analysis of Critical Control Points (HACCP) used in the food industry, may be applied to drinking water production, although WSP has not yet been directly stipulated in any European legislation.

Investigation and reporting of the food- and waterborne outbreaks at the national level in Finland have been mandatory since 1997 (http://www.ktl.fi/portal/suomi/terveyden_ammattilaisille/infektiotaudit/infektiotauditnosto/epidemioiden_selvittaminen/National Surveillance and Reporting System).

In the EU, data collection and reporting became mandatory in 2004 after the implementation of Directive 2003/99/EC of the European Parliament and the Council. Local food and public health protection authorities collect data, and make primary notices and the final investigative reports to the central public health and food safety authorities. Water companies themselves are responsible for the hygienic quality and safety of distributed drinking water and of the water treatment and distribution system.

6.2.2 Market and economic aspects

The total cost of a waterborne outbreak is difficult to estimate. However, the costs include, among others, those of the patients in the health care systems and medication, sick leave for employees and employers, and extra work costs for public health protection, as well as maintenance and reparation costs at the water plant. Further costs may arise if there are food or beverage industries in the affected area, where the direct and indirect costs can be high. An important consequence of an outbreak is also the loss of public confidence in water quality and safety, which may require considerable time and effort to recover.

6.3 The case history continues

A total of ten groundwater wells located in three different aquifers were used as raw water sources for the drinking water in the community. Water from separate wells was collected and combined in a water tower and distributed further to the consumers without any treatment or disinfection. Previously, in the routine sampling the water had fulfilled the quality criteria stated in the Drinking Water Directive. Routine monitoring on 8 October showed no indication of contamination. Extensive water sampling started on 23 October. This sampling included, among others, samples of 8–10 l from each groundwater well to detect and identify the well(s) that might have been contaminated with *C. jejuni*. In addition, water samples were studied for *Escherichia coli* and coliforms to detect possible faecal contamination.

Both *C. jejuni* and low counts of coliforms were identified in well no. 2. Furthermore, another well (well no. 1) contained low counts of *E. coli* and coliforms, but no *C. jejuni*. These two wells were located in the same aquifer and further microbiological studies were focused on these two wells. Well no. 1 was later found positive for indicator bacteria as well as for *C. coli*. However, all further samples from well no. 2 were negative for *C. jejuni* and for coliforms as indicator bacteria.

Further investigation of the contamination history of wells no. 1 and no. 2 revealed that they were located close to a golf course. In this area construction work had resulted in changes to the collection and runoff of rainwater, directing the runoff in the direction of these particular wells. Furthermore, there were especially heavy rains in September in the region and the rainwater formed a pond close to the wells. The investigation also revealed that there were animal farms close to the wells and there were runoff water dikes running from the farms in the direction of these wells. Both *C. jejuni* and *C. coli* were detected in the dike water samples. It is thus possible that surface water runoff could have contaminated the wells. However, none of the distributed tap water samples from these wells were positive for *C. jejuni* or coliforms. A schematic time-associated description of the outbreak is presented in Fig 6.1.

The typing of ten *C. jejuni* isolates from the patients by Penner serotyping and pulsed field gel electrophoresis (PFGE) revealed that all isolates had an identical serotype (Pen1) and KpnI PFGE genotype. However, the *C. jejuni* isolate from well no. 2 was serotype Pen 55 and its PFGE genotype was different from that of the patients' isolates. The PFGE genotypes from the dike water samples were also unique and differed from that of the patients' isolates.

A total of 1000 people (5% of the total population) were estimated to have acquired campylobacteriosis from contaminated tap water in this outbreak during October–November.

6.4 Management of the outbreak and source tracing

6.4.1 How should a waterborne outbreak be managed?

In general, close cooperation at the local level between different stakeholders, i.e. public health authorities, medical care, the water supplier and the laboratory performing water analysis, is required for effective management of the outbreak. A common strategy and a clear plan of what to do, how to delegate tasks and duties, and how to inform the public and communicate with the media are needed. A municipality should establish a local organization to handle such foodborne or waterborne outbreak situations. This organization must include those with medical expertise, public health authorities, water treatment managers and laboratory experts.

The main goal of an outbreak investigation is to end the spread of the outbreak. One of the first practical interventions is to inform the public about the outbreak and contamination incident. It is very important to advise the public to stop using tap water unless boiled for drinking or in food preparation. If there are food or beverage industries in the affected region, they need to have specific information about the episode, treatment requirements and distribution breaks.

Further measures include extensive water sampling for microbiological analysis of raw water sources, the water treatment plant and the distribution network, with samples taken from taps at various locations. In addition, data collection for preliminary epidemiological analysis to identify the potential contamination source, including geographical information on the water distribution network and location of the patients within the network, as well as data on the emergence of the first cases, is needed.

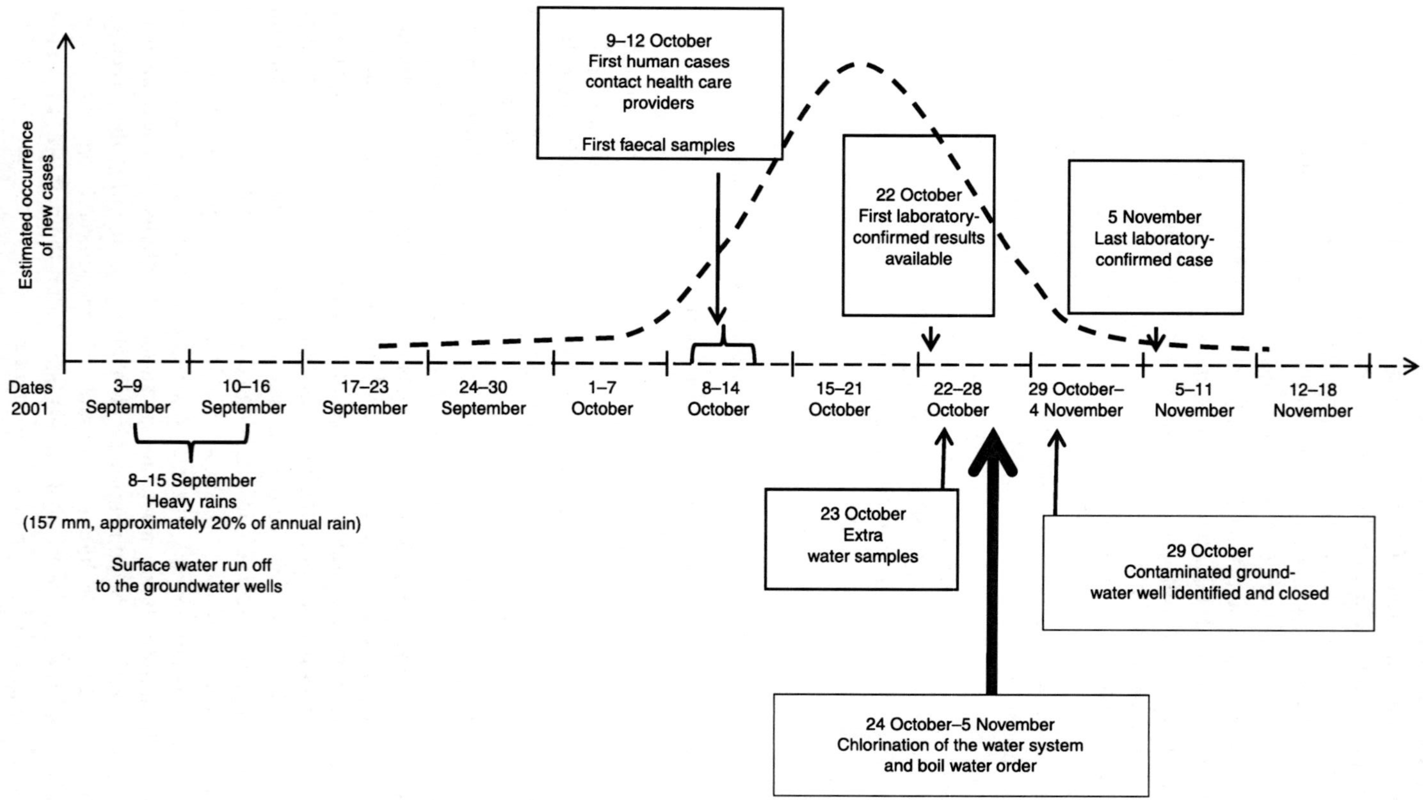

Fig. 6.1 Estimated epidemiological curve and events associated with the community-wide *Campylobacter jejuni* outbreak in a Finnish community in 2001. Estimated total number of cases was 1000, of whom 220 contacted healthcare providers; 12 were hospitalized.

An essential part of outbreak management is the active, reliable and continuous supply of information to the public and the media, for example facts relating to the outbreak. This information should also include advice on how to treat water, as well as information on alternative drinking water supplies. The information should be given regularly and should start as soon as the outbreak or contamination incident is discovered, and should continue until the episode is over and the normal situation has been restored.

After the contamination source has been detected and identified, corrective actions to repair the failures and to clean and disinfect affected plants and infrastructure, including shock chlorination of the distribution network, should be done. Furthermore, an appropriate intensified monitoring programme and control measures should be in place until the water safety criteria have been demonstrated to be fulfilled once again.

6.4.2 Data that should be collected for source tracing during an outbreak investigation

1. Patient data including symptoms, their onset and the results of the microbiological analysis of the faecal samples (which should include both pathogen isolation and characterization). There should be a descriptive epidemiological study (when, who, where) and, if possible, an analytical epidemiological (case–control, or cohort) study.
2. Data from environmental analysis: of the water treatment process, raw water source(s), water distribution network, geography, weather conditions or other exceptional incidents that could be associated with the contamination.
3. Data from analysis of the water samples: microbiological analysis for faecal indicator bacteria and pathogens from several points: raw water, water treatment plant, distribution network and tap water.
4. Results of comparison of the microbial strains: strains isolated from patients, source water and distributed tap water should be compared, as well as those from faecal samples from wild or farm animals whose faeces could have contaminated the water (association studies).

6.5 Conclusions of the outbreak and attempts to identify its source

In this particular case, the waterborne outbreak continued for a relatively long period of approximately one month, but the exact starting time for the outbreak remained unclear. The first patients were laboratory-confirmed on 22 October. Most probably there were already some cases before 22 October, because the heavy rains suspected of causing the contamination of the groundwater wells occurred in mid-September. One confusing factor in linking potential cases to the outbreak in the early phase was that patients from the affected area contacted two different health care centres.

Wells no. 1 and no. 2 remained faecally contaminated by surface water runoffs, according to the microbiological analysis, for a longer period after the heavy rain had passed. Another factor prolonging the outbreak detection was linked to the pathogenesis of campylobacteriosis. The incubation time for the disease can be rather long: up to 7 days. Furthermore, the analysis of faecal samples may require 1 week or even more. There was a 1-month period between the heavy rains with flooding and diagnosis of the first *Campylobacter* patients. The advice to boil water was given and chlorination was initiated on 24 October. The epidemic was regarded as being over in early November.

All patient isolates showed identical sero/genotypes, confirming that there was a common source. However, the fact that water sampling began a while after the water contamination incident might explain the difference between the *C. jejuni* genotypes from the water samples and those from the patients, with the *C. jejuni* sero/genotype infecting the patients no longer being detected in the water samples. The detection of *E. coli*, coliforms, *C. jejuni* and *C. coli* isolates, as well as diversity of the *C. jejuni* sero/genotypes, all indicated that the groundwater wells were faecally contaminated, most probably from the farm(s) located in the vicinity of the wells.

A crucial factor in water sampling and analysis is the sampling volume. Although in routine analysis a 100 ml sample is analysed, a sample volume of up to 10 l may be needed before the presence of bacterial waterborne pathogens can be confirmed. The increased sample volume is necessary due to the low concentration of a pathogen and uneven distribution of the bacteria in water samples. In this particular outbreak, *C. jejuni* was only detected in the water samples of 10 l.

Epidemiological analysis based on a questionnaire would have improved the strength of the evidence for association of the outbreak with drinking water. However, a communitywide questionnaire study was not conducted due to a lack of resources and since reasonable evidence of the possible source of the outbreak already existed. Further evidence for the drinking water as a source of the outbreak was received because, after the advice to boil water order and chlorination of the water distribution network, no new *Campylobacter* patients appeared in the community.

Quite often in a real-life outbreak investigation, and also in this case, not all requirements for the highest category of the strength of evidence are fulfilled. However, there was enough evidence to conclude that the drinking water was faecally contaminated and contained *Campylobacter* spp. both before and during the campylobacteriosis outbreak in the community, and this was the most obvious source of the epidemic.

6.6 Commentary: how can safety be improved and what constitutes an outbreak?

To improve water safety and set public health targets at local, national and EU levels surveillance data should be collected. Once analysed, data collected relating to contamination incidents and outbreaks can also help to provide information that can be used to prevent further outbreaks and improve drinking water safety both in general and at the individual water plant level.

To refer to an occurrence of cases as a microbial waterborne outbreak, an epidemiologic link between the cases must be established. Evidence for a link includes exposure to a common source, exposure to the source during a clearly defined time period, and symptoms caused by common enteric pathogens such as bacteria, parasites and viruses. A common route of exposure to waterborne pathogens is swallowing contaminated water. Epidemiological evidence must implicate water as the probable source of the illness for the outbreak to be deemed waterborne.

Assessment of the strength of the evidence implicating water as the source of the outbreak can be performed at the time of the final report on the outbreak, when all data have been collected and all analytical results are available. An evaluation report assessing the strength of the evidence should include:

- Patient data: details of the increased numbers of patients contacting local medical health care facilities with gastroenteric symptoms typical of a particular pathogen (evidence of 'spatial and temporal association')
- Faecal samples of patients testing positive for a causative pathogen.
- The interviews (questionnaires) implying drinking water as a potential common source.
- Analytical epidemiology, e.g. case–control or cohort studies confirming whether there is a statistically significant association between consumption of the water and the cases.
- Environmental analysis of the drinking water supply and distribution showing contributing factor(s). This should contain records of exceptional incidents, e.g. heavy rains before the outbreak, maintenance work on pipelines or construction work close to supplies, and potential reservoirs, e.g. cattle or poultry farming close to the raw water sources or groundwater wells.
- Microbiological analysis of drinking water samples: indicator bacteria or a particular causative pathogen detected.
- Typing of the patient and the water microbial isolates: evidence of whether the same types of the causative pathogen have been identified in patients and in water samples (whether or not there is an 'indistinguishable causative agent').

The EFSA divides the strength of the microbiological or the epidemiological evidence for an outbreak into two categories: strong and weak (EFSA 2011b). Strong microbiological evidence exists when an indistinguishable causative agent has been identified in an outbreak case and from food or water consumed by a case. Further, a strong epidemiological evidence requires a statistically significant association between the cases and the source in an analytical epidemiological study. The evidence may be considered weak when a causative pathogen has been identified in patients, but the link between the source and the patients remains unclear (weak epidemiological evidence) or the microbiological analysis has not been able to identify the causative agent in the food or water consumed (weak microbiological evidence).

6.7 Critical questions for discussion

1. How can we guard against outbreaks associated with untreated drinking water?
2. What needs to be done to prevent waterborne outbreaks in general?
3. Why is *C. jejuni* a common causative bacterium in waterborne outbreaks in some countries?
4. How should a drinking water system associated with a waterborne outbreak be sampled?
5. How should surveillance data be used?
6. How can evidence-based risk assessments be strengthened?
7. What is the definition of a waterborne disease outbreak?
8. What is meant by the term 'waterborne disease burden'?
9. What is the basic concept of the Water Safety Plan proposed by the WHO?
10. What is meant by the term 'water-related environmental tracking'?

6.8 Further reading

HÄNNINEN M.-L., HAAJANEN H., PUMMI T., WERMUDSEN K., KATILA M-L. *et al.* 2003. Detection and typing of *Campylobacter jejuni* and *Campylobacter coli* and analysis of indicator organisms in three waterborne outbreaks. *Appl. Env. Microb.*, **69**: 1391–6.

NATIONAL INSTITUTE FOR HEALTH AND WELFARE. *National Surveillance and Reporting System for Food and Waterborne Outbreaks*. Available at www.ktl.fi/portal/suomi/terveyden_ammattilaisille/infektiotaudit/infektiotauditnosto/epidemioiden_selvittaminen.

WORLD HEALTH ORGANIZATION (WHO). 2010. Environment and health information system (ENHIS). In *Health and Environment in Europe: Progress Assessment*. WHO Press, Geneva, Switzerland.

WORLD HEALTH ORGANIZATION (WHO). 2011. *Guidelines for Drinking-Water Quality* (4th ed.). World Health Organization, Geneva, Switzerland. Available at http://whqlibdoc.who.int/publications/2011/9789241548151_eng.pdf.

6.9 References

EUROPEAN FOOD SAFETY AUTHORITY (EFSA). 2011a. The European Union summary report on trends and sources of zoonoses, zoonotic agents and food-borne outbreaks in 2009. *EFSA Journal*, **9**: 2090. Available at www.efsa.europa.eu/efsajournal.

EUROPEAN FOOD SAFETY AUTHORITY (EFSA). 2011b. Updated technical specification for harmonised reporting of food-borne outbreaks through the European Union reporting system. *EFSA Journal*, **9**: 2101. Available at www.efsa.europa.eu/efsajournal.

7

One pathogen but two different types of foodborne outbreak: *Bacillus cereus* in catering facilities in Germany

M. Ehling-Schulz, University of Veterinary Medicine Vienna, Austria and U. Messelhäusser, Bavarian Health and Food Safety Authority, Germany

Abstract: Changes in lifestyle and eating habits have raised the importance of catering and ready-to-eat foods. However, mass catering facilities such as canteens and large commercial kitchens require certain hygienic measures, especially if they provide food for nurseries, schools, hospitals and retirement homes. These facilities are at special risk for contamination with spore-forming bacteria and can pose a public health risk, especially if these spore-formers are capable of producing toxins. The following case studies describe two typical foodborne outbreaks linked to mass catering facilities and provide guidelines for the prevention of similar cases.

Key words: catering, endospore-forming bacteria, *Bacillus cereus*, spores, toxins, emetic syndrome, diarrhea.

7.1 Introduction to the cases

Endospore-forming bacteria are frequent contaminants in food production, food processing and mass catering facilities. Mostly these bacteria are introduced into the food production process via raw materials, e.g. pasta, rice or spices. Due to the heat and acid resistance of their spores they cannot be eliminated by pasteurization or normal sanitation processes and therefore represent a major risk in mass catering facilities, if the temperature management of ready-to-eat food fails.

Among the endospore-formers, *Bacillus cereus* is gaining increasing attention as a foodborne pathogen. This aerobic, Gram-positive rod, which also grows well anaerobically, can cause two distinct forms of food poisoning: vomiting (emesis) and diarrhea. The small heat- and acid-stable peptide toxin cereulide is responsible for the emetic syndrome, whilst different heat-labile enterotoxins are linked to the diarrheal form of the disease. The emetic type of disease is the result of intoxication by cereulide-contaminated food, whereas the diarrheal type is thought to be the consequence of an infection with enterotoxic *B. cereus* (Ehling-Schulz *et al.*, 2011; Stenfors Arnesen *et al.*, 2008). As *B. cereus* is mainly a problem of catering facilities, we present two cases from such settings.

7.1.1 How did the cases begin?

Engineering service agency (2010)
Half an hour after lunch time, four employees of an engineering service agency became ill with acute vomiting. The lunch had been served from a catering facility. It was prepared in a large commercial kitchen that provided a delivery service to different agencies. The transport time to the engineer service agency was about half an hour. Because of an important meeting, the lunch was not served immediately but was stored at 8 °C in the fridge for 1 hour and afterwards reheated in the microwave. The lunch comprised Chana masala (cooked chickpea) with baked potatoes in curry sauce, cooked rice and mixed salad.

Nursery (2011)
Five hours after serving lunch in a nursery, 20 children aged 3–5 years and two nurses became ill with acute diarrhea; four children had to be taken to hospital. The lunch consisted of baked pasta with bacon, cheese and eggs and an additional green salad. The pasta was prepared the day before in a big casserole, then cooled down for 2 hours at room temperature in the kitchen and stored overnight at 7 °C in the refrigerator. Thirty minutes before lunch, the pasta was mildly heated to consumption temperature, but not completely baked again.

7.1.2 Infective dose of *B. cereus*

The human risk profile highly depends on the type of *B. cereus*. Enterotoxin-producing bacteria causing diarrhea are usually found in higher numbers ($>10^4$) in contaminated food remnants, whereas the bacteria themselves might even be absent in the case of emetic intoxications caused by cereulide-producing *B. cereus* strains. Spores from enterotoxigenic *B. cereus* strains are ingested with contaminated food and pass to the intestine where they can re-grow and start to produce enterotoxins. In contrast, the emetic *B. cereus* toxin cereulide is preformed in food during the food production process and ingested with the contaminated food. Unlike the depsipeptide toxin cereulide, enterotoxins are heat- and acid-labile and digested by pepsin and trypsin. The infective doses for both scenarios are currently unclear. The European Food Safety Agency (EFSA) has assessed that 10^5–10^8 enterotoxigenic *B. cereus* cells or spores are normally necessary to cause diarrhea. Consequently, foods containing more than 10^4 viable cells or spores per gram can pose a risk (EFSA, 2005). Deduced from animal feeding studies and outbreaks, the intoxication dose for the cereulide toxin is thought to be around 0.03–13.3 μg toxin/g food. However, because of the large differences in the capacity of toxin production among the different *B. cereus* strains, there can be major variations in the infective dose, and the numbers provided above can only be seen as a rough estimate.

7.1.3 Sources of *B. cereus* in food

B. cereus spores are ubiquitous in the environment and can be introduced via soil-grown vegetables or spices to large kitchens, as well as by other raw materials such as rice or pasta. The ability to form biofilms and to persist on the surfaces of processing equipment can lead to additional cross-contamination during the food production process (Andersson *et al.*, 1995).

7.1.4 Different diagnostic methods

In the case of *B. cereus* contamination, a mixture of strains is usually found. It is therefore not sufficient to analyze one isolate from the suspicious food for its ability to produce toxin. Emetic strains in particular are often accompanied by non-emetic strains, making it necessary to test the food directly by molecular methods for the ability for toxin production or for the toxin itself (Fricker *et al.*, 2007) (Table 7.1).

Table 7.1 Methods for identification and differentiation of *Bacillus cereus*

	Enterotoxigenic *B. cereus*	Emetic *B. cereus*
Qualitative and quantitative cultural detection of presumptive *B. cereus* (without any differentiation of the emetic or enterotoxigenic potential)	ISO 7932:2004: Microbiology of food and animal feeding stuffs – Horizontal method for the enumeration of presumptive *B. cereus* – Colony-count technique at 30 °C For a review see Ehling-Schulz *et al.*, 2011	
Determination of the toxigenic profile (PCR)	Guinebretière *et al.*, 2002 Ehling-Schulz *et al.*, 2006	Ehling-Schulz *et al.*, 2006
Determination of the toxigenic profile (real-time PCR)	Wehrle *et al.*, 2010	Fricker *et al.*, 2007 Wehrle *et al.*, 2010
Quantitative real-time PCR	Yang *et al.*, 2007	
Detection of enterotoxins (ELISA)	Wehrle *et al.*, 2009	
Detection and quantification of cereulide (mass spectrometry)		Häggblom *et al.*, 2002
Detection and quantification of cereulide with internal standard (stable isotope dilution assay)		Bauer *et al.*, 2010

PCR, polymerase chain reaction; ELISA, enzyme-linked immunosorbent assay.

In the case of emetic syndrome, sometimes only the heat-stable toxin can be detected in the processed food but not the bacteria themselves. Therefore, in suspicious cases, a toxin test should be performed as well, if the microbiological tests are negative. In contrast, in outbreak situations that are putatively linked to diarrhea-causing enterotoxic *B. cereus* strains, a colony count method in combination with polymerase chain reaction (PCR)-based toxin gene profiling methods and/or enzyme-linked immunosorbent assay (ELISA) techniques is recommended for diagnosis. In the latter case, a toxin test directly from foodstuffs is not informative. Enterotoxin production in food is possible but of low importance, since the enterotoxins are mostly digested by pepsin and trypsin during the gastrointestinal passage in the human host. The enterotoxins relevant to the human illness are produced after outgrowth of *B. cereus* spores in the human small intestine.

7.2 Significance of the cases

B. cereus is a spore-forming, facultatively anaerobic growing, Gram-positive bacterium. The strains are closely related to *B. anthracis* and to *B. thuringensis*, which is the reason why the three species are also subsumed under the term '*Bacillus* cereus group' or '*Bacillus cereus* sensu lato'. In recent years, this bacterium has been increasingly recognized as an important foodborne pathogen, particularly in mass-catering facilities. For instance, it is assumed that in the years 1998–2000 *B. cereus* was involved in up to 60% of foodborne intoxications resulting from mass-catering facilities in Germany, and in this timeframe it was also the most frequently isolated microorganism from samples taken from large commercial kitchens. Generally, reports on *B. cereus* intoxications and infections are steadily increasing throughout Europe (EFSA, 2010) and the USA (Scallen *et al.*, 2011).

The symptoms of *B. cereus* intoxications and infections are very similar to gastrointestinal illness caused by other bacteria. Considering only the clinical symptoms, the diarrheal syndrome caused by *B. cereus* cannot be distinguished from a *Clostridium perfringens* infection. The same problem exists when comparing the emetic type of a *B. cereus* food poisoning and *Staphylococcus aureus* intoxication. This is one of the reasons why the true incidence of infections and intoxications caused by *B. cereus* is still unknown. However, in recent years many new diagnostic methods for the characterization and typing of *B. cereus* strains have been developed (Ehling-Schulz *et al.*, 2011) and are increasingly implemented in routine diagnostics. Therefore, it is likely that it will be possible to assess the true incidence of food poisonings caused by *B. cereus* during coming years and obtain more accurate statistics on the real contribution of *B. cereus* to the total burden of foodborne infections and intoxications.

7.2.1 Regulatory aspects

The general rules for food safety in the EU are laid down in Regulation (EC) 178/2002. This regulation does not differentiate between, for example, chemical or microbiological risks, but assures comprehensive consumer protection. Article 14 of Regulation (EC) 178/2002 requires that food shall be not placed on the market if it is unsafe. This is, for example, the case if the food is considered to constitute a risk to the consumer.

When evaluating if a food is unsafe or not, other aspects such as, for instance, eating habits or usual conditions of intake must be included in the risk assessment. In both cases described above, ready-to-eat-foods, which caused illness, were served to a large number of customers. These are typical examples of unsafe foods according to Article 14 paragraph 2a of Regulation (EC) 178/2002. It is therefore the obligation of the food manufacturer to remove the whole batch of the unsafe food from the market. The relevance of such a case as regards criminal law or private law varies according to the national laws of the different EU countries. In Germany, for example, it is a criminal offence to place unsafe foods on the market.

7.2.2 Economic and market aspects

The number of food poisonings caused by bacterial toxins has been steadily increasing in Europe throughout the last years. In 2008, bacterial toxins were involved in about 15% of the human foodborne diseases in Europe. EFSA also reported an increase of 18% in the number of foodborne outbreaks caused by *B. cereus* toxins in the European countries in comparison to the year 2007. This corresponds with a reporting rate of 0.03 cases/100 000 inhabitants (EFSA, 2010).

The major challenge to a proper risk assessment model for *B. cereus* is the high variability in toxigenic potential observed among *B. cereus* strains. In order to avoid substantial economic losses by destroying products that may just carry non-toxigenic or only mildly toxigenic strains, it is of utmost importance to know the toxic potential of the strains found in the different food matrices. However, such a specific diagnosis is complex and is currently not yet available for routine laboratory diagnostics. Furthermore, because of the ubiquitous occurrence of *B. cereus* in the environment, it is not possible to completely avoid *B. cereus* contamination in most food categories. Correct production management and an effective hazard analysis and critical control points (HACCP) system are of utmost importance to ensure safer food production and avoid food poisoning, especially severe cases involving hospitalization and even deaths.

7.3 The case histories continue

In both cases described here the sources of the outbreaks were unknown when the police and food inspectors contacted the engineering agency (case one) and the nursery (case two). Severely ill patients had already been admitted to the hospital. The retained food samples or the remnants from the lunch and samples of the raw components used for meal production were sent for investigation to a governmental microbiological laboratory.

7.3.1 Emetic syndrome in the agency

The symptoms (vomiting half an hour after lunch) described to the food inspectors indicated a foodborne intoxication presumably caused by *S. aureus* enterotoxins or by the emetic *B. cereus* toxin cereulide. Since no food remnants were left from the lunch, the food inspectors took retained samples and the raw materials (raw rice, raw vegetables and chickpeas) from the kitchen of the catering facility. The retained samples were checked for the presence of *S. aureus* enterotoxins by ELISA, and real-time PCR was performed as described by Fricker *et al.* (2007) to screen the retained samples as well as the raw materials for the presence of emetic *B. cereus* strains. *S. aureus* enterotoxins were not detected, but the real-time PCR provided a positive result for emetic *B. cereus*. Therefore, a cultural quantitative detection of *B. cereus* was carried out. The cooked rice sample contained 2.8×10^4 colony-forming units (cfu) *B. cereus*/g sample material. All *B. cereus* isolates from the plate were tested positive by real-time PCR for the ability to produce the cereulide toxin. In the cooked chickpea sample, as well as in raw rice used for the production of lunch, emetic *B. cereus* was also detected, but the number was lower than 10 cfu/g sample material. All other raw materials tested negative for the presence of *B. cereus*.

7.3.2 Diarrhea in the nursery

Diarrhea is a common symptom in foodborne infections caused by *Salmonella* spp., thermotolerant *Campylobacter* spp., *C. perfringens* or *B. cereus*. Therefore the remnants of the pasta found in the nursery were checked for all of these pathogens and/or toxins using both cultural and molecular methods. The real-time PCR provided negative results for *Salmonella* spp. and thermotolerant *Campylobacter* spp. after enrichment as well as for *C. perfringens*, but provided positive results for genes encoding the non-hemolytic enterotoxin (NHE) and also the hemolysin BL (HBL) toxin of *B. cereus*. Quantitative cultural methods revealed a contamination level of 2.2×10^6 cfu/g of *B. cereus* in the pasta. All *B. cereus* isolates tested positive for the presence of the enterotoxin genes *nhe* and *hbl* by real-time PCR. In the raw pasta from the nursery kitchen, the diarrheal strains of *B. cereus* were found in a range of between 1 and 10 cfu/g.

7.4 Resolution and outcomes

In both cases described above, the raw materials (rice/pasta) were contaminated with *B. cereus* spores, but only in low numbers (1–10 spores per gram). If handled correctly in the kitchen, the ready-to-eat food produced with the contaminated ingredients would have been unlikely to cause illness. The real problem was most likely improper temperature management in both kitchens. Due to the production of a large amount of food that was not cooled down immediately, the *B. cereus* spores, which survived the cooking temperature, germinated and grew very well without competitive bacterial flora. In the first case (engineering agency), the bacteria even produced a toxin.

To avoid foodborne intoxications caused by emetic or diarrheal *B. cereus* strains, appropriate temperature management is of utmost importance. Due to the ubiquitous presence of *B. cereus* spores in the environment it is often not possible to prevent the contamination of raw food material with a low number of spores. It is possible, however, to prevent the germination of the spores in order to avoid high concentrations of *B. cereus* or the production of toxin until consumption. Therefore, it is necessary either to keep food not eaten immediately after cooking hot (>65 °C) or to cool it down quickly and keep it at temperatures below 8 °C. One should always bear in mind that the extremely stable emetic toxin cereulide can hardly be inactivated once it has been produced. Reheating of precooked food after inadequate temperature management will not destroy potentially preformed cereulide and is therefore not an appropriate measure to prevent intoxications caused by cereulide.

In addition, avoiding cross-contamination in the kitchen, especially from raw material to ready-cooked food, can be seen as the next most important measure to prevent *B. cereus*-associated food poisonings. To support this conclusion, the detection of emetic *B. cereus* strains in the cooked chickpeas (raw chickpeas and spices having been negative for *B. cereus*) as reported in case one showed that, if there is a lack of hygiene in the kitchen, ready-to-eat food can easily become contaminated with *B. cereus*.

7.5 Commentary and personal reflections

It is necessary to develop rapid and cost-effective detection methods for routine use to estimate the toxigenic potential of the different *B. cereus* strains. This will enable a decision taken into consideration whether a suspicious food is 'only' unfit for human consumption (as mentioned by Article 14 para. 2b of Regulation (EC) 178/2002), or if it really poses a health risk.

7.6 Critical questions for discussion

1. In which cases is food called 'unsafe' and what are the legal consequences?
2. What are the differences between a foodborne infection and foodborne intoxication?
3. Why is *B. cereus* such a common problem in mass-catering facilities?
4. Which other microbial contaminants can cause foodborne intoxications? Please mention at least three pathogens.
5. The true incidence of cases of foodborne intoxication is unknown. What are the reasons for this?
6. The EFSA in its annual summary reports consistently mentions a large number of 'unverified foodborne outbreaks'. What is meant by this and why might it be impossible to verify the outbreaks?
7. Does an exclusively cultural detection of *B. cereus* in the context of food safety make sense, and which alternative or additional methods could be used?

8. Is it possible to find products completely free from *B. cereus* spores and vegetative cells? Please explain the reasoning behind your answer.
9. What is the HACCP concept, and which legal regulations in the European Union prescribe the concept?
10. Which storage times and conditions should be applied to heated or cooled ready-to-eat food in catering facilities in order to avoid an increase in numbers of bacterial contaminants?

7.7 References

ANDERSSON, A., RONNER, U. and GRANUM, P.E. (1995). What problems does the food industry have with the spore-forming pathogens *Bacillus cereus* and *Clostridium perfringens*? *Int J Food Microbiol*, **28**:145–55.

BAUER, T., STARK, T., HOFMANN, T. and EHLING-SCHULZ, M. (2010). Development of a stable isotope dilution analysis for the quantification of the *Bacillus cereus* toxin cereulide in foods. *J Agric Food Chem*, **58**:1420–8.

EFSA (2005). Opinion of the scientific panel on biological hazards on *Bacillus cereus* and other *Bacillus* spp in foodstuffs. *EFSA J*, **175**:1–48.

EFSA (2010). The community summary report on trends and sources of zoonoses, zoonotic agents and food-borne outbreaks in the European Union in 2008. *EFSA J*, **8**(1):1496.

EHLING-SCHULZ, M., GUINEBRETIERE, M.H., MONTHAN, A., BERGE, O., FRICKER, M. *et al.* (2006). Toxin gene profiling of enterotoxic and emetic *Bacillus cereus*. *FEMS Microbiol Lett*, **260**:232–40.

EHLING-SCHULZ, M., MESSELHÄUSSER, U. and GRANUM, P.E. (2011). *Bacillus cereus* in milk and dairy production. In Hoorfar J. (ed.), *Rapid Detection, Characterization and Enumeration of Food-Borne Pathogens*, pp. 275–89. ASM Press. Washington DC.

FRICKER, M., MESSELHÄUSSER, U., BUSCH, U., SCHERER, S. and EHLING-SCHULZ, M. (2007). Diagnostic real-time PCR assays for the detection of emetic *Bacillus cereus* strains in foods and recent food-borne outbreaks. *Appl Environ Microbiol*, **73**:1892–8.

GUINEBRETIERE, M.H., BROUSSOLLE, V. and NGUYEN-THE, C. (2002). Enterotoxigenic profiles of food-poisoning and food-borne *Bacillus cereus* strains. *J Clin Microbiol*, **40**:3053–6.

HÄGGBLOM, M.M., APETROAIE, C., ANDERSSON, M.A. and SALKINOJA-SALONEN, M.S. (2002). Quantitative analysis of cereulide, the emetic toxin of *Bacillus cereus*, produced under various conditions. *Appl Environ Microbiol*, **68**:2479–83.

SCALLEN, E., HOEKSTRA, R.M., ANGULO, F.J., TAUXE, R.V., WIDDOWSON, M.A. *et al.* (2011). Foodborne illness acquired in the United States – major pathogens. *Emerg Infect Dis*, **17**(1): 7–15.

STENFORS ARNESEN, L.P., FAGERLUND, A. and GRANUM, P.E. (2008). From soil to gut: *Bacillus cereus* and its food poisoning toxins. *FEMS Microbiol Rev*, **32**:579–606.

WEHRLE, E., DIDIER, A., MORAVEK, M., DIETRICH, R. and MÄRTLBAUER, E. (2010). Detection of *Bacillus cereus* with enteropathogenic potential by multiplex real-time PCR based on SYBR Green I. *Mol Cell Probes*, **24**:124–30.

WEHRLE, E., MORAVEK, M., DIETRICH, R., BÜRK, C., DIDIER, A. *et al.* (2009). Comparison of multiplex PCR, enzyme immunoassay and cell culture methods for the detection of enterotoxinogenic *Bacillus cereus*. *J Microbiol Methods*, **78**:265–70.

YANG, I.C., SHIH, D.Y., WANG, J.Y. and PANI, T.M. (2007). Development of rapid real-time PCR and most-probable-number real-time PCR assays to quantify enterotoxigenic strains of the species in the *Bacillus cereus* group. *J Food Prot*, **70**:2774–81.

8

Investigating a suspected case of mycotoxin poisoning in South Africa

M. F. Dutton, University of KwaZulu Natal, South Africa

Abstract: In 1988, the fungal/mycotoxin analytical unit at what was then known as the University of Natal, Pietermaritzburg, South Africa, was contacted by one of the large agricultural co-operatives in what was then the Transvaal with a problem affecting farm animals on a farm in the Kinross area, and the Director of the Unit went to investigate. The case describes the actions taken to find the cause of the problem, the outcomes and lessons learned.

Key words: farm animal, production loss, fungus, mycotoxin, analysis.

8.1 Introduction to the case

Mycotoxins are toxic fungal secondary metabolites and their persistent appearance in food is of major concern to all agricultural and related government ministries, food companies, legislators, health bodies and not least the consumers themselves (Fitzpatrick 1990). Once the mycotoxin is in the food it is very difficult to remove it; therefore prevention is better than cure, and efforts have to be made at the agricultural level to minimize the introduction of the producing organism, i.e. the fungus, and its mycotoxin. Although over 300 mycotoxins are known (Cole and Cox 1981) the problem is made a little easier by recognizing that there are only a few of these that are regarded as being the main contaminants of agricultural commodities (Frisvad *et al.* 2006). These are the notorious aflatoxins, deoxynivalenol, a *Fusarium* trichothecene, the fumonisins, ochratoxins and zearalenone. These contaminate mainly cereals and oil seeds and nuts (Sangare-Tigore *et al.* 2006; Perkowski *et al.* 1995). Further, the other mycotoxins commonly found in food are generally specific to certain commodities, e.g. patulin, which is found in apples (Berretta *et al.* 2000), ochratoxin A, which is found in grapes/wine (Otteneder and Majerus 2000) and *Claviceps* spp. (ergot) in rye and wheat and related cereals

(Young 1981). Mycotoxins are generally introduced into foods at the agricultural commodity production level, either through direct infection of the crop in the field (e.g. with *Fusarium verticillioides*, producer of fumonisin in maize (Dutton 2009)) or during storage where the more saprophytic organisms can infect the commodity, if the conditions are right (e.g. *Aspergillus flavus*, which is ubiquitous, can produce the aflatoxins B_1 and B_2 when the moisture content is over 15% (Toteja *et al.* 2006).

8.1.1 Commercial and control issues

In order to assist local farmers, feed millers and other agencies, a laboratory was set up at the University of Natal, Pietermaritzburg by Professor Dutton in 1985, under the auspices of the Department of Animal and Poultry Science with one technologist, Mrs Anne Kinsey, to screen samples for fungi and mycotoxins. Because of the number of mycotoxins known, only up to 20 were looked for (unless specific requests were made and then specific methods for that toxin were used). The thin layer chromatography (TLC) method of Patterson and Roberts (1969) was used, which involves a dialysis clean-up step and two-dimensional chromatography on 10 × 10 cm aluminum-backed plates (Merck) with various spray reagents where necessary, e.g. chromotropic acid reagent (Baxter *et al.* 1983) for trichothecenes. While this method has its critics because of poor recoveries (it is a multi-mycotoxin screen) and hence low sensitivity, it is a practical approach, as farmers are not worried about low and precise levels of mycotoxins, rather about those that are over the legal limit or are going to produce symptoms in their animals. Where necessary, methods for specific mycotoxins were also applied using high-performance liquid chromatography (HPLC), which became available to the laboratory in 1988. In the hands of a meticulous and experienced operator, acceptable results can be achieved and Mrs Kinsey admirably fitted these criteria. Fungal contamination screening was done using a restriction agar (Kaufman *et al.* 1963) with serial dilution so the level of contamination as well as the contaminants could be determined. After some time with isolates being sent to experts in mycological taxonomy, Mrs Kinsey became expert herself in fungal identification, certainly down to the genus level and in most cases the species. To say that in ten years of operation we screened over 8000 samples shows the need for this type of service.

Most of the compound feed produced in South Africa is for local consumption and amounts to a total of nearly 9 million tonnes annually. Generally this is produced by commercial feed manufacturers, although a certain supplementary amount may be produced by the farmers themselves. Raw commodities such as maize are drawn from South African production itself, often from established farm co-operatives but with certain components sourced from the rest of the world, particularly protein sources such as fish meal, cotton seed meal and soybean.

8.1.2 How did the problem begin?

So this was the situation when a phone call was received from a Transvaal farm co-operative in South Africa in 1988 with respect to a farmer having difficulty with a herd of cattle in the Kinross area (Fig. 8.1). In most of these cases the client wants an instant response, but this was not always possible because the laboratory was not a commercial enterprise and there was plenty of other work to do. In any case, fungal and mycotoxin screening takes a finite period of time. On the other hand 'time is of the essence' since, in the mycotoxin game, the culprit mycotoxin with its producing fungus has long since gone into the animal once symptoms have appeared. As this client had supported the laboratory, Professor Dutton drove to the farm at Kinross later that week to inspect the situation in person and take samples of everything that the animals were consuming, apart from the water, which had been tested and found to be negative for contaminants. It is very important for this type of inspection to take place. When farmers ring through with problems they are not fully aware of the dangerous circumstances that can arise on the farm in terms of fungal contamination. Generally these are of a storage nature on the farm but many scenarios can greet the mycotoxicologist, e.g. poor poultry housing conditions for both broilers and layers with damp floor litter (due to drinking water being spilt), silage pits being opened to the air to allow animals to feed *ad lib.*, commodities stored in leaking silos and barns, etc.

8.1.3 The situation on the ground

The first step was to visit the co-operative's headquarters in the nearest town to the farm, where the deputy director was contacted, and then to visit the farm. The

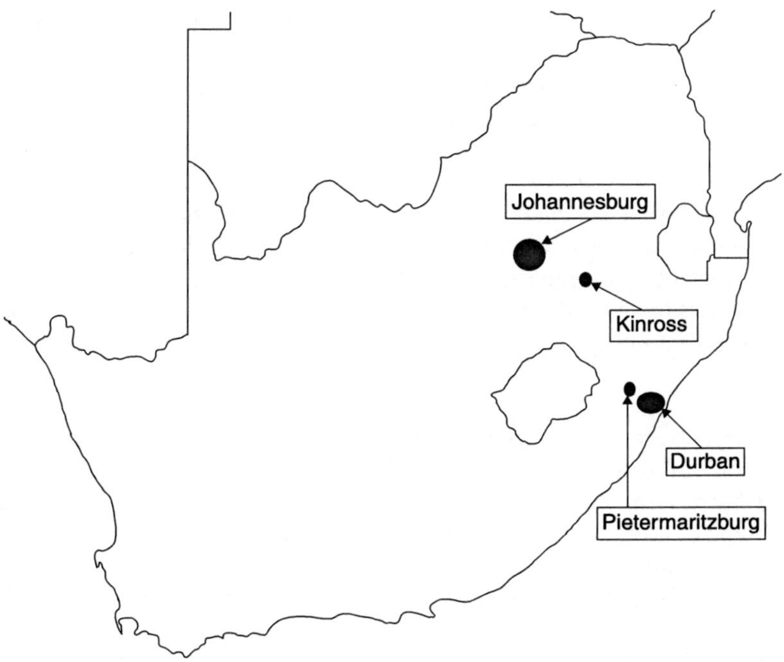

Fig. 8.1 Map of South Africa showing places mentioned in the case study.

farmer, as might be expected, was very concerned with respect to the condition of his herd. He had nearly 100 milking cows, a prize herd, and which until late in the previous year had provided record amounts of top-quality raw milk. The farm was approximately 100 ha and was mainly pasture together with some other crops such as maize and lucerne (alfalfa), which supported the herd and other animals when there had been rain for it to grow. For no apparent reason the milk yield had gone down and the herd had begun to look progressively unresponsive and unwell. The veterinarian who was called to check them through could not find anything obviously wrong; there were no bacterial or other microbial diseases, e.g. mastitis or viral conditions. As is the norm with these causes of unknown etiology, the diet comes under suspicion, with the possibility of the presence of mycotoxins or other toxic materials in the feeds.

The rest of the morning and early afternoon was taken up by going around the farm, inspecting the condition of the premises and taking samples where appropriate. This included compound feed stored on the farm and materials such as chopped lucerne and silage. None of the materials looked at all suspicious, although only laboratory analysis would prove this for sure. The immaculate condition of the farm and the obvious meticulous standards of the farmer and his staff made the poor condition of the herd a complete mystery.

8.1.4 Laboratory work

All the samples taken, including the lucerne, were screened for fungi and the mycotoxins that were normally screened for. Apart from the odd small amount of deoxynivalenol (DON) in the compound feed and patulin in the silage, there was nothing that gave a clear clue as what it might be, if it were a mycotoxin problem. This was further confirmed from the fungal screen of the compound feed, which showed low counts of *Fusarium* spp., some *Mucor* and *Rhizopus*. Under normal circumstances these samples would have been reported as being negative for dangerous organisms and mycotoxins. As this was before the laboratory introduced a methyl thiazol tetrazolium (MTT) cytotoxicity screening test (Mwanza *et al.* 2009) to detect toxins in the extracts, we had not found anything suspicious as far as the farmer's problems were concerned.

8.1.5 What were the challenges involved?

It was rather disappointing that nothing had been found to affect the herd, as it seemed fairly clear that the herd had access to something that was upsetting them. Some cases in later years were found to be due to the ingestion of silage, which was found to be toxic using the cytotoxicity screen, and this had been traced back to the farmer spraying his field with pesticide prior to cutting it for silage. However, for the Kinross case the cytotoxicity test was not yet available. It did seem, however, that some dietary factor was involved in this problem: but what?

8.2 Significance of the case

Apart from the farmer losing his livelihood, the presence of unknown factors on a prize farm affecting its output and commercial viability was of great concern, particularly as there are various contagious diseases of cattle in Africa that are not easy to diagnose. So it was of paramount importance for the source of the sickness to be determined. Consequently, further investigations, both our own and those of veterinarians, were required.

8.3 The case continues

8.3.1 Second farm visit

It was, therefore, determined that a second visit would be made to the farm at Kinross for a further look at the situation and arrangements were made accordingly. The herd was still obviously 'under the weather' with the milk yield way down. The farm was re-inspected, the farmer and his staff were interrogated, and even the pasture was inspected. Nothing seemed amiss, although it was noticed that the pasture had been cut and had not grown very much due to rather dry weather for the time of year.

On preparing for the long drive back to Pietermaritzburg, Professor Dutton happened to look across the farm yard. Some of the cattle were sauntering over to a feeding pen. This was an obvious point of trouble that had been missed. In the feed lot, which just looked like an innocent area fenced off with metal bars, was a large pile of hay. The cattle stuck their head through the bars and ate freely. Samples were immediately taken and these looked moldy. On inquiry as to from where the hay had originated, the stockman, who had just brought some in a large barrow, said that it was 'big bale' hay. On going to a field behind the barns, there were big rolled bales of hay. On hitting them hard with a convenient plank of wood, a large cloud of what turned out to be fungal spores went up, which looked like very fine dust. There were about a dozen of these bales all uncovered and sitting in the corner of the field where they had been cut. Ironically they were quite dry, so fungal growth had taken place sometime previously when they were presumably moist from rain, and then when drying began the fungus, being stressed, started producing spores and possibly mycotoxins. All the bales were sampled and all gave off spores when hit. All the hay samples from the feed lot and the field plus samples of residual dried herbage, the grass plants and the soil from where grass had been harvested were taken to the laboratory to find a source of fungal infective agents (Fig. 8.2).

8.3.2 Further laboratory work

All the samples were cut up and mixed. Some pieces of infected straw were placed on potato dextrose agar and the other samples screened for fungi and mycotoxins. This process when the matrix is straw/hay is quite difficult to handle, as it absorbs

Fig. 8.2 Dairy cattle at the farm at Kinross eating large bale hay contaminated with *Aspergillus versicolor*.

water and other solvents very readily. To extract for mycotoxins, therefore, extra extraction mixture (potassium chloride solution/acetonitrile) has to be added and then a known fraction of this is recovered by filtration with suction. The same happens in the serial dilution for fungal counting. However, this was not a problem as the hay was so heavily infected it was uncountable. What this means is that the plate of the 10^{-6} dilution was uncountable; in other words there were viable fungal propagules in excess of 10^6 per gram of hay. Not all the bales were this contaminated, but no samples of hay were recovered at $<10^5$ propagules per gram. The identification of the fungi was easier than was expected. It was an almost pure culture of *Aspergillus versicolor*.

8.3.3 Significance of the finding

The isolation of *Aspergillus versicolor* was surprising as we had never had more than low infection of material with this species in other samples analyzed. From earlier experiences with *A. versicolor* it had been found that it grows more slowly and has more particular substrate requirements than other *Aspergillus* spp., e.g. the ubiquitous *A. flavus*, hence the surprise. It is well known that this species produces the mycotoxin sterigmatocystin (ST) (Cole and Cox 1981), which is metabolically and chemically related to the aflatoxin B_1 but because of its relative rarity is not as well studied. It has a xanthone ring in place of the coumarin ring found in aflatoxins, but an important point is that it retains the *bis*-dihydrofurane ring which gives AFB1 its carcinogenic properties, and hence ST is also carcinogenic, although at a lower potency (Hendricks *et al.* 1980). The mechanism seems to be the same as that of AFB1, as ST can also form adducts with DNA

(Essigmann *et al.* 1979). ST can also exhibit acute toxicity (Sreemannarayana *et al.* 1987) and has been shown *in vitro* to cause apoptosis (Sun *et al.* 2002) and inhibit HLA-1 expression in human mononuclear cells (Xing *et al.* 2005), which suggests that it may interfere with the immuno-response.

The next question was – did the *Aspergillus versicolor* identified produce ST? The answer was a definite yes; in fact all isolates when cultured in yeast extract/sucrose medium (Davis *et al.* 1966) produced over 1 ppm (mg/kg) and in some cases more. Clearly the conditions that allowed this fungus to infest the rolled hay were also such as to trigger toxin production. Furthermore, it also produced orange hydroxyanthraquinones, e.g. versicolorin A (Dutton and Anderson 1982), some of which are precursors of AFB1 and sterigmatocystin itself (Hsieh *et al.* 1973). It was such a good producer that it was used in our metabolic studies on AFB1 biosynthesis, which was one of our main research thrusts at the time. The other samples taken of soil and hay gleaning and the actual grass plants showed no *A. versicolor* infection.

8.4 Outcomes

From the above results, it can confidently be said that the dairy herd were being fed on a daily basis a dose of *A. versicolor* metabolites, including ST and anthraquinone pigments. Presumably this gave the hay an interesting flavor for the cattle, as they seemed to like it and in no way exhibited feed refusal symptoms. Sterigmatocystin has been reported in dairy feed in the US (Vesonder and Horn 1985) where it caused bloody diarrhea and death in the dairy cattle. The farmer was advised to dispose of his rolled hay and not to feed it anymore. This seemed to do the trick as his herd improved and got back to almost the level they had been at before the problem started. He was also advised to store hay in a dry barn or, better still, ensile it either in a pit or by the big plastic bag technique.

8.5 Commentary

The lessons learned from this case were that there is no substitute for personally examining the scene of the problem, that care has to be taken in examining all aspects of animal feeding, and that one has to be prepared to visit the scene of the 'crime' at least several times. It was really sheer luck that the problem was solved by the chance observation of cattle eating hay not mentioned by the farmer or any of his staff or, of course, the cows themselves. The cause of the problem, i.e. *A.versicolor* and ST, was a complete surprise, which shows that nature can pull unexpected tricks and does not always listen to dogma spouted by mycotoxicologists and mycologists on what one should find.

Although from a commercial point of view the investigation was a success and saved the co-operative a hefty insurance claim, several questions were left unanswered.

- Does ST pass through the cow's rumen and become hydroxy-sterigmatocystin (its methyl derivative is known as aspertoxin), the analogue of aflatoxin M_1, the product found in the milk derived from AFB1? If so, what effect does it have on the consumer?
- What were the conditions that so well favored the growth of *A. versicolor*? There was always difficulty in producing it in the laboratory and the best substrate that was found for producing the labeled material was moistened shredded wheat.
- What was the source of the infection? It may have been simply the natural flora on the grass, but if that was the case then the conditions it met during the incubation must favor it, almost to the exclusion of other microorganisms on grass.
- Was the problem of *Aspergillus versicolor* in damp hay just local to this farm or was the potential for this problem to appear present on surrounding farms?

None of these questions were investigated due to lack of funding, once the problem of the ill-thrift herd had been solved.

8.6 Critical questions for discussion

1. Is thin-layer chromatography (TLC) adequate for screening feed and feed commodities for mycotoxin contamination?
2. Does screening for fungi and cytotoxicity in feeds assist or add anything to mycotoxin screening?
3. Is there any point in analysing feed for mycotoxins after the animals have developed a possible mycotoxicosis?
4. What is the role, if any, of non-government organizations analyzing food and feeds for mycotoxins?
5. Why do open silage silos constitute a hazard in animal feeding?
6. How should cut grass be handled to produce safe hay for cattle feed?
7. Should veterinarians include mycotoxicoses in their primary disease cause assessment rather than as a last resort?

8.7 Further reading

DIAZ, D. (2005). *Mycotoxin Blue Book.* Nottingham University Press, Nottingham.
VAN EGMOND, H.P. (1989). *Mycotoxins in Dairy Products*. Elsevier, London.

8.8 References

BAXTER J.A., TERHUNE S.J., and QURESHI S.A. (1983). Use of chromotropic acid for improved thin-layer chromatographic visualisation of trichothecene mycotoxins. *J. Chromatog.*, **261**:130–3.

BERETTA B., GAIASCHI A., GALLI C.L., and RESTANI P. (2000). Patulin in apple-based foods: occurrence and safety evaluation. *Food Addit. Contam.*, **17**:399–406.

COLE R.J., and COX R.H. (1981). *Handbook of Toxic Fungal Metabolites*. Academic Press, New York.

DAVIS N.D., DIENER U.L., and ELDRIDGE D.W. (1966). Production of aflatoxin B_1 and G_1 by *Aspergillus flavus* in a semi-synthetic medium. *Appl. Microbiol.*, **14**:378–80.

DUTTON M.F. (2009). The African Fusarium/maize disease. *Mycotoxin Res.*, **25**:29–39.

DUTTON M.F., and ANDERSON M.S. (1982). Role of versicolorin A and its derivatives in aflatoxin biosynthesis. *Appl. Environ. Microbiol.*, **43**:548–51.

ESSIGMANN J.M., BARKER L.J., FOWLER K.W., FRANCISCO M.A., REINHOLD V.N. *et al.* (1979). Sterigmatocystin–DNA interactions: identification of a major adduct formed after metabolic activation *in vitro*. *Proc Natl. Acad. Sci. U.S.A.*, **76**:179–83.

FITZPATRICK D.W. (1990). Mycotoxins in the food chain: Nutritional and toxicological considerations. *Can. J. Physiol.*, **68**:979–81.

FRISVAD J.C., THRANE U., SAMSON R.A., and PITT J.I. (2006). Important mycotoxins and the fungi which produce them. *Adv. Experiment. Med. Biol.*, **571**:3–31.

HENDRICKS J.D., SINNHUBER R.O., WALES J.H., STACK M.E., and HSIEH D.P.H. (1980). Hepatocarcinogenicity of sterigmatocystin and versicolorin A to rainbow trout (*Salmon gairdneri*) embryos. *J. Natl. Cancer Inst.*, **64**:1503–9.

HSIEH D.P.H., LIN M.T., and YAO R.C. (1973). Conversion of sterigmatocystin to aflatoxin B_1 by *Aspergillus parasiticus*. *Biochem. Biophys. Res. Comm.*, **52**:992–7.

KAUFMAN D.D., WILLIAMS L.E., and SUMNER, C.B. (1963). Effect of plating medium and incubation temperature on growth of fungi in soil dilution plates. *Can. J. Microbiol.*, **9**:741–51.

MWANZA M., KAMETLER L., BONAI A., RAJLI V., KOVACS M. *et al.* (2009). The cytotoxic effect of fumonisin B_1 and ochratoxin A on human and pig lymphocytes using the methyl thiazol tetrazolium (MTT) assay. *Mycotoxin Res.*, **25**:233–8.

OTTENEDER H., and MAJERUS P. (2000). Occurrence of ochratoxin A in wines: influence of the type of wine and its geographical origin. *Food Addit. Contam.*, **17**:793–8.

PATTERSON D.S., and ROBERTS B.A. (1979). Mycotoxins in animal feedstuffs: Sensitive thin layer chromatographic detection of aflatoxin, ochratoxin A, sterigmatocystin, zearalenone and T-2 toxin. *JAOAC*, **62**:1265–7.

PERKOWSKI J., MIEDANER T., GEIGER H.H., MULLER H.M., and CHELKOWSKI J. (1995). Occurrence of deoxynivalenol (DON), 3-acetyl DON, zearalenone and ergosterol in winter rye inoculated with *Fusarium culmorum*. *Cereal Chem.*, **72**:205–9.

SANGARE-TIGORE B., MOUKHU S., KOUADIO H.J., BETBEDER A.M., DANO D.S. *et al.* (2006). Co-occurrence of aflatoxin B_1, fumonisin B_1, ochratoxin A and zearalenone in cereals and peanuts from Cote d'Ivorie. *Food Addit. Contam.*, **23**:1000–7.

SREEMANNARAYANA O., FROHLICH A.A., and MARQUARDT R.R. (1987). Acute toxicity of sterigmatocystin to chicks. *Mycopath.*, **97**:51–59.

SUN X.M., ZHANG X.H., WANG H.Y., CAO W.J., YAN X. *et al.* (2002). Effects of sterigmatocystin, deoxynivalenol and aflatoxin G_1 on apoptosis of human peripheral blood lymphocytes *in vitro*. *Biomed. Environ. Sci.*, **15**:145–52.

TOTEJA G.S., MUKHERJEE A., DIWAKAR S., SINGH P., SAXENA B.N. *et al.* (2006). Aflatoxin B_1 contamination in wheat grain samples collected from different geographical regions of India: A multicenter study. *J. Food Protect.*, **69**:1463–7.

VESONDER R.F., and HORN B.W. (1985). Sterigmatocystin in dairy cattle feed contaminated with *Aspergillus versicolor*. *Appl. Environ. Microbiol.*, **49**:234–5.

XING L.X., ZHANG X.H., LI Y.H., YAN X., WANG J.J. *et al.* (2005). Effects of sterigmatocystin on HLA-1 expression of human peripheral blood mononuclear cells *in vitro*. *Wei Sheng Yan Jiu*, **34**:454–6. (Abstract in English on PubMed PMID 16229275).

YOUNG J.C. (1981). Variability in the content and composition of alkaloids found in Canadian ergot I Rye. *J. Environ. Sci. Health B*, **16**:83–111.

9

Investigations of foodborne outbreaks caused by a parasite in Canada

B. Dixon, Health Canada, Canada, L. Landry, Public Health Agency of Canada, Canada and F. Milord, Direction de la Santé Publique de la Montérégie, Canada

Abstract: Two well-documented foodborne outbreaks of diarrheal illness associated with infection with *Cyclospora cayetanensis* occurred in Canada in the spring and early summer of 2005. One outbreak occurred in Ontario and the other in Quebec. This chapter describes these outbreaks and the epidemiological and traceback investigations that followed, as well as the risk management decisions that were made. A number of surveillance studies have reported the presence of this parasite in a variety of vegetables, including leafy greens in Nepal, Egypt, and Costa Rica, market vegetables in Peru, water spinach in Cambodia, and herbs in Vietnam. However, there are a variety of hurdles faced in making an association between a specific food and illnesses resulting from infection with protozoan parasites.

Key words: *Cyclospora*, foodborne, outbreak, fresh produce, food testing, traceback investigation, control.

9.1 Introduction to the cases

9.1.1 How did it all begin?

Two well-documented foodborne outbreaks of diarrheal illness associated with infection with *Cyclospora cayetanensis* occurred in Canada in the spring and early summer of 2005. One outbreak occurred in Ontario and the other in Quebec.

9.1.2 Ontario outbreak

Toronto Public Health was notified of a suspected 'food poisoning' incident involving students and staff of a Toronto school attending a retreat with catered meals in April 2005. A total of 63 students and staff attended this retreat, of whom

40 met the case definition. Age and sex distribution was comparable for both cases and non-cases. A probable case was defined as any person who consumed food prepared for this retreat and became ill after attending the retreat, and experienced multiple episodes of diarrhea and/or vomiting and at least one of stomach cramps/ pain, increased gas/bloating, nausea, loss of appetite or fatigue. Sixteen cases were subsequently laboratory confirmed, and 24 others were probable cases.

The most common symptom among cases was diarrhea (95.0%), while other common symptoms included abdominal cramps (90.0%), loss of appetite (90.0%), fatigue (90%), gas/bloating (85.0%), and nausea (82.5%). The mean incubation period for onset of symptoms was 8 days, with a range of 6–13 days. The duration of illness ranged from 2–10 days, with a median of 4 days. Sixteen cases visited a physician for treatment, and one case was hospitalized.

A retrospective cohort study was conducted among the students and staff and their families to identify the source and extent of the outbreak. In this investigation, standardized questionnaires listing all menu items were distributed and were completed by 98.4% of the attendees. Attack rates and relative risks were calculated for each meal and food item served. Univariate analysis revealed that only drinking tap water and eating pasta salad were associated with significant risk of illness. Tap water was, however, ruled out because overall consumption was not associated with a risk of illness; the water source was well water that was filtered, chlorinated, and treated with UV light; water testing showed no elevation in fecal coliforms or turbidity; and no visitors to the retreat before or after the event reported illness. The risk of illness was highest in those individuals who consumed pasta salad, with an 82.2% attack rate. The pasta salad contained pesto sauce (prepared using fresh basil, fresh garlic and oil), pasta and sun-dried tomatoes cooked together, mayonnaise, olives, sugar, salt, pepper, and balsamic vinegar. The only fresh produce ingredients were basil and garlic. No foods were available for testing following this outbreak. There was no illness reported among food handlers at this event.

9.1.3 Quebec outbreak

The Public Health Office in Montérégie Region, Quebec (Fig. 9.1), was informed of gastrointestinal illness in personnel from seven schools 1 week after consuming meals at a restaurant on 29 or 30 June 2005. This restaurant provided a special menu available to school guests only. A total of 250 people were possibly exposed. Of the 160 people interviewed, only 12 did not report illness, indicating an attack rate of over 90%. Among the 148 people ill, 20 were laboratory confirmed, 122 were probable cases, and six were unclassified cases. A probable case was defined as having diarrhea along with at least one other symptom, or, in the absence of diarrhea, at least five symptoms including three gastrointestinal symptoms.

Symptoms in the cases began, on average, 7.2 days after exposure. Diarrhea was the most commonly reported symptom (96%), with a duration of 5–15 days, while nausea was reported in 88%, fatigue/weakness in 87%, and cramps in 85%. Other symptoms included fever, headache, and vomiting. There were four people hospitalized and four others kept overnight. Sixty-eight percent of cases took

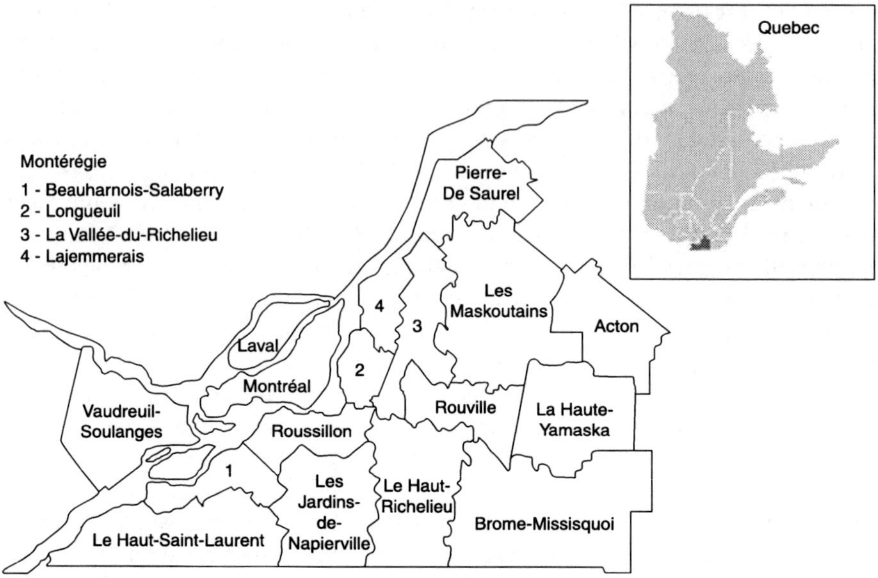

Fig. 9.1 Map of the Montérégie Region in southwestern Quebec, Canada.

antibiotics for 7–10 days, and approximately 40% noted some improvement in symptoms. Eight patients had persisting symptoms and were seen by physicians at a tropical medicine clinic.

In order to complete a retrospective cohort study, a questionnaire was distributed to attendees and was completed by 133 cases and 12 non-cases regarding the foods eaten at the meals. All cases consumed the appetizer, which consisted of bocconcini cheese and tomato with basil, while only 67% of the non-cases consumed it. The main course consisted of poultry, pork, or salmon, with cooked vegetables, and the dessert was chocolate cake with fresh fruits. Three employees of the restaurant, including a cook who prepared the appetizer and main courses, and two waiters who tasted the appetizer and dessert, also became sick a week after the event. Fresh basil used in the appetizer was considered to be the most likely cause of this outbreak (Milord *et al.*, 2012).

9.2 Significance of the cases

9.2.1 Cyclosporiasis

Cyclospora cayetanensis is a recently identified protozoan parasite responsible for enteric disease in humans worldwide, particularly in the endemic countries of Nepal, Haiti, Peru and Guatemala. The transmission stage of *C. cayetanensis*, known as an oocyst, is spherical and 8–10 μm in diameter. Unsporulated (immature) oocysts are shed with the feces of infected individuals into the environment where they undergo sporulation and become infectious within 7–15 days (Ortega and Sanchez, 2010). The infectious dose of oocysts is thought to be relatively low, and the incubation period is approximately one week. The most common symptom of infection is diarrhea, with a variety of other abdominal complaints also reported, along with fatigue and fever. If left untreated, these symptoms may persist anywhere from a few days to several weeks. These symptoms may be more severe and long-lasting in very young or old people, or immunocompromised individuals. Those at greatest risk of infection include travelers to endemic countries. In developed countries, most people are susceptible to infection, while in endemic countries susceptible populations are restricted to the very young and the very old (Ortega and Sanchez, 2010).

Cyclosporiasis is generally diagnosed by microscopic examination of stools for the presence of oocysts using a variety of methods including wet-mount, modified iron hematoxylin, and/or modified acid-fast. In addition, *C. cayetanensis* oocysts are auto-fluorescent and appear as bright blue circles when wet-mounts are examined under UV light (Fig. 9.2). Other methods available for the detection

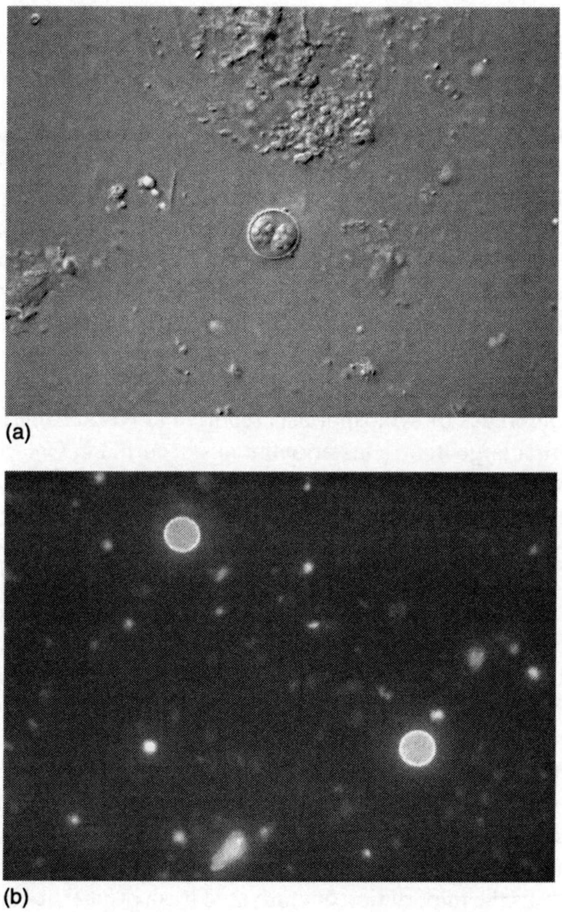

Fig. 9.2 *Cyclospora cayetanensis* oocysts. (a) Differential interference contrast (DIC) microscopy; (b) epifluorescence microscopy.

of this pathogen in stools include the polymerase chain reaction (PCR) (Ortega and Sanchez, 2010) and flow cytometry (Dixon *et al.*, 2005). Cyclosporiasis is effectively treated with the antibiotic bactrim (trimethoprim-sulfamethoxazole).

9.2.2 Transmission of illness

Due to the sporulation period in the life cycle of *C. cayetanensis*, person-to-person transmission of cyclosporiasis through the fecal-oral route is considered to be very unlikely. Similarly, zoonotic transmission is also unlikely as *C. cayetanensis* appears to be largely restricted to human hosts. While cyclosporiasis was originally recognized as a waterborne illness, numerous foodborne outbreaks have occurred in Canada and the US since 1995 (Ortega and Sanchez, 2010; Dixon *et al.*, 2011). Cyclosporiasis became a nationally notifiable disease in Canada in 2000.

9.3 The background to the cases: previous outbreaks and challenges associated with detection and control

9.3.1 Epidemiological aspects of foodborne outbreaks of cyclosporiasis in North America

It has been estimated that 90–99% of cyclosporiasis cases in the US are foodborne (Mead *et al.*, 1999; Scallan *et al.*, 2011). Epidemiological and traceback studies have implicated a variety of foods, particularly fresh imported produce, in these outbreaks. There is a distinct seasonality in foodborne outbreaks of cyclosporiasis, whereby most have been reported in the spring to early summer (Herwaldt, 2000). The reasons for this are not entirely clear and are only partially explained by environmental factors such as temperature and rainfall.

Foodborne outbreaks of cyclosporiasis reported in North America are listed in Table 9.1. The first large multi-cluster outbreaks occurred in Ontario, Canada and in the US in the spring of 1996 and 1997, and were associated with the consumption of fresh raspberries, and possibly blackberries, imported from Guatemala (Dixon *et al.*, 2011). Mesclun lettuce was also implicated in an outbreak in Florida in early 1997. In 1998, another outbreak associated with Guatemalan raspberries was reported in Ontario. While fresh raspberries from Guatemala were not accepted for importation into either Canada or the US in 1999, outbreaks were still reported in both Ontario and Florida, with the Canadian outbreak possibly being associated with fresh Guatemalan blackberries, and the US outbreak associated with fresh berries of unknown origin. A qualitative risk assessment on *Cyclospora* on raspberries from Guatemala was developed by Health Canada in 1999 to assess the health risks to consumers in Canada associated with this commodity, and to guide subsequent risk management decisions. A policy on fresh raspberries and blackberries imported from Guatemala was drafted in 2000 and revised in 2009. This policy permits the importation of cultivated fresh raspberries and blackberries between 15 August and 14 March of each year, and restricts their importation between 15 March and 14 August, which corresponds to a high-risk period when

Guatemalan fresh raspberries and blackberries could be contaminated with *Cyclospora*. Subsequent outbreaks in Canada and the US were associated with a variety of fresh produce items including basil, Thai basil, raspberries, cilantro and snow peas originating from a number of countries (Dixon *et al.*, 2011).

Table 9.1 Foodborne outbreaks of cyclosporiasis in North America

Date	Location	No. of cases	Implicated food	Reference
May 1995	Florida	38	Raspberries (Guatemala?)	Koumans *et al.*, 1998
May–June 1995	New York	32	Fruit?	Herwaldt, 2000
May 1996	Boston, MA	57	Berries	Fleming *et al.*, 1998
May 1996	Charleston, SC	38	Raspberries (Guatemala)	Cáceres *et al.*, 1998
May–June 1996	US/Ontario	1465	Raspberries (Guatemala)	Herwaldt *et al.*, 1997; Manuel *et al.*, 1999; 2000
March–April 1997	Florida	220	Mesclun (Peru)	Herwaldt, 2000
April–May 1997	US/Ontario	1012	Raspberries (Guatemala)	DeGraw *et al.*, 1997; Herwaldt *et al.*, 1999
June–July 1997	Washington, DC	260	Basil	Pritchett *et al.*, 1997
September 1997	Virginia	21	Fruit plate	Herwaldt, 2000
December 1997	Florida	12	Mesclun (Peru)	Herwaldt, 2000
May 1998	Ontario	192	Raspberries (Guatemala)	Toronto Public Health *et al.*, 1998
May 1999	Ontario	104	Blackberries?	Herwaldt, 2000
May 1999	Florida	94	Berries	Herwaldt, 2000
July 1999	Missouri	62	Basil (Mexico or US)	Lopez *et al.*, 2001
July 1999	Vancouver, BC	15	Unknown	BC Centre for Disease Control, 2002 (unpublished data)
May 2000	Atlanta, GA	21	Raspberries (Guatemala)	Jack Guzewich, USFDA, 2002 (pers. comm.)
June 2000	Philadelphia, PA	54	Raspberries (Guatemala)	Ho *et al.*, 2002

(*Continued*)

Table 9.1 Continued

Date	Location	No. of cases	Implicated food	Reference
May 2001	Vancouver, BC	17	Thai basil (US)	Hoang *et al.*, 2005
January 2002	Vermont	22	Raspberries (Chile?)	Jack Guzewich, USFDA, 2002 (pers. comm.)
June–July 2003	Vancouver, BC	10	Cilantro?	BC Centre for Disease Control, 2003 (unpublished data)
February 2004	Texas/Illinois	95	Unknown	Ortega and Sanchez, 2010
June–July 2004	Pennsylvania	96	Snow peas (Guatemala)	Crist *et al.*, 2004
April 2005	Florida	592	Basil	Ortega and Sanchez, 2010
April 2005	Ontario	40	Basil (Latin America)	Public Health Agency of Canada, 2005 (unpublished data)
June 2005	Quebec	148	Basil (Mexico)	Milord *et al.*, 2012

9.3.2 Food testing

A number of surveillance studies have reported the presence of this parasite in a variety of vegetables, including leafy greens in Nepal, Egypt, and Costa Rica, market vegetables in Peru, water spinach in Cambodia, and herbs in Vietnam (Dixon *et al.*, 2011). However, there are a variety of hurdles faced in making an association between a specific food and illnesses resulting from infection with protozoan parasites. With the relatively long incubation times of many of these parasites, and the difficulties in diagnosing human cases, illnesses are generally not reported and outbreaks are not recognized or investigated until well after the exposure event. Most foods, particularly fresh produce, are either consumed or disposed of due to their short shelf lives, and little or no material is left to test. Further, with the exception of *Toxoplasma gondii* in meats, there are currently no standardized methods available for the detection of protozoan parasites in foods, and published methods often have relatively low recovery efficiencies.

The elution and concentration of protozoan parasites from foods generally involves agitation of a food sample in a buffer with detergent, followed by centrifugation. However, parasitic cysts and oocysts can be very difficult to wash off produce due to the porous nature or rough surfaces of some fruits and vegetables. Further, the number of parasites present in or on foods is generally much lower than that found in fecal material, and, since protozoans will only multiply in the host, enrichment steps such as those used in the detection of bacterial pathogens in foods are not generally available.

Once oocysts are eluted, a variety of microscopical methods can be used for the detection of *C. cayetanensis* in foods. PCR methods developed for clinical diagnoses have also been useful in detecting the presence of this pathogen in foods.

As a result of the challenges in testing foods for the presence of *C. cayetanensis*, an association between illnesses and a specific food or foods is generally based on epidemiological data alone rather than actual detection of parasites. However, *C. cayetanensis* has been detected in implicated foods by means of PCR in two separate outbreaks in the US (Lopez *et al.*, 2001; Ho *et al.*, 2002).

9.3.3 Control measures

The contamination of produce and other foods with *C. cayetanensis* oocysts may occur directly by means of infected pickers, sorters, inspectors, or food handlers, and is due to poor personal hygiene, namely lack of hand washing. The use of 'night soil' to fertilize crops may also play a role. Contamination may also occur indirectly through contaminated water used for irrigation, mixing pesticides, or washing hands and equipment. Therefore, control measures to reduce the risk of foodborne cyclosporiasis include farm-level controls, post-harvest disinfection, and consumer-level controls.

There is little information available regarding environmental resistance of *C. cayetanensis* oocysts, although they are considered to be more susceptible to adverse environmental conditions than *Cryptosporidium* spp. For example, sporulation may be retarded in oocysts stored at either 4 °C or 37 °C for 14 days (Smith *et al.*, 1997). These findings could have important implications regarding risk control during production, storage and transportation. Desiccation inactivates oocysts on produce, although the moist, irregular surfaces of some fruits and vegetables likely provide some protection. While most of the work done on chemical and physical disinfection of protozoans in foods has involved *Cryptosporidium* spp., studies have been done on the effectiveness of commercial freezing, irradiation, ultraviolet light, and high hydrostatic pressure against *C. cayetanensis*, or *Eimeria acervulina* as a surrogate organism (Kniel *et al.*, 2007; Ortega and Sanchez, 2010). As with other protozoan parasites, *C. cayetanensis* oocysts are thought to be very resistant to the chemical disinfectants commonly used in the food industry as well as in water treatment.

There are also some important control points at the consumer level. As with all enteric pathogens, good personal hygiene, particularly hand washing, is of considerable importance in reducing the risk of transmission. However, as discussed, person-to-person transmission of cyclosporiasis is unlikely, and oocysts contaminating foods are not immediately infective. While rinsing of fresh fruits and vegetables with water is recommended for reducing the risk of transmission of pathogens, it is unlikely that this practice will remove all *C. cayetanensis* oocysts from the foods. Furthermore, delicate fruits such as raspberries will not tolerate anything more than a gentle rinse. Other treatments such as cooking and freezing may be used as final barriers against transmission, although most of the produce items implicated in outbreaks of cyclosporiasis are typically consumed raw. Further, there are few data available regarding the susceptibility of *C. cayetanensis* to cooking and freezing temperatures. Recent studies, however, suggest that *C. cayetanensis* oocysts are relatively resistant to household freezing and to short exposures to high temperature (Ortega and Sanchez, 2010).

9.4 The case histories continue: traceback investigations and follow-up

Both Canadian outbreaks described in the present chapter were associated with the consumption of imported fresh basil. Based on the results of the epidemiological investigations, the Canadian Food Inspection Agency (CFIA) conducted traceback studies on the fresh basil implicated in each of these outbreaks.

Basil consumed during the Ontario outbreak was traced to a single farm in an unnamed Latin American country, which was also implicated in US outbreaks during the same period. Following the Quebec outbreak, an investigation conducted by the CFIA, along with the Ministère de l'Agriculture, des Pêcheries et de l'Alimentation du Québec (MAPAQ), led to the conclusion that fresh basil imported from a farm in Mexico was the likely source (Milord *et al.*, 2012).

9.5 Resolution and outcomes

Based on recommendations from authorities in Canada and the US following the Ontario outbreak, the farm of origin installed a water filtration system and additional field bathrooms and locker rooms with showers, initiated stool examinations for new employees and an education program for employees on hygiene, and began using colour-coded uniforms distinguishing different working areas. No import restrictions were deemed necessary in Canada.

Based on the results of the traceback study following the Quebec outbreak, an import restriction was imposed on herb-growing farms in Mexico until such time as members of the CFIA were able to visit the region. All aspects of basil production were examined at the farm of origin, including soil type, use of organic fertilizer, irrigation water quality, manual harvesting, packing and coding, and worker health and hygiene. Good Agricultural Practices and Good Management Practices were observed during the CFIA visit and, as a result, the import restriction was lifted.

By the time traceback studies were completed for the Quebec outbreak, most foods containing fresh basil from the original shipment would have been consumed or disposed of. However, the basil remaining in the marketplace at the time was recalled by the supplier, which led to subsequent recalls involving distributors, processors, retailers and restaurants. In addition to the fresh basil, the recall was extended to include some commercial pesto, which is a long shelf-life product.

9.6 Commentary

With consumer demands for fresh produce year-round and the increase in international trade, foodborne illness outbreaks and cases associated with protozoan parasites such as *C. cayetanensis* may continue to be reported in Canada and other countries. Short of continued import restrictions, however, there are numerous control measures available from the farm level through to the consumer, which may be implemented for the prevention, removal or destruction of protozoan parasites in foods. These measures may involve hazard analysis and critical control points (HACCP) and risk assessment approaches, as well as educational programs. There is also a need for continued international cooperation in terms of tracebacks of implicated produce items, and a need for standardized detection methods for foods, as well as further studies on environmental resistance and disinfection.

9.7 Acknowledgments

The authors would like to thank Toronto Public Health, Toronto, Ontario, for their collaboration during the investigation of the Ontario outbreak described in this chapter.

9.8 Critical questions for discussion

1. What is an 'attack rate' in relation to infectious diseases?
2. Why do we use the word 'associated' and not 'caused' when we relate a certain food consumption to an outbreak, such as the one described in this chapter?
3. What is meant by 'the cases were laboratory confirmed'?
4. How is it possible to investigate the source of a foodborne outbreak when no food is left to be tested after consumption of the suspect meal?
5. What is a retrospective cohort study and how is it designed?
6. What is a univariate analysis, and why is it used in outbreak investigations?
7. Why is person-to-person and zoonotic transmission of *C. cayetanensis* unlikely?
8. What is meant in food testing by 'published methods often have relatively low recovery efficiencies'?
9. What is the main difference between HACCP and Good Agriculture Practice?
10. Is there an increase in vegans (people eating raw foods) in your country, and are they more frequently sick due to food poisoning than the average population?

9.9 References

CÁCERES, V.M., BALL, R.T., SOMERFELDT, S.A., MACKEY, R.L., NICHOLS, S.E. *et al.* 1998. A foodborne outbreak of cyclosporiasis caused by imported raspberries. *J. Family Practice*, **47**:231–4.

CRIST, A., MORNINGSTAR, C., CHAMBERS, R., FITZGERALD, T., STOOPS, D. *et al.* 2004. Outbreak of cyclosporiasis associated with snow peas – Pennsylvania, 2004. *Morb. Mortal. Wkly Rep.*, **53**:876–8.

DEGRAW, E., HEBER, S., and ROWAN, A. 1997. Update: Outbreaks of cyclosporiasis – United States and Canada, 1997. *Morb. Mortal. Wkly Rep.*, **46**:521–3.

DIXON, B.R., BUSSEY, J.M., PARRINGTON L.J., and PARENTEAU, M. 2005. Detection of *Cyclospora cayetanensis* oocysts in human fecal specimens by flow cytometry. *J. Clin. Microbiol.*, **43**:2375–9.

DIXON, B.R., FAYER, R., SANTIN, M., HILL, D.E., and DUBEY, J.P. 2011. Protozoan parasites: *Cryptosporidium, Giardia, Cyclospora*, and *Toxoplasma*. In Hoorfar, J. (ed.), *Rapid Detection, Characterization and Enumeration of Food-Borne Pathogens*, pp. 349–70. ASM Press, Washington, DC.

FLEMING, C.A., CARON, D., GUNN, J.E., and BARRY, M.A. 1998. A foodborne outbreak of *Cyclospora cayetanensis* at a wedding. *Arch. Intern. Med.*, **158**:1121–5.

HERWALDT, B.L. 2000. *Cyclospora cayetanensis*: a review, focusing on the outbreaks of cyclosporiasis in the 1990s. *Clin. Infect. Dis.*, **31**:1040–57.

HERWALDT, B.L., ACKERS, M.-L., and the *Cyclospora* Working Group. 1997. An outbreak in 1996 of cyclosporiasis associated with imported raspberries. *New Engl. J. Med.*, **336**:1548–56.

HERWALDT, B.L., BEACH M.J., and the *Cyclospora* Working Group. 1999. The return of *Cyclospora* in 1997: Another outbreak of cyclosporiasis in North America associated with imported raspberries. *Ann. Int. Med.*, **130**:210–20.

HO, A.Y., LOPEZ, A.S., EBERHART, M.G., LEVENSON, R., FINKEL, B.S.J. *et al.* 2002. An outbreak of cyclosporiasis associated with imported raspberries in Pennsylvania in 2000. *Emerg. Infect. Dis.*, **8**:783–8.

HOANG, L.M.N., FYFE, M., ONG, C., HARB, J., CHAMPAGNE, S. *et al.* 2005. Outbreak of cyclosporiasis in British Columbia associated with imported Thai basil. *Epidemiol. Infect.*, **133**:23–27.

KNIEL, K.E., SHEARER, A.E.H., CASCARINO, J.L., WILKINS, G.C., and JENKINS, M.C. 2007. High hydrostatic pressure and UV light treatment of produce contaminated with *Eimeria acervulina* as a *Cyclospora cayetanensis* surrogate. *J. Food Prot.*, **70**:2837–42.

KOUMANS, E.H.A., KATZ, D.J., MALECKI, J.M., KUMAR, S., WAHLQUIST, S.P. *et al.* 1998. An outbreak of cyclosporiasis in Florida in 1995: A harbinger of multistate outbreaks in 1996 and 1997. *Am. J. Trop. Med. Hyg.*, **59**:235–42.

LOPEZ, A.S., DODSON, D.R., ARROWOOD, M.J., ORLANDI, P.A.J., DA SILVA, A. *et al.* 2001. Outbreak of cyclosporiasis associated with basil in Missouri in 1999. *Clin. Infect. Dis.*, **32**:1010–7.

MANUEL, D.G., SHAHIN, R., LEE, W., and GRMUSA, M. 1999. The first reported cluster of food-borne cyclosporiasis in Canada. *Can. J. Public Health*, **90**:399–402.

MANUEL, D.G., NEAMATULLAH, S., SHAHIN, R., REYMOND, D., KEYSTONE, J. *et al.* 2000. An outbreak of cyclosporiasis in 1996 associated with consumption of fresh berries – Ontario. *Can. J. Infect. Dis.*, **11**:86–92.

MEAD, P.S., SLUTSKER, L., DIETZ, V., MCCAIG, L.F., BRESEE, J.S. *et al.* 1999. Food-related illness and death in the United States. *Emerg. Infect. Dis.*, **5**:607–25.

MILORD, F., LAMPRON-GOULET, E., ST-AMOUR, M., LEVAC, E., and RAMSAY, D. 2012. *Cyclospora cayetanensis*: a description of clinical aspects of an outbreak in Quebec, Canada. *Epidemiol. Infect.*, **140**:626–32.

ORTEGA, Y.R., and SANCHEZ, R. 2010. Update on *Cyclospora cayetanensis*, a food-borne and waterborne parasite. *Clin. Microbiol. Rev.*, **23**:218–34.

PRITCHETT, R., GOSSMAN, C., RADKE, V., MOORE, J., BUSENLEHNER, E. *et al.* 1997. Outbreak of cyclosporiasis – northern Virginia – Washington, DC – Baltimore, Maryland, metropolitan area, 1997. *J. Am. Med. Assoc.*, **278**:538–9.

SCALLAN, E., HOEKSTRA, R.M., ANGULO, F.J., TAUXE, R.V., WIDDOWSON, M.-A. *et al.* 2011. Foodborne illness acquired in the United States—major pathogens. *Emerg. Infect. Dis.*, **17**:7–15.

SMITH, H.V., PATON, C.A., MTAMBO, M.M.A., and GIRDWOOD, R.W.A. 1997. Sporulation of *Cyclospora* sp. oocysts. *Appl. Environ. Microbiol.*, **63**:1631–2.

TORONTO PUBLIC HEALTH *et al.* 1998. Outbreak of cyclosporiasis – Ontario, Canada, May 1998. *Morb. Mortal. Wkly Rep.* **47**:806–9.

Part II

Source tracing

10

Troubleshooting the environmental source of contamination with *Listeria monocytogenes* in a typical small food manufacturing plant in Ireland

K. N. Jordan, K. Hunt and E. M. Fox, Teagasc Food Research Centre, Ireland

Abstract: Ready-to-eat artisan food products are an important part of the food manufacturing industry in Ireland. However, such food products are susceptible to contamination with *Listeria monocytogenes*, a foodborne pathogen that causes listeriosis. Food contamination with *L. monocytogenes* is potentially a serious public health risk that is taken seriously by the food manufacturing industry. The following case study describes how an artisan food manufacturer addressed an occurrence of *L. monocytogenes* in the processing facility.

Key words: Ready-to-eat dairy foods, *Listeria*, typing, tracing.

10.1 Introduction to the case

In Ireland, the primary measure of food safety is currently end-product analysis. As part of this end-product analysis, food from one of the producers (a small-scale manufacturer producing <10 tonnes per annum) was regularly positive for *Listeria monocytogenes*. Because *L. monocytogenes* is regularly found in the environment, its complete absence from food products, especially artisanal products, is unreasonable. Issues with *L. monocytogenes* are very serious for small producers as large quantities of product may need to be destroyed, resulting in heavy losses. Because of the serious potential public health implications of contaminated end-product, and the potential commercial loss involved, this situation was investigated further.

The Irish speciality food industry consists of many small, family-owned manufacturers that are very important to maintaining employment in rural areas.

They also export their product and contribute to the image of Ireland as the 'Food Island'. Food safety and quality are extremely important to them, and the responsibility for food safety is taken seriously.

10.1.1 How did it all begin?

As part of a sampling strategy at one production facility, the regular occurrence of *L. monocytogenes* warranted further investigation. The initial analysis involved an assessment of the number of contaminating *L. monocytogenes*, the ability of the product to support growth of *L. monocytogenes*, their location either in or on the product and the strain similarity of the isolates obtained. Such analysis helps to assess the type of issues involved.

10.1.2 Numbers

As reflected by the EU regulation, the numbers of *L. monocytogenes* on a sample is an important issue – greater than or less than 100 cfu/g. In the current situation, 90% of the samples had numbers <100 cfu/g.

10.1.3 Growth

The ability of a product to support the growth of *L. monocytogenes* is important. If the product does not support growth, and the numbers are <100 cfu/g, then the human risk profile is reduced as the numbers will stay <100 cfu/g. However, if growth is supported, the risk profile is increased dramatically as the numbers (if present) will increase to levels dependent on growth rate, temperature, etc. In the current situation, the product did support the growth of *L. monocytogenes* at a rate that could result in numbers of 10^6 cfu/g by the end of storage, if the numbers were initially 10^2 cfu/g. This was a high-risk scenario.

10.1.4 Location

Bacteria can be located in a food product, indicating contamination during the product manufacturing process, on the surface of the product, indicating environmental contamination, or both. Analysis of the current product, dividing it into surface, core or composite of both surface and core, showed that the contamination was exclusively on the surface.

10.1.5 Strain similarity

Pulsed field gel electrophoresis (PFGE) is a genotyping method used to cut the DNA from bacteria into several fragments using a specific enzyme (AscI for *L. monocytogenes*). The fragments are then separated by electrophoresis. The fragment patterns from different strains can be compared (using computer software) and, if the patterns are the same, the implication is that the strains are indistinguishable. In order to provide further evidence that strains are indistinguishable, cutting the DNA with a second specific enzyme (Apa1) is undertaken.

PFGE can be used to sub-type bacterial isolates, compare strains and therefore determine potential routes of transmission. Of the 156 isolates obtained from the current analysis, >90% of them were indistinguishable. This indicates the continuous recycling of a strain within the processing environment – contamination from the food to the environment, from the environment to the food, and so on.

10.1.6 What were the challenges involved?

From these results, we determined that the contaminating numbers were generally low but that the product could support growth. Contamination was on the surface of the product, rather than in the core of the product. This indicates environmental contamination during storage, rather than raw material contamination during the manufacturing process. However, other possibilities, such as raw material contamination, or continuous contamination of the environment from the outside, could not be ruled out. In the follow-up analysis, all three possibilities had to be considered, but the focus was on environmental contamination as the initial results indicated that this was the highest probability.

10.2 Significance of the case

Within the bacterial group '*Listeria*' there are several subspecies, of which *L. monocytogenes* is the most significant. It causes listeriosis, a foodborne illness with a mortality rate of about 30%, and a low but increasing incidence. It is thought to have a high infectious dose and is potentially harmful only to a particular risk group of young, old and immunocompromised. The organism is ubiquitous in the environment and, therefore, the regulations regarding *L. monocytogenes* are quite complex (see below).

10.2.1 Regulatory aspects

European Regulation 2073/2005 sets the microbiological criteria for foods in Europe. Microbiological criteria are designed to control the concentration and prevalence of bacteria in foods. They specify the microorganisms of concern, the analytical methods for their detection/quantification, a sampling plan defining the number and size of samples to be taken, the microbiological limits deemed appropriate at a specified point in the food chain, and the number of analytical units that should conform to those limits. More specifically, the regulation places responsibility for food safety on the food manufacturer that produces the food. The specifications for *L. monocytogenes* are quite complex. In general, absence from foods is required, although, in certain foods able to support growth of *L. monocytogenes*, manufacturers must have sufficient controls in place to ensure the concentration of *L. monocytogenes* remains below 100 cfu/g throughout its shelf life until the point of consumption.

10.2.2 Market and economic issues

In Ireland there are over 100 manufacturers of artisan food products, such as dairy products, meats and fish. Annual production is less than 1500 tonnes. Most of the producers are small, concentrated in the south and east and supplying a local market. However, particularly with regard to cheesemakers, a few have capitalized on the export market. The volumes of export of speciality artisan products are small, less than 50 tonnes, but the value of these exports to small rural economies outweighs the volume.

Currently, the number of manufacturers of artisan products is increasing with an increase in the potential market for such products. From a consumer perspective, locally produced artisan food products are highly regarded, are considered safe, and their consumption is increasing.

Issues such as that described in the current case are very serious, not only for the manufacturer involved, but for the entire artisan food industry. An outbreak of listeriosis resulting from consumption of an artisan product would have a long-term negative effect on the industry. From the point of view of the manufacturer, disposal of contaminated product has the immediate effect of financial loss for that product (raw materials, staff costs, storage costs, etc.), but also a long-term effect of not being able to supply markets, thus losing market share.

10.3 The case history continues

As the cleaning and hygiene procedures in the processing facility were very good, the indications were that the isolate was persistent in the environment. This was confirmed by showing that the PFGE patterns of the current isolates were indistinguishable from those of isolates taken 10 years previously (Fig. 10.1).

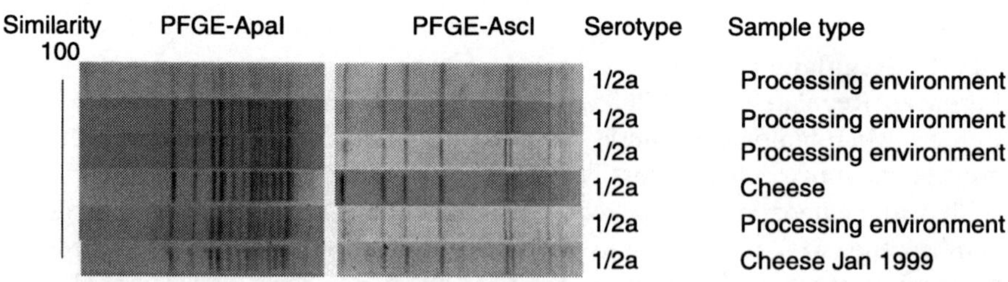

Fig. 10.1 Pulsed field gel electrophoresis (PFGE) and serotype analysis of isolates from the processing facility, showing 100% similarity with the isolate from 10 years previously (row labelled Cheese Jan 1999). Identical fragment patterns (as seen by the vertical bands) can be seen for all isolates. In addition, all strains had the same serotype. Serotyping is another method of differentiating *L. monocytogenes* strains, although in general its power of discrimination is less than PFGE.

Therefore, the probability was that the *L. monocytogenes* isolate had remained in the processing environment for 10 years, controlled but not eliminated by good hygiene practices. However, as stated previously, raw material or external re-contamination could not be ruled out entirely.

As the likely cause of the contamination was environmental, swabbing of surfaces in the manufacturing and storage areas was undertaken. With a view to determining the source of contamination, raw materials and external areas were also analysed. The results showed that the primary areas of contamination were surface areas and drains in the storage rooms. While raw materials and external areas yielded the occasional isolate, the PFGE profile was different from that of the commonly found persistent strain. This result again confirmed the probability of internal recycling of the contaminant. From the swabbing analysis results, it was obvious that the best solution to addressing the issue was to shut down manufacturing for a week and undertake a thorough cleaning of the facility. In association with this, batches of product that were identified as being contaminated were destroyed. Despite these measures, contamination was still evident and further investigation was required.

10.3.1 Troubleshooting

Not all the product was contaminated and so the uncontaminated product needed to be stored on the premises. In order to do this, all uncontaminated product was placed in one storage room. The manufacturing area, other store rooms, the packing area and general purpose areas were thoroughly cleaned. The product was then moved to one of the clean rooms while that room was cleaned.

While this was being done, workflows within the dairy facility were examined. In a small manufacturing facility, it is not practical to have specific staff for specific tasks and it is necessary for all the staff to be involved in the different activities of manufacture, preparation, packing, cleaning, etc. Therefore, proper workflows are difficult to maintain as staff move from one task to the next. Good procedures of hygiene, footwear changing and hand-washing when moving from one task to another can reduce the risks involved.

To make the situation more complex, many small processing facilities are not purpose-built factories, but have developed as the needs of the manufacturer grew. Hence, the workflows tend to be haphazard. A plan for workflow changes and staff movement restrictions was drawn up. This was implemented once manufacturing recommenced.

On recommencement of manufacture and implementation of altered workflows, environmental swabbing continued. Unfortunately, the problem was not resolved. Both end-product analysis and environmental swabs continued to show the presence of *L. monocytogenes*. In an effort to identify the source of the contamination, areas likely to harbour *L. monocytogenes*, such as air conditioning units, drains and crevices, were targeted for swabbing. While more isolates of the common PFGE profile were found, no source was identified. In particular, no isolates were obtained from the air or the air conditioning units. Drains were commonly contaminated and the manufacturer installed a new drainage system with back-flow prevention. The cleaning detergent in use at the facility was changed in case the strain associated with the facility had acquired resistance.

The problem continued. A further week of closure and cleaning followed. Further environmental swabbing, raw material analysis and external environment analysis were undertaken. No isolates were obtained from the raw materials or external samples. Occasional positive swabs were obtained, particularly from the walls and floors of the storage rooms.

10.4 Resolution and outcomes

During this time, end-product analysis continued. Only product that was negative for *L. monocytogenes* was distributed during this period. Manufacturing recommenced at about 50% of capacity. When manufacturing was resumed, environmental swabbing continued. When a positive swab was found, this signalled a targeting of that area for specific cleaning attention. No source of the *L. monocytogenes* contamination was found. Analysis of environmental surface swabs continues as a means of pre-empting product contamination. The issue is not solved, it is controlled.

10.5 Commentary and personal reflections

This case shows the importance of environmental analysis in assessing contamination in a processing facility. Comparison of isolates by PFGE is a valuable tool in understanding the issues and designing a suitable plan to address them. Continuous monitoring can help to pre-empt a re-occurrence. Further research on the persistent strain from this facility is ongoing. Resistance of cleaning detergent and analysis of the transcriptome (the genes transcribed) under sub-lethal detergent conditions are being studied. This will help to determine if any specific genes that facilitate persistence and survival are expressed. Hopefully, this research will contribute to the prevention of future incidences with *L. monocytogenes*.

10.6 Critical questions for discussion

1. Why is *Listeria* contamination of product so important?
2. What is a 'mortality rate'?
3. What is an 'infectious dose' and how is it defined?
4. What other microbial contaminants of ready-to-eat foods can be a threat to human health?
5. What is the difference between 'microbiological criteria' and 'food safety objectives'?

6. Does raw milk pose a bigger risk for cheese production than pasteurized milk, and what do international regulations say about using raw milk?
7. Are certain food types more at risk of *Listeria* contamination than others?
8. Are small producers at more risk of food contamination than larger producers?
9. What are microbial genotyping methods and what is the main advantage of genotyping methods in comparison to phenotyping methods?
10. What is the science of transcriptomics and why is it important to microbial food safety?
11. In the current situation, what other measures could have been taken to address the issue?

10.7 Further reading

FOX E., HUNT K., O'BRIEN M., and JORDAN K. 2011. *Listeria monocytogenes* in Irish farmhouse cheese processing environments. *Int. J. Food Microbiol.*, **145**:S39–S45.

OLIVE M. and BEAN P. 1999. Principles and applications of methods for DNA-based typing of microbial organisms. *J. Clin. Microbiol.*, **37**:1661–9.

RYSER E.T., and MARTH E.H. 2007. Listeria, *Listeriosis and Food Safety*, 3rd ed. CRC Press, Boca Raton, Florida.

TENOVER F.C., ARBEIT R.D., GOERING R.V., MICKELSEN P.A, MURRAY B.E. *et al.* 1995. Interpreting chromosomal DNA restriction patterns produced by Pulse-Field-Gel-Electrophoresis: criteria for bacterial strain typing. *J. Clin. Microbiol.*, **33**:2233–9.

TODAR K. 2011. *Listeria monocytogenes* and listeriosis. In *Online Textbook of Bacteriology*. Available at http://www.textbookofbacteriology.net/Listeria.html. Accessed 13 February 2011.

11

Tracing the source of persistent contamination of ready-to-eat chicken products with *Listeria monocytogenes* in Italy

V. Biguzzi (retired), G. Pezzotti, Istituto Zooprofilattico Sperimentale dell'Umbria e le Marche, Italy and M. V. Gianfranceschi and D. De Medici, Istituto Superiore di Sanità, Italy

Abstract: *Listeria monocytogenes* is a Gram-positive, facultative intracellular bacterial pathogen that can cause a severe invasive disease (listeriosis), mainly in immune-compromised, elderly individuals and pregnant women. Its control in the poultry chain requires strict hygiene practices. We report the case of an Italian poultry meat production plant, where high numbers of chicken wurstels (a frankfurter sausage made from chicken) were contaminated with *L. monocytogenes*. The chapter describes the significance of the problem and the actions taken to resolve it. The important points that the case demonstrates are also discussed.

Key words: Listeria, poultry, wurstel, hygiene.

11.1 Introduction to the case

Listeria monocytogenes (*L. monocytogenes*) is an important microorganism in the poultry chain, whose control requires strict hygienic practices. In some cases its presence in products may be linked to the persistence of the microorganism in processing plants. The first cases of *L. monocytogenes* contamination in chicken wurstel (a kind of frankfurter sausage made from chicken) in Italy were reported in the mid-1980s. Increasing attention by the public health authorities followed, leading to several recalls of contaminated lots from the market.

Before these incidents, *L. monocytogenes* had never been considered a significant risk in poultry products, either by public health authorities or by food business operators (FBO). The unexpected cases demonstrated, though, that poultry could represent a potential vehicle for *L. monocytogenes* and caused great concern among producers due to the high persistence of the microorganism in the environment and its potential to resist or replicate in fresh products, even when stored at low temperatures. Thus, special efforts were made to prevent *L. monocytogenes* along the whole production chain, leading, in some cases, to the modification of the processes and introduction of new effective control measures.

In the meantime, in order to clarify the legal measures to be adopted in case of contamination and to guarantee the protection of the consumers, the Italian Health Authorities laid down a set of microbiological criteria for products intended to be cooked before consumption, such as raw, frozen, pre-cooked and pasteurized foods. An additional label 'to be properly cooked before consumption' was also introduced and used on a large scale for some pre-cooked and/or pasteurized products, including chicken wurstel.

11.2 Significance of the case

L. monocytogenes is a Gram-positive, facultative intracellular bacterial pathogen that can cause a severe invasive disease (listeriosis), mainly in immune-compromised, elderly individuals and pregnant women, characterized by sepsis, meningitis, meningoencephalitis and abortion (Slutsker and Schuchat 2007). It is a ubiquitous organism that is widely distributed in the environment, including the food production environment (Gianfranceschi *et al.* 2003, Busani *et al.* 2005). Listeriosis is of great concern for public health due to its clinical severity and high mortality rate (20–30%), despite its low incidence, which is 0.36 cases per 100 000 population in Europe (EFSA/ECDC 2011). Based on the reported fatality rates and the total number of reported confirmed cases, it is estimated that in 2009 approximately 270 people died in the EU due to listeriosis (EFSA/ECDC 2011). Both sporadic and epidemic human listeriosis cases are associated with the consumption of contaminated foods (McLauchlin *et al.* 2004). A significantly increasing trend in listeriosis notification rates from 2005–2009 was noted in Austria, Denmark, Hungary, Italy, Spain and Sweden (EFSA/ECDC 2011).

Since 1993, the reporting of listeriosis is mandatory in Italy, and from 1996 epidemiological data are available on the Ministry of Health's database (www.sanita.it). An average of 38 cases/year (ranging from 32 to 59 cases/year) (0.06 cases per 100 000 population) were reported from 1996 to 2006. During 2007–2008 there was an increase of notifications with an average of 104 cases/year (ranging from 89 to 118 cases/year) (0.2 cases per 100 000 population) (provisional data for 2009 confirm this trend).

11.2.1 Regulatory aspects

Currently, in Italy, there are legal criteria for *L. monocytogenes* in food imposed by National Law (Ordinanza Ministeriale 7 December 1993) and by Commission

Regulation (EC) 2073:2005 (and its amendments). The National Law is applied to non-ready to eat (RTE) foods, including raw meat products and any other products labelled 'to be cooked before consumption', such as foods frozen, pre-cooked or pasteurized before dispatch from food production plants. The sampling plan of this National Law is a three-class attributes plan that incorporates two limits, *m* and *M*, and which distinguishes three classes of sampling results: the number of sample units to be drawn by the lot *n*, and the maximum number of sample units *c* that are allowed to fall into the region between *m* and *M*. Table 11.1 reports the different tolerance limits applied to different types of food, depending on processing treatments. The method is based on the most probable number (MPN) approach for the enumeration of *L. monocytogenes*.

The *L. monocytogenes* criteria laid down by Regulation No (EC) 2073/2005 cover, instead, primarily RTE food products, and stipulate:

- Zero tolerance in products intended for infants and for special medical purposes (*L. monocytogenes* must not be present in 25 g).
- That *L. monocytogenes* must not be present at levels greater than 100 cfu/g during the shelf life of those RTE products both able and unable to support the growth of *L. monocytogenes* (other than products intended for infants and for special medical purposes, for which the rule above applies).
- That, for RTE foods that support the growth of the bacterium, *L. monocytogenes* may not be present in 25 g at the time of leaving the production plant. However, if the producer can demonstrate to the competent authority that the 100 cfu/g limit will not be exceeded throughout the shelf life, this criterion does not apply.

The detection and enumeration method in the Commission Regulation is based on ISO 11290 part 1 (detection) and part 2 (enumeration). Regulation No (EC) 2073/2005 (*Article 5*) establishes also that 'Food Business Operators manufacturing ready-to-eat foods, which may pose a *L. monocytogenes* risk for public health,

Table 11.1 Italian criteria for the tolerable presence of *L. monocytogenes* in bulk and packaged food products or intended for using after cooking, or products which are labelled 'to be cooked before consumption' (excluding milk and dairy products)

Food product	Analytical unit		Microbiological limit (/g)	
	n	*c*	*m*	*M*
Raw food	3	2	11	110
Frozen food	5	3	11	110
Pre-cooking and pasteurized food	5	1	11	110

n = number of sample units; *c* = maximum number of sample units that are allowed to fall between *m* and *M*. *m* = the threshold value for the number of bacteria; the result is considered satisfactory if the number of bacteria in all sample units do not exceed this value. *M* = the maximum value for the number of bacteria; the result is considered unsatisfactory if the number of bacteria in one or more sample units is equal to or greater than this value.

shall sample the processing areas and equipment for *L. monocytogenes* as part of their sampling scheme'.

11.2.2 Economic and market aspects

Poultry products in Italy are a significant portion of the meat market; according to the Italian National Poultry Association (Unione Nazionale dell'Avicoltura; www.unionenazionaleavicoltura.it). In 2010 the total amount of poultry meat produced was 1 221 700 tonnes (+0.6% in respect to 2009), in particular:

- 780 400 tonnes of chicken meat,
- 88 000 tonnes of hens' meat,
- 279 300 tonnes of turkey meat,
- 74 000 tonnes of the meat of other poultry species.

In respect of chicken meat, 15% is consumed as whole carcasses, 64% as cuts (breast, legs, wings, etc.) and 21% as processed products (wurstel, roasted meat, hamburgers, etc.). Only 2% of turkey meat is commercialized as whole carcasses, 79% as cuts and 19% as processed products.

In Italy, integrated companies directly manage all components of the production chain (parental stocks, commercial farms, hatcheries, feedmills, slaughterhouses and processing plants) producing more than 90% of the poultry meat. This integrated system has been consolidated throughout the years, applying effective control procedures along the entire food chain, in a real 'farm to fork approach'.

However, in the mid-1980s, the detection of a high number of poultry or poultry products (particularly in chicken wurstel) contaminated with *L. monocytogenes* during official controls carried out by the Italian Veterinary Laboratories caused serious economic damage to this sector. The story was featured strongly by the mass media and there was a dramatic reduction in the consumption of poultry meat. This situation forced the major companies to multiply their efforts to identify effective control measures to eliminate, or reduce to an acceptable level, the presence of *L. monocytogenes* in products.

11.3 The case history continues

In this Italian chicken producing company, routine laboratory controls reported, as the weeks went by, an increasing level of contamination by *L. monocytogenes* in packaged chicken wurstel, with a real risk that levels would exceed the criteria laid down in national regulations. Thus it was necessary to investigate the process and improve control measures to reduce the contamination.

Firstly, it was hypothesized that the mechanically deboned meat (MDM) used as a major raw ingredient for wurstel could be the source of contamination; in this matrix, in fact, high rates of contamination were repeatedly observed. A more rigid application of good manufacturing practices (GMPs) and good hygienic practices (GHPs) was adopted in this part of the process; however, no significant reduction of *L. monocytogenes* in MDM was observed. On the contrary, the percentage of samples and the level of contamination, in terms of counts per gram, kept on increasing as the weeks went by (Figs 11.1, 11.2). This could make the cooking step in wurstel production ineffective at maintaining the level of contamination reasonably below the legal criteria. The same feature was observed

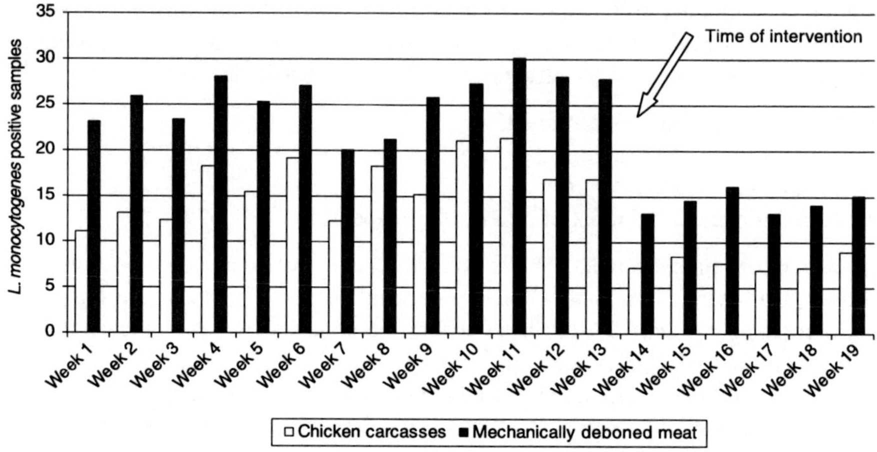

Fig. 11.1 Percentage of *Listeria monocytogenes* positive samples detected at the end of each week of production before and after the intervention (the figure does not represent the real situation at the time of the case described here, but was created by the authors to clarify the situation for teaching purposes).

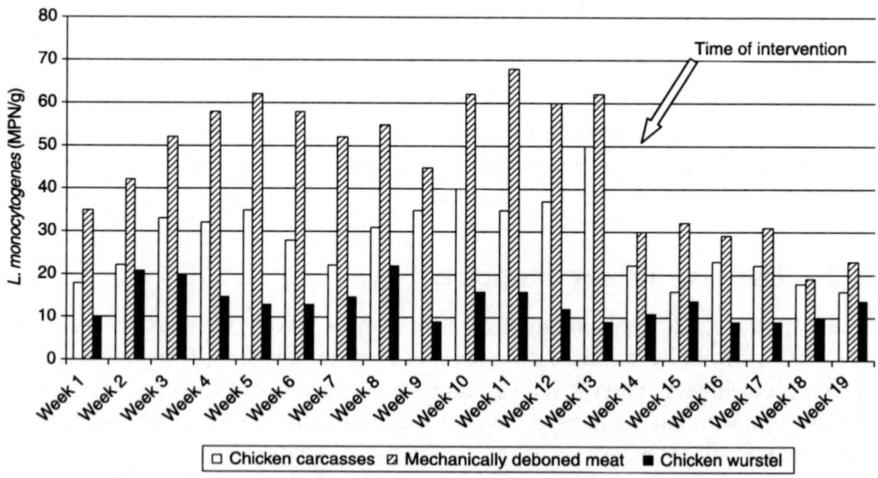

Fig. 11.2 *L. monocytogenes* enumerations (MPN/g) in chicken carcasses, mechanically deboned meat (MDM) and chicken wurstel at the end of each week of production, before and after the intervention (the figure does not represent the real situation at the time of the case described here, but was created by the authors to clarify the situation for teaching purposes).

in chicken carcasses, with a rising number of contaminated samples. At this point, it was clear that the root of the problem could be in the slaughtering plant.

The possibility of environmental contamination in the slaughtering plant was thus taken into consideration. Routine controls demonstrated that the hygienic conditions of the plant at the beginning of each working day were acceptable, but the Quality Control Unit decided anyway to proceed with further investigations and verify if there were significant differences in levels of carcass contamination during the day and at different times of the week. To evaluate the variation in behaviour of *L. monocytogenes* with time, samples were collected at the beginning (Monday) and at the end (Friday) of a week; the first lot from each day's production was sampled and tested after 2 days of storage at 0 °C in order to simulate the average storage time before distribution. From each lot (approximately 3500 chicken carcasses) 35 pools of breast skin were sampled from about 100 chicken carcasses. The results showed that 23% and 45% of the samples collected on Monday and Friday respectively were contaminated by *L. monocytogenes*.

To investigate what happened during the process, at different times in the working day, carcasses were then sampled (taking 25 g of breast skin) at different steps of the chain:

1. at the end of the slaughtering line, where carcass temperature was 37–39 °C,
2. at the end of the drying–cooling tunnel, where carcass temperature was 0–4 °C,
3. during the handling of the product for packaging (the entire procedure lasted about 20 minutes, carried out at 12 °C and 55% relative humidity),
4. after 2 days of storage at 0 °C (average time of storage of the chicken carcasses before distribution or further processing, including mechanical deboning).

For each step, carcasses were sampled at the beginning and at the end of each working day. The following results were observed:

1. At the end of the slaughtering line (before the carcasses were introduced into the tunnel) the percentage of *L. monocytogenes* contaminated carcasses never exceeded 7% and *L. monocytogenes* enumeration was always below 10 cfu/g. No differences were observed during the day (first or last working hour).
2. At the end of the tunnel, the percentage of *L. monocytogenes* in pooled samples taken in the first working hour of the day ranged from 10 to 25%, with an average level of 10–50 cfu/g, and the percentages from pooled samples taken during the last working hour of the day ranged from 20 to 40%, with an average level of 25–100 cfu/g.
3. No significant differences were observed during handling of the product for packaging, yielding similar results to those obtained after the tunnel.
4. No significant differences were observed after 2 days of storage at 0°C.

11.4 Resolution and outcomes

These tests confirmed that the source of contamination was the drying–cooling tunnel, in spite of the regular checks carried out (controls on temperature, relative humidity, defrosting, etc.) and the cleaning and disinfecting procedures routinely adopted. At the end of the working day, after routine cleaning and disinfection, the Quality Control Unit carried out a more accurate and time-consuming inspection of the tunnel, particularly of the evaporative cooling pads where residues of small feathers and plumes, released from carcasses by the strong air circulation inside the tunnel, were found. These residual materials could not be observed during routine checks due to the abundant presence of ice, but could be observed after the complete defrosting of the equipment.

The problem was solved, in part, by introducing new and more accurate cleaning and disinfection procedures. These included cleaning and disinfection of evaporative cooling pads, front and rear, and fan coolers, at the beginning and at the end of each working day. After these interventions a contamination average of 7% or below, with a maximum count of 10 cfu/g, was found on carcasses. The slaughtering process was thus considered again under control but the problem of contamination of the chicken wurstel was still occasionally present, reasonably considered to be a consequence of the re-contamination of the product during handling at the final packaging stage.

The objective, therefore, became the guarantee of total absence of *L. monocytogenes* in products, both at the time of distribution and until the end of the shelf-life. Application of GMPs and GHPs, although very rigid, was not considered sufficient to achieve this result. The process was therefore modified, including a pasteurization step (72 °C for 15 min) after the final packaging of the product. In addition, to prevent any potential risk to consumers due to thermal abuse during the shelf-life, it was also recommended on the label to cook the wurstel before consumption. After these additional precautionary actions, *L. monocytogenes* was detected in wurstel samples, but at a level that complied with the microbiological criteria set by the Italian regulations.

11.5 Commentary

This episode demonstrated that regular monitoring of the processes and products is essential for risk assessment. The solution required, other than laboratory tests, additional quality control measures based on direct inspection of the entire production chain to identify the critical points. Correct control procedures could then be applied. Sometimes, however, these measures are not sufficient to guarantee the elimination or reduction of the risk to an acceptable level. This happened in the episode described here, where only after the introduction of a final pasteurization step (pasteurization of the packaged product) was the problem solved.

11.6 Acknowledgements

This work was supported by the EU 6th Framework Programme under the project BIOTRACER, project number 036272.

11.7 Critical questions for discussion

1. How should a HACCP (hazard analysis and critical control points) team perform on-site review of the operation to verify the accuracy and completeness of the flow diagram?
2. Why is Listeria contamination a very difficult problem to solve for the food industry?
3. What temperature and relative humidity should be applied in a room normally used for handling muscle foods?
4. How can cleaning procedures be improved to reduce the contamination of *L. monocytogenes* in the food industry? What type of disinfectant should be selected?
5. How, for RTE foods that support the growth of *L. monocytogenes*, can the producer demonstrate that this microorganism will not multiply to levels of 100 cfu/g, or above, during the entire shelf life of the product?

11.8 References

BUSANI, L., CIGLIANO, A., TAIOLI, E., CALIGIURI, V., CHIAVACCI, L. *et al.* (2005). 'Prevalence of *Salmonella enterica* and *Listeria monocytogenes* contamination in foods of animal origin in Italy.' *J Food Prot*, **68**(8):1729–33.

EFSA/ECDC (2011). 'The European Union summary report on trends and sources of zoonoses and zoonotic agents and food-borne outbreaks in the European Union in 2009.' *EFSA Journal*, **9**(3):2090. Available at http://www.efsa.europa.eu/en/efsajournal/doc/2090.pdf.

GIANFRANCESCHI, M., GATTUSO, A., TARTARO, S. and AURELI, P. (2003). 'Incidence of *Listeria monocytogenes* in food and environmental samples in Italy between 1990 and 1999: serotype distribution in food, environmental and clinical samples.' *Eur J Epidemiol*, **18**(10):1001–6.

MCLAUCHLIN, J., MITCHELL, R.T., SMERDON, W.J. and JEWELL, K. (2004). '*Listeria monocytogenes* and listeriosis: A review of hazard characterisation for use in microbiological risk assessment of foods.' *Int J Food Microbiol*, **92**(1):15–33.

SLUTSKER, L. and SCHUCHAT, A. (2007). *Listeriosis in Humans*. In E. T. Ryser and E. H. Marth (eds) Listeria, *Listeriosis, and Food Safety*. CRC Press, Boca Raton.

12

An unforeseen source of aflatoxins in milk in Sweden

E. Nordkvist, National Veterinary Institute, Sweden

Abstract: Aflatoxin M1 at elevated but low levels in consumers' milk was observed in routine checks in the south of Sweden in February 2006. Up to 257 ng aflatoxin M1 per kg raw milk was found. This case describes the investigation that followed and the actions that were taken to protect the consumer. The source of the contamination was eventually traced and its route of entry to the food chain determined.

Key words: mycotoxin, aflatoxin, milk, compound feed.

12.1 Introduction to the case

In February 2006 traces of aflatoxin M1 were observed within the control program for consumer milk conducted by the Swedish Dairy Association. Aflatoxins are secondary metabolites commonly produced by the moulds *Aspergillus flavus* and *Aspergillus parasiticus*. Aflatoxins B1, B2, G1 and G2 are the most important, with B1 being the most toxic and carcinogenic in humans (International Agency for Research on Cancer, 2008). Aflatoxin M1 is a metabolite produced from aflatoxin B1 in the liver of mammals. It is known that milk can be contaminated by aflatoxin M1 if the dairy cow feed contains aflatoxin B1. Due to their high toxicity and proven carcinogenic effects, Sweden has established a rigorous control system for aflatoxins in feeds. Since 20 years ago, imported protein-rich feed materials (products of babassu, coconut, maize, palm kernel, ground nuts, shea nuts and soy beans) have been analyzed for aflatoxins.

12.1.1 How did it all begin?

Within the control program of the Swedish Dairy Association, packaged consumers' milk is analyzed with respect to aflatoxin M1 five times per year per dairy. Within this program, aflatoxin M1 was detected in early February 2006 in consumer milk from a dairy in the south of Sweden (county of Skåne) by the dairy's own control program. The levels were low, 12 ng and 13 ng aflatoxin M1 per kg milk, compared with the legal limit, which is 50 ng/kg. According to the control program, all samples with a content of 8 ng/kg or higher should be followed up in order to trace the source of contamination. In this process a total of 333 milk samples were analyzed for aflatoxin M1. During this trace-back process, incoming milk from several collection lines was found to contain elevated levels of aflatoxin M1. The common factor for the farms with contaminated milk was the use of compound feed from the same company. A compound feed is a feed produced by combining different raw materials and additives.

12.1.2 What were the challenges involved?

The very strict rules of the dairy company and the Swedish Dairy Organisation result in the investigation of every case where more than 8 ng/kg of aflatoxin M1 is detected in consumer packed milk. Such investigations invoke a series of important decisions and actions. In this case there was a need for high capacities in sampling and analysis of milk. Both analyses to detect non-compliant milk (i.e. milk with levels of aflatoxin M1 higher than 50 ng/kg) and analyses to trace the source of contamination were needed. A need for rapid methods of analysis for aflatoxins in feed and feed ingredients could also be foreseen by those managing the investigation. In addition, the investigation managers needed to be prepared to take action and ban farms producing contaminated milk from delivering to the dairy, and all of these actions and decisions had to be taken under time pressure.

12.2 Significance of the case

12.2.1 Economic aspects

Milk and dairy products are important food commodities in Sweden. Large amounts of milk products are consumed daily, especially by children and young people. About 350 000 dairy cows produce annually around 2.8 million tonnes of milk. Drinking milk, fermented milk and cream constitute 45% of the dairy products, while 33% of the milk is used for cheese production. In Sweden the annual consumption of drinking milk is close to 100 kg per capita; in addition to this about 70 kg fermented products, cheese, cream and butter are consumed. A major incident of milk contamination with aflatoxin M1 can put pressure on the whole food supply chain. For the farmer, being banned from delivery of milk causes severe economic damage. If the ban continues, disposal of the milk can cause environmental problems.

12.2.2 Regulatory aspects

Due the highly toxic and carcinogenic nature of aflatoxins, regulations containing maximum permitted levels exist for both feed and food. For milk the maximum level of aflatoxin M1 is 50 ng/kg according to Regulation (EC) No 1881/2006 (last amended in Regulation (EU) 165/2010) (European Commission 2010). The maximum levels for aflatoxin B1 in feed are stated in Directive 2002/32/EC (last amended by Regulation (EU) No 574/2011) (European Union 2011). The maximum level is 5 μg/kg for compound feed intended for dairy cows. Official control of the feed/food chain is described in Regulation (EC) No 882/2004 of the European Parliament and of the Council (European Commission 2004).

12.3 The case continues

As a result of the trace-back analyses of milk, 68 dairy farms that had been using feed from the same producer were banned from delivering milk to the dairy. Among these, one farm had 257 ng aflatoxin M1 per kg raw milk. A few more had levels above the legal limit for human consumption (50 ng/kg). All banned farms had to discard the compound feed and clean their silos and feeding systems before restocking them with uncontaminated compound feed.

The first step in tracing the source of contamination of the feed was to check with the Swedish import control if any contaminated feed had been observed. The answer was no. The feed producers and the Board of Agriculture (the official control authority) then sampled feed both at the factories and at contaminated farms. Thirty-eight samples of feed and all feed ingredients that had been used during the actual period were analyzed.

At contaminated farms, 22 feed samples or feed ingredients were tested. Only compound dairy feed from one single feed mill showed levels of aflatoxin B1 higher than the limit of quantification. In addition, contaminated feed was found in one feed mill, where ingredients including distiller's grain, rapeseed meal, sugar beet pellets, biscuit meal, soy bean and rice feed meal were subsequently analyzed (Table 12.1).

Only one commodity, rice feed meal (a by-product from the preparation of Basmati rice by a factory in the south of Sweden), showed elevated levels of aflatoxin B1. The content of aflatoxin B1 was 56 μg/kg. The rice feed meal consists mainly of the outer parts of the rice grain, containing significant levels of fibers and protein. When tracing this product to the producer of Basmati rice for human consumption, aflatoxin contents of up to 154 μg/kg were detected.

After the farms' silos and feeding systems had been process cleaned, milk samples were analyzed and, if the results were satisfactory, milk deliveries were resumed. The metabolism of aflatoxins in the body of mammals is fairly rapid. Four to five days after discontinued consumption the blood (or milk) levels of aflatoxin metabolites are negligible. There is a linear relationship between aflatoxin B1 ingested to aflatoxin M1 in milk, and the carry-over rate has been estimated to be 1–3% in cows (van Eijkeren *et al.* 2006). More exact estimates are difficult to obtain. However, it has been shown that the composition of the diet and source of aflatoxin influence the conversion rate. In the present case cattle were fed with new uncontaminated feed for a period of 3 days before new milk samples were taken.

Table 12.1 Content of aflatoxin B1 in different feeds sampled at the factory and at farm level in the trace-back study. It can be seen that only feeds from one factory were contaminated

Feed	Aflatoxin B1 (μg/kg)
Pooled dairy feed from delivery at feed mill	3.6
Pooled dairy feed from January to March 2006	7.2
Pooled pig feed from January to March 2006	26
Compound dairy feed A at feed mill	22
Compound dairy feed B at feed mill	0.6
Compound dairy feed A at farm level	26
Compound dairy feed B at farm level	0.6
Straw for feeding	0.6
Compound dairy feed from a different feed mill	<0.2

The importance of thorough cleaning in conjunction with the change of feed was demonstrated clearly at the follow-up sampling stage. At the farm with the highest contamination level of milk (257 ng/kg), milk samples taken 3 days after the feed change showed contents of aflatoxin M1 of less than 2 ng/kg (the limit of detection). On the other hand, there were examples of farms with lower levels of contamination that still had elevated, albeit below the legal limit, levels (2–30 ng/kg) of aflatoxin M1. These farms had apparently been less careful when cleaning their silos and other parts of their feeding systems.

12.4 Resolution and outcomes

The present case was well handled by the organizations and authorities. Rapid trace-back with potent actions and follow-up in order to maintain food safety was taken. No milk with contamination of aflatoxin M1 above the legal limit was put on the market. Still, the case demonstrates the impact of feed on the food chain.

12.5 Commentary and personal reflections

Sweden has several control systems in place to prevent aflatoxins from entering the food chain. Firstly, there is the import control program that lays down rules for sampling and analysis of feed ingredients assessed to pose a risk for aflatoxin contamination. These commodities are analyzed before they are allowed to enter the country. A second level of control is the official control of food and feed according to Regulation 882/2004 (EC), where the authorities for both feed and food conduct controls of products for consumption by animals or humans, respectively. And finally, a control program is conducted by the dairy organization. This might seem overcautious, but in the present case only one of the systems reacted.

The lesson from the present incident is that the relationship between feed and food is not as linear as one might think in the first place. The pathway to food of animal origin is not always the simple direction field→animal→food. It is more complicated; several agricultural products of vegetable origin, intended for human consumption, undergo various processing steps such as rinsing, cleaning, polishing etc., which separate the commodities into a fraction for human consumption and a residual fraction. This latter fraction is often used as an ingredient in animal feed. Other than the Basmati rice implicated in this case, examples of agricultural products processed in this way are brewer's and distiller's grain, soy bean meal, rape seed meal, bran and similar cereal fractions, among many others.

In many of these cases, the processing consists of removing the outer parts of kernels or similar. In those cases one should be aware that contaminants, both natural (Jayaraman and Kalyanaundaram 1990) and man-made, such as pesticides, often are concentrated at or near the surface of the kernel and will therefore accumulate in the residual fraction often used as feed ingredient. Also other by-products, including residues from fat/oil processing, dairies, or from the food industry, among others, may be used for feeding purposes. In such cases, it is important to be aware of potential contaminants of natural origin such as mycotoxins or those from industrial processes.

12.6 Acknowledgments

Per Häggblom and Anna Stepinska are gratefully acknowledged for invaluable discussions and superb technical skills.

12.7 Critical questions for discussion

1. What are the food safety implications of aflatoxins?
2. Why are aflatoxins of interest in animal feed?
3. What is the carry-over rate for aflatoxins from food to feed?
4. Why in the above case was a period of 3 days applied before new milk samples were analyzed after changing the feed?
5. What was the action when contaminated milk was found at farm level?
6. Which control systems are in place to prevent the consumption of contaminated milk by the public in your country?
7. Why wasn't the rice feed meal analyzed within the import control program?
8. Discuss the possibility of there being hitherto unknown pathways of contamination of feed that might have adverse effects on food quality and safety.

12.8 References

EUROPEAN COMMISSION (2004). Regulation (EC) No 882/2004 of the European Parliament and of the Council.

EUROPEAN COMMISSION (2010). Regulation (EC) No 1881/2006 (last amended in Regulation (EU) 165/2010).

EUROPEAN UNION (2011). Commission Regulation No 574/2011.

INTERNATIONAL AGENCY FOR RESEARCH ON CANCER (IARC) (2008). *Summaries & Evaluations, Aflatoxins (Group 1)*. Available at http://www.inchem.org/documents/iarc/vol82/82-04.html. Accessed 6 November 2008.

JAYARAMAN, P. and KALYANAUNDARAM, I. (1990). Natural occurrence of toxigenic fungi and mycotoxins in rice bran. *Mycopathologica*, **110**:81–5.

VAN EIJKEREN, J.C.H., BAKKER, M.I., and ZEILMAKER, M.J. (2006). A simple steady-state model for carry-over of aflatoxins from feed to cow's milk. *Food Addit Contam*, **8**:23, 833–8.

13

Investigations prompted by the discovery of high levels of *Salmonella* in fresh herbs in the UK

N. C. Elviss and F. Jørgensen, Health Protection Agency (HPA), UK

Abstract: During a UK survey in 2007, 1.6% of the 3760 fresh herbs on retail sale tested were found to be of unsatisfactory microbiological quality. *Salmonella* species were detected in 18 of the samples and the serotypes were identified. Thirty-two human cases of *S.* Senftenberg were also identified from March 2007 in England and Wales and a further 19 in Scotland, Denmark, The Netherlands and the USA. This is a case study illustrating how vigilant surveillance, rapid strain typing and prompt liaison between public health authorities led to the tracking down of the source of an international outbreak.

Key words: *Salmonella*, herbs, basil, outbreak investigation.

13.1 Introduction

Fresh herb sales had increased considerably in the UK in years prior to 2007 with the largest market share being for coriander, basil and parsley (www.mintel.com). The nature of herb production alongside their tendency to be eaten raw makes them challenging in terms of food safety. In 2005 the first Rapid Alert System for Food and Feed (RASFF) notifications reporting detection of *Salmonella* in fresh herbs appeared (European Commission website). In that year the London Port Health Authority, liaising with the UK Health Protection Agency (HPA)'s London Food, Water and Environmental Microbiology Laboratory, decided to investigate the microbiological quality of imported fresh herbs as part of routine import sampling requirements. Of 244 samples examined, 13% were positive for Salmonellas, finding different serotypes, but no cases of salmonellosis could be attributed to these contaminated herbs. *Salmonella*-positive samples included four varieties of basil (sweet, holy, lemon, and tree basil) grown in Thailand. Finding such a high prevalence in a product likely to be eaten without further processing prompted a further investigation, this time looking at the retail sale of fresh herbs. In this 2006 pan-London study of fresh herbs, 1.7% of samples were contaminated with different serotypes of *Salmonella*. This finding provided further evidence for fresh herbs being potential sources of *Salmonella* infection in the UK, particularly considering their widespread use as an ingredient in ready-to-eat foods (Surman-Lee *et al.* 2008). It was then decided to undertake a UK-wide survey to establish the microbiological quality of fresh herbs at retail sale and this study was co-ordinated by the HPA in liaison with the Local Authority Coordinators of Regulatory Services (LACORS, now Local Government Regulation). As this study started in May 2007, surveillance of human salmonellosis in the UK had identified an increase in the number of *Salmonella* Senftenberg cases. In the following we describe how public health professionals liaised and identified the source of an outbreak as a result of these investigations.

13.2 Significance of the case

In 2007, *Salmonella* was responsible for ~14 000 reported cases of infection in England and Wales. There is a considerable burden associated with *Salmonella* cases in terms of morbidity and costs, ranging from lost work time to extended time in hospital. Transmission occurs by eating contaminated food, mainly of animal origin, or by faecal contamination from an infected person or animal. The infectious dose can be low and legislation is in place in relation to the presence of *Salmonella* in foods (see below).

13.2.1 Regulatory aspects

Ready-to-eat food is food intended by the producer or the manufacturer for direct human consumption without the need for cooking or other processing effective to eliminate or reduce to an acceptable level micro-organisms of concern (Regulation (EC) No. 2073/2005).

EC (Regulation) number 2073/2005 (as amended) states that detection of *Salmonella* in ready-to-eat pre-cut fruit and vegetables, including fresh herbs, placed on the market is unsatisfactory. An unsatisfactory result would require the product or batch of foodstuffs to be withdrawn or recalled, for the food business operator to take corrective action and to find the cause of the unsatisfactory result in order to prevent the recurrence of the unacceptable microbiological contamination. One of the challenges during outbreak and food incident investigations is traceability. The complexity of the food supply chain and the lack of identifying markers on fruit and vegetables can make it extremely difficult to track produce back to its origin. EU legislation (Regulation (EC) 178/2002) governing full traceability came into force on 1 January 2005, and requires that a food business operator can, at any point in the food chain, identify its suppliers and the businesses that it has supplied with ingredients or final food product, i.e.

traceability is determined on a one step forward, one step back basis. Nonetheless, labelling of some fresh vegetables, particularly those sold wholesale, has been shown not to be sufficient to allow proper tracing of some products. This has implications for public health, since food alerts and product withdrawals are contingent on accurate identification of the suspect product. However, the samples of retail pre-packed basil in which *S.* Senftenberg was detected in 2007 did have sufficient identifying markers to enable the original grower to be identified.

13.2.2 Economic and market aspects

The value of fresh herbs sales in the UK had reached ~£46 million in 2007 and sales were predicted to grow further. The short shelf-life of the batch from which *S.* Senftenberg was identified meant that there was no product recall undertaken as by the time the contamination was identified the shelf-life of the batch had expired. Investigation of samples taken from the grower's farm identified no further positive samples. While it is believed that farm production did not cease trading, affected UK retailers did, at least temporarily, suspend supply from this producer. As the grower was based in Israel, there is no mandate for EU authorities to investigate suppliers outside the EU. The effect on sales of fresh herbs is uncertain but there was a limited interest from the media in relation to this outbreak.

13.3 The case history continues

During the 2007 UK survey, 1.6% of the 3760 fresh herbs on retail sale tested were found to be of unsatisfactory microbiological quality. *Salmonella* species were detected in some of the samples. The first *Salmonella* strains were isolated from fresh basil in early May 2007 and immediately sent to the Salmonella Reference Laboratory (SRL) at the HPA Microbiology Services, Colindale, London. Since the SRL identified some of the isolates as *Salmonella enterica* sero-variant Senftenberg, this prompted further rapid strain typing to be undertaken as the laboratory was investigating an increase in the number of human *S.* Senftenberg cases at this time.

The *S.* Senftenberg isolates from the basil had a pulsed field gel electrophoresis (PFGE) profile designated as SSFTXB.0014, indistinguishable from the profiles obtained from the strains isolated from the human *S.* Senftenberg cases. Acting on this finding, the HPA alerted authorities and on 25 May 2007 the UK retailers, including three major supermarket chains, publicly recalled product and the Food Standards Agency (FSA) advised consumers who might have bought certain batches of packaged fresh basil that these should not be consumed (FSA 2007). The retailers withdrew all potentially affected basil products with a use-by date up to and including 28 May 2007. All of the *S.* Senftenberg strains were isolated in May 2007 (Fig. 13.1). In total, 32 human cases of illness involving this outbreak strain of *S.* Senftenberg were identified in England and Wales from 5 March to 6 June 2007 (Pezzoli *et al.* 2007).

In the LACORS HPA survey of fresh herbs *Salmonella* spp. was detected in a total of 18 of 3760 (0.5%) samples that were collected at retail in the UK from

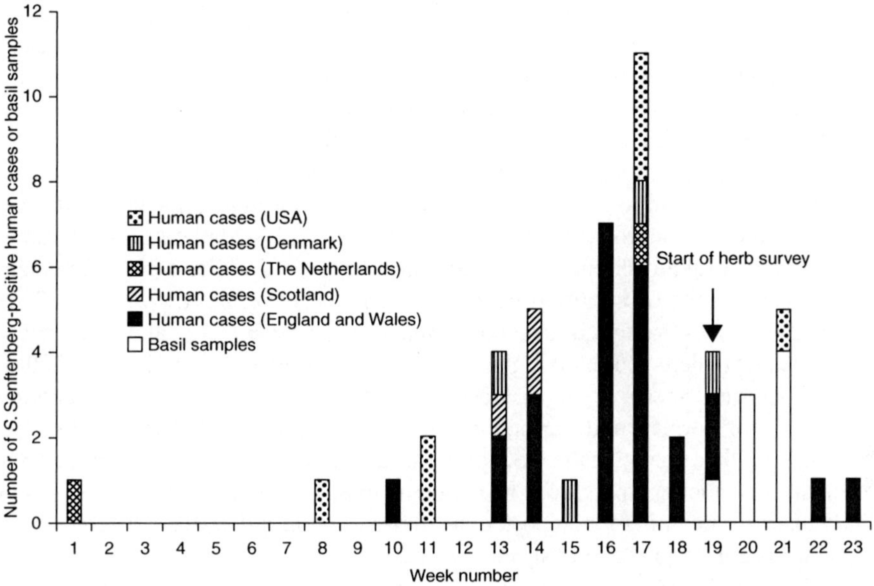

Fig. 13.1 Weekly detection of *Salmonella* Senftenberg PFGE designated type SSFTXB.0014 in samples obtained from 41 human cases (source: Pezzoli *et al.* 2008) and in basil samples collected as part of a UK coordinated study in 2007. Arrow indicates start of herb survey (source: Elviss *et al.* 2009).

May to October 2007. Ten different *Salmonella* serotypes were obtained from the 18 contaminated herb samples. The serotypes identified in the contaminated samples were Senftenberg ($n = 8$), Agona ($n = 2$), Anatum ($n = 1$), Durban ($n = 1$), Javiana ($n = 1$), Mgulani ($n = 1$), Montevideo ($n = 1$), Unnamed (II 16:g, t: z42) ($n = 1$), Virchow ($n = 1$) and mixed Newport & Virchow ($n = 1$). These serotypes were obtained from ten samples of different herbs (coriander (3), curry leaves (2), mint (1), parsley (2), sweet basil (1) and walleria (1)) (Elviss *et al.* 2009). The study also found that 54% (1665/3103) of the fresh herbs sampled were imported products, with 64% of these being sourced from Israel. *Salmonella* was isolated from 13 herb samples grown in Spain, Israel and Sri Lanka, but the study was unable to establish country of origin for five of the *Salmonella* positive samples. Most fresh herbs were displayed at >8 °C and all *Salmonella* positive samples were collected from displays with temperatures between 10 and 22 °C (Elviss *et al.* 2009). In all cases, the retailer and the UK FSA were immediately informed and remedial action was taken. These findings resulted in a number of EC Rapid Alert System for Food and Feed notifications being issued to EU member states.

Cases outside the UK were also discovered and, based on *S.* Senftenberg PFGE profile and plasmid carriage, altogether 55 primary human cases in England and Wales, Scotland, Denmark and the Netherlands were linked with these isolates. The USA also reported on 19 cases infected with the outbreak strain of *S.* Senftenberg between January and June 2007. Investigations into the source of the fresh basil contaminated with *Salmonella* linked it to a single grower in Israel, who exported fresh basil to the USA, North Europe and Russia. Eight samples of pre-packed/pre-cut basil from the UK survey were linked to this grower. Full investigations that were undertaken in response revealed no further contaminated product (FSA 2007; HPA 2007b; Pezzoli *et al.* 2008).

Around the same time fresh herbs were also linked to one other outbreak. In Denmark in 2006, imported basil used to make pesto was implicated in an outbreak of *S.* Anatum and enterotoxigenic *E. coli* associated with school dinners in Greater Copenhagen (Pakalniskiene *et al.* 2008). In Norway in 2005 and 2007, 28% and 15%, respectively, of imported fresh pre-cut herbs (basil, mint, coriander) from South East Asia were found to be contaminated with *Salmonella*, and 18 different serotypes were detected (Norwegian Scientific Committee for Food Safety 2008). Data from the USA surveying domestic fresh produce in 2000 and 2001 reported that 1.1% of parsley and coriander were contaminated with *Salmonella* spp., whereas another survey in the USA of imported fresh produce in 1999 found 8.8% of fresh coriander and parsley to be contaminated. The different detection rates of *Salmonella* contamination reported in the studies might reflect differences in conditions during pre- and post-harvest operations.

13.4 Resolution and outcomes

While the source of this *Salmonella* outbreak strains is most likely to be of animal origin, herbs can also be cross-contaminated through poor handling, and contamination can be exacerbated through poor storage practices. Application of good basic food hygiene would greatly reduce the risk of transmission via cross-contamination or infected food handlers. One particular challenge has recently been highlighted as there is evidence that *Salmonella* can invade plant tissues, where it may be protected from processing (e.g. washing with potable/chlorinated water) (Golberg *et al.* 2011). The current study, however, did contribute to an increased awareness of pathogens in fresh herbs and highlighted the need for standards to be set out in industry guidelines for the herb sector (as is available for produce suppliers to chilled food manufacturers) to help minimize contamination and/or bacterial growth in these products. Awareness of pathogen risks in fresh herbs also increased in retailers, possibly prompting a review of control measures, reflecting a clear recognition of the product as falling into the definition of a ready-to-eat food. Finally, there is now legislation in place which stipulates that all consignments originating from Thailand of basil (holy, sweet), coriander and mint must be checked for *Salmonella* on embarkation to EU countries.

13.5 Commentary

This outbreak highlights the importance of international surveillance systems, such as the European Centre for Disease Prevention and Control Epidemic Intelligence

Information System (EPIS), which can be vital mechanisms in their recognition and investigation. Investigations need to be timely, well designed and properly conducted, and investigators should bear in mind that the distribution of contaminated product may not be widespread or at high levels. Contaminated products are not always associated with bad practices. Responsible companies that have implemented hazard analysis and critical control point (HACCP) systems may supply contaminated product as the controls in place may not be sufficient to prevent poor-quality items being supplied to consumers. Washing is not always able to remove all pathogens, and additional control measures are needed to prevent contamination of seeds and crops in the field, during harvesting, and during processing and distribution of the fresh produce. The absence of microbiological evidence of contamination in the field or along the distribution chain is not the same as evidence of absence of product contamination. The outbreak also highlights how targeted microbiological food surveillance can be critical for identifying the source and prompt increased scrutiny of the effectiveness of food safety procedures and control of food safety for products intended to be eaten without further processing. Sharing of information was timely and detailed and important for verifying this outbreak. While this outbreak was short-lived, outbreaks can sometimes be prolonged and reach global proportions; it is difficult to assess how widespread an outbreak will become at the early stages when the source is unknown, and prompt investigation can help deliver the most cost-effective public health intervention. For fresh produce producers and retailers it is important to realize that the best systems cannot guarantee microbiological safety. Government agencies and industry need to work together to ensure effective collaboration and response to outbreaks. Finally, it should be noted that the use of analytical epidemiology to investigate outbreaks can be hampered by the number of people affected, complex distribution of products through different outlets or the widespread use of a contaminated product as an ingredient in a variety of foods. The application of a combination of food surveillance, molecular microbiology and epidemiology coordinated through established national and international networks means that information can be gathered quickly and communicated effectively.

13.6 Acknowledgements

We would like to thank Dr Christine Little of the Health Protection Agency for her critical comments.

13.7 Critical questions for discussion

1. Describe current EU legislation relevant to products such as fresh herbs and import controls.
2. What is the shelf-life of products such as fresh herbs, and how many days could pass between harvest and last day of use/shelf-life?

3. What level of primary contamination could be tolerated considering the potential for growth and death of *Salmonella* during the life of fresh herb/salad products?
4. How is primary contamination controlled?
5. Can washing of fresh herbs/salad remove all pathogens?
6. What would be the implications for investigation of samples and cases if several *Salmonella* types were involved in an outbreak?
7. Who can stop an import or ask for a recall?
8. Are some herbs more likely to contain human pathogens than others, and why?
9. Could *Salmonella* persist in environments where herbs are packed?

13.8 References

ELVISS, N.C., LITTLE, C.L., HUCKLESBY, L., SAGOO, S., SURMAN-LEE, S. *et al.*, Food, Water and Environmental Surveillance Network (2009). Microbiological study of fresh herbs from retail premises uncovers an international outbreak of Salmonellosis. *International Journal of Food Microbiology*, **134**:83–8.

EUROPEAN COMMISSION (2007). *Rapid Alert System for Food and Feed (RASFF) – Overview Week 22 2007*. Available at http://ec.europa.eu/food/food/rapidalert/reports/week22-2007_en.pdf. Accessed 21 July 2011.

FOOD STANDARDS AGENCY (2007). *Agency issues warning on* Salmonella *in basil*. Available at http://www.food.gov.uk/news/newsarchive/2007/may/warningbasil. Accessed 21 July 2011.

GOLBERG, D., KROUPITSKI, Y., BELAUSOV, E., PINTO, R., and SELA, S. (2011). *Salmonella typhimurium* internalization is variable in leafy vegetables and fresh herbs. *International Journal of Food Microbiology*, **145**:250–7.

HEALTH PROTECTION AGENCY (2007). A national outbreak of *Salmonella* Senftenberg in England and Wales: April to May 2007. *Health Protection Report*, **1**:22. Available from: http://www.hpa.org.uk/hpr/archives/2007/hpr2207.pdf. Accessed 21 July 2011.

NORWEGIAN SCIENTIFIC COMMITTEE FOR FOOD SAFETY (2008). *Risk assessment of import and dissemination of intestinal pathogenic bacteria in fresh herbs and leafy vegetables from South-East Asia*. Available at http://www.vkm.no/dav/22a5fd4852.pdf.

PAKALNISKIENE, J., FALKENHORST, G., LISBY, M., MADSEN, S.B., OLSEN, K.E. *et al.* (2009). A foodborne outbreak of enterotoxigenic *E. coli* and *Salmonella* Anatum infection after a high school dinner in Denmark, November 2006. *Epidemiology and Infection*, **137**:396–401.

PEZZOLI, L., ELSON, R., LITTLE, C., FISHER, I., YIP, H. *et al.* (2007). International outbreak of *Salmonella* Senftenberg in 2007. *Euro Surveill.* **12**:pii=3218. Available at: http://www.eurosurveillance.org/ew/2007/070614.asp#3. Accessed 21 July 2011.

PEZZOLI, L., ELSON, R., LITTLE, C. L., YIP, H., FISHER, I. *et al.* (2008). Packed with *Salmonella* – Investigation of an international outbreak of *Salmonella* Senftenberg infection linked to contamination of prepacked basil in 2007. *Foodborne Pathogens and Disease*, **5**:661–8.

SURMAN-LEE, S., MURPHY, N., PATHAK, K., CLEMENTS, J., and DE PINNA, E. (2008). A pan-London study of the microbiological quality of fresh herbs. Abstract 258 at *Health Protection 2008*. Available at www.hpa-events.org.uk/hp2008. Accessed 21 July 2011.

14

Source tracing in Thailand following contamination of dog chew toys with *Salmonella*

S. D. Pillai, Texas A&M University, USA

Abstract: Dog chew toys are growing in popularity around the world. They are manufactured primarily in Asia, in countries such as Thailand and China, and exported worldwide. The ingredients used in the manufacture and the manufacturing process can vary widely. Very often they are either pig ears or raw hide. The concerns associated with pathogen-contaminated chew toys have caused countries to consider banning their import. Such actions can cause major financial ramifications for the exporting countries. The following case study describes the challenges of microbial pathogen contamination of dog chew toys.

Key words: pathogen, dog chew toys, contamination, HACCP (hazard analysis and critical control points).

14.1 Introduction to the case

In September 1999, FDA learned that 30 cases of *Salmonella infantis* in Canada were strongly associated with pig ear dog treats. All affected individuals were dog owners, indicating that these infections were in people who handled dog chew toys. Many of the cases were children. The contaminated products were traced back to a Canadian company that manufactured the chew toys.

Pig ears and raw hide leather (untanned leather of cattle) are some of the main ingredients involved in the manufacture of dog chew toys. The trace back of the contamination was verified using phage typing and pulsed field gel electrophoresis (PFGE) analysis. Based on the reported infections in Canada, the US Food and Drug Administration (FDA) issued warnings about dog chew toys and started paying close attention to imported dog chew toys, especially those made from pig ears.

14.2 Significance of the case

Salmonella infantis is one of the ten most common *Salmonella* serovars. It is often found associated with farm animals and feedlot run-off. Though *S. infantis* infections are not life-threatening in healthy individuals, these infections can be life-threatening in immune-compromised individuals such as young children.

14.2.1 Regulatory, economic and market aspects

Currently, about 62% of US households have a pet, and 46% of these pets are dogs. The pet industry is valued at around $48 billion per year in the US (American Pet Products Association, 2011). Play toys especially designed for use by dogs are commercially available around the world. These toys are shaped in an extraordinary number of sizes, shapes, colors, textures and even tastes. Thailand, China and Brazil are among the top countries in the world that are involved in exported dog chew toys. On average a US household spends approximately $45 per year on dog toys.

Though the infections in Canada were linked to a manufacturer in Canada, a number of dog chew manufacturers in Thailand and other parts of Asia were faced with a potential ban of their products in Canada and the US due to their links with human infections. The ramifications of these actions were especially felt in Thailand, one of the world's largest exporters of dog chew toys.

Companies were faced with product recalls, order cancellations, and having to resort to gamma irradiation to sterilize the chew toys, which added to their costs. The US FDA has a current import alert in place, which is the Agency's guidance to FDA field personnel at the ports of entry to potentially detain pet treats, especially those that contain pig ears from companies that are on a 'Red List' (US FDA, 2011).

If companies wish to be removed from the Red List then they should provide, among other items, evidence that they have identified the source of the problem and taken remedial steps to prevent the occurrence of the problem.

14.3 Case history continues

The author was invited by one of the progressive dog chew toy manufacturers in Thailand in early 2000 to provide technical assistance to solve the *Salmonella* contamination issue. Though this facility was not directly implicated in any infections, quality assurance/quality control (QA/QC) tests performed by the company revealed that their products did harbor *Salmonella*. The author spent 3 weeks in Thailand to analyze the production practices, provide recommendations to prevent *Salmonella* contamination, provide guidance on developing a hazard analysis and critical control points (HACCP) plan, identify critical control points for controlling or eliminating *Salmonella* contamination, and provide guidance for the establishment of a QA/QC laboratory.

14.3.1 The facility and production process as it existed

The manufacturing facility was located on a long, narrow plot of land. Though the manufacturing facility was covered by corrugated metal roof, the sides were partially open. The processing areas could be separated into:

- raw material handling area,
- cutting/twisting/folding area,
- oven-drying area,
- quality control area,
- warehousing area,
- packaging area.

Different products were being manufactured in the facility. An interesting observation was the location of the warehouse in the middle of the facility and certain post-oven processing operations such as the band-saw (cutting/trimming) operations at the sides of the facility. The location of the warehouse in the middle of the process was interrupting the process flow of products from start to finish. Also, locating the cutting/trimming operations at the sides of the facility (which was not adequately protected from the weather and possible external contamination) was a risk factor. Oven baking is one of the critical control points in reducing/eliminating potential pathogens. Thus, taking products from the oven and moving them to the sides of the facility increases the vulnerability of these products to contamination.

The company purchased both locally-obtained raw hide and imported raw hide (from Germany). The company preferred the imported raw hide because of its uniform shape and size, which allowed for minimal waste. However, the imported material had to be washed to remove the salt, and then sun-dried.

Both the premium imported raw hide and the local raw hide were exposed to the weather, rodents, insects and birds. The company officials were not aware that rodents, insects and birds could harbor *Salmonella*. Neither were they aware that moisture on raw leather could allow the proliferation of potential pathogens, which could in turn result in cross-contamination, nor that excess rainfall (since the raw materials were not protected from the weather) could cause these organisms to leach from the raw hide and flow into the surrounding production areas. Finished products were stored right next to the raw material handling areas. Some of the packaging of the finished products was also found to be torn and damaged.

Depending on the desired product, the raw material was used in either the dry process or the wet process. In the dry process, the raw hide is either cut, punched, hand-woven or soaked and pressed into bone-shaped products. The wet process involved the mixing of the crushed raw hide material with dye, cooked rice and other ingredients.

The company, due to customer demands, had introduced a line of chicken jerky chew toys. These toys had a sliver of marinated and dried chicken meat within the toy. Raw chicken was brought into the facility and cleaned/washed on-site using large buckets and containers. The drains were not functioning and so the waste water either stagnated or flowed towards the wall separating the chicken-washing area from the raw hide storage area. A majority of workers were not wearing any protective foot wear, gloves or apparel. Many of these workers also worked in other parts of the manufacturing process. The workers 'disinfected' the floor by pouring bleach over the work area.

There was a significant amount of noise and dust where the raw hide was either cut or ground up. Depending on the product, the material was soaked in a bleach solution (for disinfection) prior to being shaped into defined forms. There were no records of the chlorine concentration in these bleach tanks, there was no understanding of the concept of chlorine demand and it was not clear whether there was any standard operation procedure (SOP) for the routine cleaning of the bleach tanks.

Products were moved from one section to another using baskets. However, these baskets were not color coded and so it was impossible for the workers to know whether these baskets were used to transport clean or unclean material. Workers were found to handle wet raw materials with their bare hands and were not wearing any protective clothing over their street clothes.

All the products were oven-dried at 55–60 °C for 100 hours to reduce the moisture content to around 11–12%. Some of the ovens had temperature monitors that were capable of monitoring the temperature at the front, middle and back of the oven. A review of the data indicated that the temperature within the oven was not uniform.

Some of the products were packaged after the drying process on wooden tables. Though the workers at the packaging station wore gloves and caps (to prevent contamination), there was no indication that there were procedures in place to disinfect or clean the tables during or after work shifts. Some of the plastic bins at the packaging areas were the same as those that were seen in the raw material storage areas.

14.3.2 Microbiological analysis of materials

The raw hide material, the product just prior to oven drying and the products following oven drying were analyzed for the presence of *Salmonella* spp. and Enterobacteriaceae. Similarly, equipment and personnel swabs were collected and analyzed for the presence of *Salmonella* spp. The laboratory tests proved that the raw hide was contaminated with *Salmonella* and Enterobacteriaceae. Moreover, even the post-oven samples were positive for *Salmonella*, albeit at lower levels of incidence.

14.4 Resolution and outcomes

Laboratory tests confirmed the assumption that the raw hide material harbored *Salmonella*. Though the imported raw hide had lower numbers of pathogens and Enterobacteriaceae, this material was exposed to the domestic material in different areas such as the bleach soaking steps, which are supposed to contain 2% chlorine. There was no validation that the bleach solution was capable of disinfecting the raw hide. Based on the detection of *Salmonella* on the raw hide and the lack of validation of the bleach solution, one recommendation was that all raw hide in the manufacturing process should be considered to be contaminated.

There were only two steps in the entire process that could potentially control *Salmonella*, namely the initial oven drying (55 °C for 100 min) and a disinfection drying cycle (85 °C for 20 min). The small number of post-oven samples that were positive for *Salmonella* and Enterobacteriaceae indicated that the two heating cycles were controlling *Salmonella* incidence rate in the products. The low numbers of contaminated samples suggested that the contamination levels were low to begin with and/or sporadic in nature. If the manufacturing process had been heavily contaminated then most of the post-oven products, equipment, and personnel swabs would have showed greater incidence of *Salmonella*.

The issue of *Salmonella* contamination was a key concern to the company management. The management had already invested a significant amount of resources, both monetary and personnel, trying to solve the problem. The company management was also well aware that for the company to produce *Salmonella*-free products, and have the confidence that *Salmonella*-free products were being exported, additional monetary as well as human resource investments were needed. The author made a total of about 75 specific recommendations that covered different areas of manufacturing such as raw material handling, packaging, equipment, personnel, plant sanitation, suggested QA/QC tests, and developing a HACCP plan. These recommendations included:

- developing SOP for each product and process,
- encouraging supervisory control to promote employee adherence to sanitation and good manufacturing practice (GMP),
- installing proper drainage,
- physically separating raw materials from finished product, and color coding bins and carriers that were used at different part of the facility,
- products that contained chicken slivers should be considered as harboring pathogens and they needed to be handled with extra special safeguards such as frequent disinfection of work areas,
- validating and monitoring disinfection cycle conditions,
- monitoring cleaning effectiveness,
- routine disinfection of equipment,
- experimental validation of *Salmonella* inactivation under varying time–temperature–humidity combinations,
- assembling and empowering a HACCP team.

14.5 Commentary and personal reflections

This case history shows the importance of good manufacturing practice and standard operation procedures in controlling pathogen contamination of a product.

During a 6-month follow-up discussion with a company representative, I was informed that my suggestion of separating the finished products from the unfinished products was being instituted and my recommendation of using color-coded bins was being adopted. However, it was surprising to find that the company's concerns had switched from *Salmonella* to BSE. The company feared that BSE could be associated with the imported raw hides from Europe! It was surprising that the company had to start to deal with another emerging issue even before they could fully address the previous threat. At the time of writing, this company continues to be one of the leading dog chew toy exporters of Thailand.

14.6 Acknowledgements

This work was supported by Hatch grant H 8708 of the Texas AgriLife Research and the National Center for Electron Beam Research.

14.7 Critical questions for discussion

1. What are the different exposure pathways for pathogens from dog chew toys to humans?
2. Though the initial infections were associated with dog chew toys manufactured in Canada, why would Asian countries worry about their products?
3. Do you think that the introduction of chicken slivers into dog chew toys was responsible for *Salmonella* contamination at this plant? What could be the reasons why? Or why not?
4. If you were the owner of a dog chew toy company and you detected *Salmonella* positives in a voluntary monitoring program, would you spend resources to identify the root cause of the contamination or would you employ irradiation as a clean-up technology for the contaminated products?
5. Would you or would you not recommend the building and equipping of a microbiology testing lab in a dog chew facility? What is the reason for your answer?
6. What specific hygienic recommendations would you make about dog chew toy manufacturing?
7. Are dog chew toys a high pathogen risk commodity?
8. Would a dry dog chew toy have a higher probability of harboring pathogens compared with a moist dog chew toy?
9. If you had to employ temperature to eliminate *Salmonella* from dog chew toys, would you employ a temperature with a long holding time, a higher temperature with a shorter holding time, or moist heat for disinfection?
10. Would ionizing radiation be a suitable application for dog chew toys? Do you know of any food commodity that is regularly radiated?

14.8 References

AMERICAN PET PRODUCTS ASSOCIATION (2011). *National Pet Owners Survey 2011–2012.* American Pet Products Association Inc., Greenwich, CT.

US FOOD AND DRUG ADMINISTRATION (2011). Import Alert # 72–03. *Detention Without Physical Examination and Intensified Coverage of Pig Ears and Other Pet Treats due to the Presence of* Salmonella. Available at http://www.accessdata.fda.gov/cms_ia/importalert_218.html.

15

Analysing tap-water from households of patients with inflammatory bowel disease in Norway

K. Rudi, Hedmark University College and The Norwegian University for Life Sciences, Norway, P. Ricanek, Akershus University Hospital, Norway, T. Tannæs and S. Brackmann, University of Oslo, Norway, G. Perminow, Oslo University Hospital, Norway and M. H. Vatn, University of Oslo, Norway and Oslo University Hospital, Norway

Abstract: Tap-water has recently been identified as an important epidemiological parameter in inflammatory bowel disease (IBD). Whether bacteria present in tap-water are involved in IBD still remains unknown. To address this question, the dominant microbiota in tap-water samples from households of patients with IBD, and from those of non-IBD control patients, were screened. Tap-water from both pediatric patients (<18 years) and adults with ulcerative colitis (UC), Crohn's disease (CD) and IBD unclassified (IBDU) were sampled in the south-eastern part of Norway. A total of 145 patients diagnosed during 2005–2007 were included in the study.

Key words: IBD, tap-water, 16S rRNA gene, microbiota.

15.1 Introduction to the case

Tap-water represents a main vector for human bacterial exposure (Szewzyk *et al.* 2000). Despite the potential impact, our knowledge about the distribution pattern and human colonization potential of bacteria found in tap water is limited (Szewzyk *et al.* 2000; Schmeisser *et al.* 2003; Williams *et al.* 2004; Keinanen-Toivola *et al.* 2006; Rudi *et al.* 2009a; Rudi *et al.* 2009b). Recent analyses of gut mucosa from inflammatory bowel disease (IBD) patients, however, showed that bacterial phylotypes from tap-water were present in higher numbers in diseased mucosa compared with mucosa from healthy controls (Frank *et al.* 2007). Further support for this observation derives from a study showing that certain tap-water sources are correlated with increased incidences of inflammatory bowel disease (IBD) (Aamodt *et al.* 2008). No studies, though, have yet determined the correlation between tap-water microbiota and the gut bacteria present in higher numbers in IBD. It is important to conduct such studies to increase understanding of the role of environmental factors in IBD.

The aim of the present case was to determine the potential correlation between tap-water microbiota and IBD. We analysed the microbiota in the tap-waters of the Inflammatory Bowel South-Eastern Norway (IBSEN) II patient cohort, which contains patients diagnosed in 2005, 2006 and 2007. Tap-water was collected from both paediatric patients (<18 years) (Perminow *et al.* 2009) and adults (≥18 years) (Ricanek P. *et al.* 2010, unpublished data). We investigated the direct correlation between the tap-water microbiota and diagnosis of IBD, and in addition compared the tap-water microbiota with recently published IBD mucosal library data (Frank *et al.* 2007).

15.2 Significance of the case

15.2.1 Causes of IBD

Currently, the precise aetiology of IBD has not been completely elucidated. However, observations from various experimental studies suggest a multifactorial interplay involving the enteric microbiota, host genetic factor and immune function and with environmental implications. Tap-water has recently been suggested as an important environmental factor in IBD (Aamodt *et al.* 2008).

15.2.2 Regulatory aspects

In Norway, the quality of drinking water is controlled by the 'Regulations on water supply and drinking water' (FOR 2001-12-04 nr 1372). These regulations aim to ensure the supply of drinking water in adequate quantity and of satisfactory quality, including the lack of harmful pollutants of any kind. With respect to microbial contamination, the law sets out the levels of *Clostridium perfringens, Escherichia coli, Pseudomonas aeruginosa* and intestinal enterococci that are permitted. The lack of regulations for *Burkholderiales* is probably due both to methodological limitations and to the fact that this family has not yet been associated with water-borne infections. The potential correlation between tap-water and IBD represents a major regulatory challenge since IBD is a chronic disease, and the bacteria associated with this disease are common water-borne bacteria.

15.2.3 Economic and market aspects

Varying costs are reported for IBD in different countries and studies. Studies report the total economic burden of Crohn's disease (CD) to be $10–15 billion in the US and €2–16 billion in Europe (Yu *et al.* 2008). In Germany, cost diaries report a mean 4-week cost per patient of €1425 for CD and €1015 for UC. Indirect costs accounted for 64% and 54% of the figures for CD and ulcerative colitis (UC), respectively (Stark *et al.* 2006). Annually, CD patients cost ~$18 500 in the United States and €2800–6900 in other Western countries. Indirect costs accounted for 28% and 64–69% in the US and Europe, respectively (Yu *et al.* 2008). Thus, measures that lead to reduced incidence rate and severity of IBD would represent enormous savings to society.

15.3 The case history continues

15.3.1 The laboratory work

One hundred and forty-five water samples (1 l) collected throughout the south-eastern part of Norway from 2005 to 2007 were included in this study. For all samples the sampling date was recorded, and additionally, for 115, the sampling location. A total of 77 samples were from patients with diagnosis. These were composed of 23 paediatric patients ($n = 5$ UC, $n = 9$ CD, $n = 2$ IBD unclassified (IBDU) and $n = 7$ non-IBD) and 54 adults ($n = 15$ UC, $n = 14$ CD, $n = 4$ IBDU and $n = 21$ non-IBD). The non-IBD control patients were referred and investigated because their symptoms meant that IBD had been suspected, but full investigation revealed normal macroscopic and microscopic endoscopic findings. They served as symptomatic controls. Irritable bowel syndrome (IBS) or recurrent abdominal pain was the main diagnosis in these patients.

The water samples were collected in sterile bottles from the kitchen tap, or the most frequently used tap in local households. The samples were centrifuged (3000 g, 30 min), supernatants were discarded and the pellets were re-suspended in 300 µl 1.0 mM Tris HCl, pH 8.0. Acid-washed glass beads (300 mg) were added to the suspensions (Sigma-Aldrich, Steinheim, Germany), which were then homogenized using a Mixer Mill MM 300 (Retsch, Düsseldorf, Germany) (20 Hz, 45 s). DNA purification was done using the Qiagen DNeasy Tissue kit following the manufacturer's recommendations (Qiagen, Hilden, Germany).

16S rRNA gene clone libraries were constructed and analysed following the concept previously described (Rudi *et al.* 2007; Rudi *et al.* 2009b). Briefly, this involved the transformation of the sequences into multimers and compression of the redundant information by the principal component analysis (PCA) provided in the Unscrambler 9.8 software (CAMO Software Inc., Woodbridge, NJ). The clone library sequences generated were deposited in Genbank with the accession numbers HM598687–HM600721. The DNA from the water samples was amplified and sequenced using the mixed sequence approach that has previously been described (Trosvik *et al.* 2007). Classification of the dominant microbiota was carried out by searches in the Ribosomal Database Project II (RDP II) database (Cole *et al.* 2008) using the consensus sequences derived from the mixed sequences.

The comparisons between our tap-water clone library data and the IBD mucosal clone libraries (Frank *et al.* 2007) were carried out by local blast searches. The local blast database construction and searches were carried out using the CLC Genomic Workbench (CLC bio, Århus, Denmark). Statistical analyses were done using the Spotfire S+ 8.1 software (TIBCO Software Inc., Palo Alto, CA). Correlations between patient IBD type (UC, CD, IBDU) and tap-water microbiota were determined by either ANOVA or Fisher's Exact Test.

15.3.2 Results

We presented a surprisingly large overlap of the microbiota from the IBD patients, tap-water and a recently characterized mucosal microbiota library (Frank *et al.* 2007). The overlap between the tap-water and the gut mucosal microbiota was determined for five random tap-water samples. This was done through a combination of mixed sequence (Trosvik *et al.* 2007; Trosvik *et al.* 2008; Zimonja *et al.* 2008) and clone library analyses. More specifically, we compared an in-depth 16S rRNA gene cloning dataset of the tap-water microbiota with a recently published mucosal 16S rRNA gene dataset from both IBD and control patients (Frank *et al.* 2007).

First, we determined the distribution of the dominant taxonomic groups in the water samples by a search in the RDP II database using the consensus for the mixed sequences. These analyses showed that the major portion of the water samples were dominated by *Betaproteobacteria*, and that there were approximately equal portions of water samples dominated by *Alpha*- and *Gammaproteobacteria*. At the order level there was a dominance of *Burkholderiales*, while *Comamonadaceae* were the most predominant at the family level. Finally, *Pelomonas* and *Herminiimonas* were the most dominant genera (Fig. 15.1).

We then compared all the tap-water clones with a local database created for the mucosal dataset. This analysis showed that 50% of the overlapping genera had previously been identified as IBD enriched (Table 15.1). There were 17 phylotypes (2% divergence level) that were concordantly enriched in both the tap-water and the mucosal datasets. These represent 17% of tap-water clones and 5.5% of the mucosal clones; 35% of the clones in the tap-water data were related to a clone in the mucosal dataset at the >98% similarity level (Table 15.1). We identified an overlap (>98% identity) for sequences related to the genera *Tepidimonas, Sphingomonas, Rhodoferax, Polaromonas, Pelomonas, Leptothrix, Herminiimonas, Curvibacter, Catellibacterium* and *Acidovorax*. The phylotype related to *Acidovorax delafieldii* showed the highest prevalence in both datasets, being present above the 1% level in both. The overlapping phylotype with highest prevalence in the mucosal data was *Burkholderia ambifaria* (1.5% of the clones), while the most prevalent tap-water bacterium (2.4% of the clones) was related to *Undibacterium* sp.

We also found there was a potential association between specific tap-water bacteria and newly diagnosed IBD patients within the IBSEN II cohort. Using the Fisher's Exact Test we found that for paediatric patients there was a dominance of *Comamonadaceae* in the tap-water associated with UC ($p = 0.04$), while for adults there was an UC association for *Enterobacteriaceae* ($p = 0.03$). For CD we found no association for paediatric patients, while there was a weak association for an unclassified group of *Burkholderiales* for adults ($p = 0.07$).

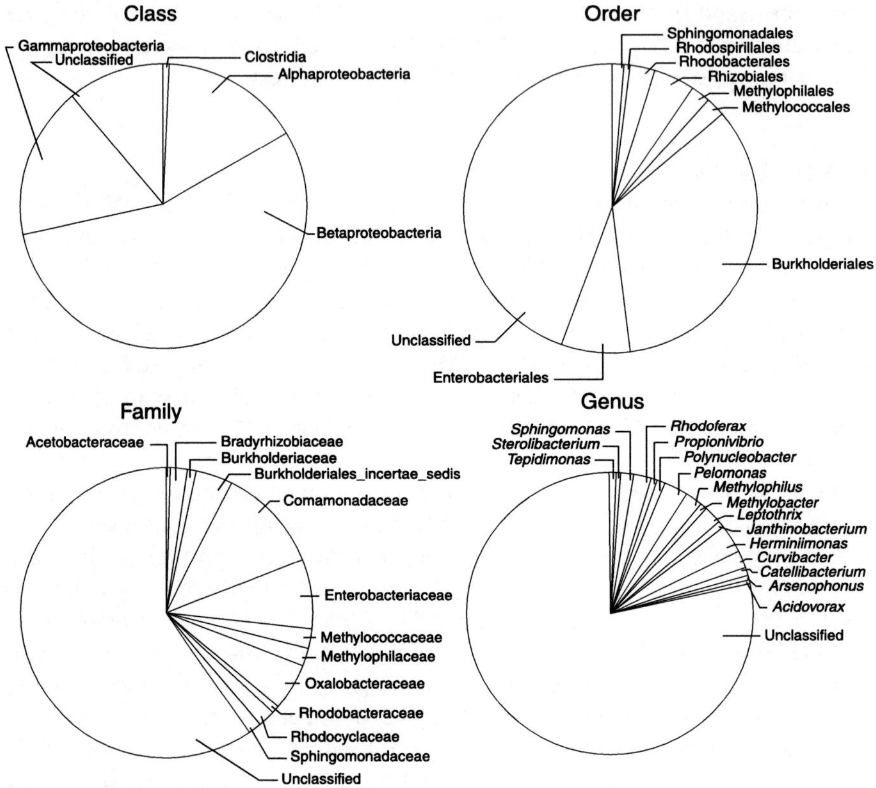

Fig. 15.1 Taxonomic assignments of the dominant tap-water microbiota for the analyses of all tap-water samples. The taxonomic assignments were done using the Ribosomal Database II (RDPII) classifier with a bootstrap cutoff of 80% for the mixed consensus sequences.

Table 15.1 Overlap between tap-water and IBD mucosal phylotypes

Species*	Identity (%)	Water (%)	Mucosa† (%)	IBD enriched‡
Janthinobacterium sp. WPCB148	100	0.48	0.06	No
Sphingomonas yanoikuyae Q1	100	0.91	0.32	Yes
Pelomonas saccharophila ATCC 15946T	99	1.90	0.16	No
Acidovorax delafieldii strain PCWCS4	100	1.30	1.01	Yes
Novosphingobium sp. K39	98	0.61	0.25	Yes
Limnobacter thiooxidans strain CS-K2	99	1.72	0.28	No
Sphingobium yanoikuyae FM3	96	1.77	0.06	No
Solirubrobacter sp. Gsoil 921	93	0.61	0.06	No
Phenylobacterium lituiforme strain Fail3	99	0.82	0.24	Yes
Burkholderia ambifaria AMMD	100	0.48	1.45	Yes

(continued)

Table 15.1 Continued

Species*	Identity (%)	Water (%)	Mucosa† (%)	IBD enriched‡
Undibacterium sp. EM 1	100	2.38	0.31	No
No cultured representatives	na	0.43	0.07	Na
Drinking water bacterium Y7	98	0.48	0.59	Yes
Drinking water bacterium MB12	100	0.99	0.07	No
Pseudomonas aeruginosa strain ATCC 27853	99	0.95	0.09	Yes
Caulobacter leidyi	100	0.69	0.07	Yes
Comamonas testosteroni CNB-2	99	0.48	0.44	No

* The species were identified as those having the highest identity by blast searches in the National Center for Biotechnology Information (NCBI) database. †Source: Frank *et al.*, 2007. ‡Match to genera enriched at the 10% level (Source: Frank *et al.*, 2007 Suppl. Table 6).

Limitations of our study, however, are that we have not matched corresponding tap-water and mucosal libraries for IBD patients, and that the techniques we have used have relatively low resolution. Therefore, we may have overseen important bacterial taxa in our work. In future, it would be highly interesting to perform a more detailed characterization of the tap-water microbiota, and to match this to the corresponding information from the patient mucosal biopsies. In this way rigorous conclusions about the effect of tap-water bacteria in IBD may be established.

15.4 Resolution and outcomes

The main outcome of this case was an increased focus on the indigenous microflora in tap-water and its relationship to chronic diseases. The large number of proteobacteria with potential for colonization of the human gut was highly surprising and could be of importance not only for IBD, but for other diseases as well. Several of the IBD-associated bacteria identified, such as *Burkholderia* and *Enterobacteriaceae*, are connected with inflammation induction (Foster *et al.* 1993; van Deventer 1999; Ergin *et al.* 2008; Ventura *et al.* 2009). These are therefore likely candidates for future validation with respect to the etiology and progression of the disease. In particular, taxa belonging to *Burkholderiales* are interesting since these are highly prevalent in both tap-water and IBD mucosa (Sim *et al.* 2010). With respect to the correlation between the tap-water microbiota and IBD, the following were also identified as potentially important: *Comamonadaceae* in UC children and an unclassified group of *Burkholderiales* for CD adults.

15.5 Commentary

As pathogens causing acute diseases are gradually being eradicated from the human food chain, a new group of harmful bacteria, which may have a role in the aetiology of certain chronic diseases, such as IBD, diabetes, rheumatism, asthma and allergy, are emerging. We believe that understanding the etiology of these diseases will be the grand challenge of the future.

15.6 Acknowledgments

We would like to give special thanks to Dung Tran for providing technical assistance. We would also like to thank Irena Szpinda and Tone Tønjum for their help with the water samples and Geir Bukholm for helpful discussions. Finally, we would like to thank Helse Sørøst, EU project # FP6-2006-036272 and Hedmark Sparebank for their financial support.

15.7 Critical questions for discussion

1. How should a test of the bacteria in the drinking water such as the one described above be implemented?
2. What do you believe are the main challenges in laboratory diagnostics of drinking-water bacteria associated with chronic diseases?
3. Do you think there can be risks associated with changing water decontamination regimes from, for example, chlorination to UV sterilization?
4. Can you think of other chronic diseases that could potentially be correlated to drinking-water bacteria?
5. Do you think there might be beneficial bacteria in drinking water, and that water decontamination could eradicate these?
6. What impact do the geological attributes of a country have on the microbial quality of drinking water (surface, ground, or treated water)?

15.8 References

AAMODT, G., BUKHOLM, G., JAHNSEN, J., MOUM, B. and VATN, M.H. (2008). The association between water supply and inflammatory bowel disease based on a 1990–1993 cohort study in southeastern Norway. *American Journal of Epidemiology*, **168**:1065–72.

COLE, J.R., WANG, Q., CARDENAS, E., FISH, J., CHAI, B. *et al.* (2008). The Ribosomal Database Project: Improved alignments and new tools for rRNA analysis. *Nucleic Acids Research*, **37**: D141–5.

ERGIN, A., ADAM, T., BUSSOW, K., THIEL, A., SIEPER, J. *et al.* (2008). Identification of the predominant antigenic epitopes in intestinal flora in IBD. *Mucosal Immunology*, **1**: Suppl. 1, S19–23.

FOSTER, S.J., MCCORMICK, L.M., NTOLOSI, B.A. and CAMPBELL, D. (1993). Production of TNF alpha by LPS-stimulated murine, rat and human blood and its pharmacological modulation. *Agents and Actions*, **38**: Spec No, C77–9.

FRANK, D.N., ST. AMAND, A.L., FELDMAN, R.A., BOEDEKER, E.C., HARPAZ, N. *et al.* (2007). Molecular-phylogenetic characterization of microbial community imbalances in human inflammatory bowel diseases. *Proceedings of the National Academy of Sciences*, **104**: 13780–5.

KEINANEN-TOIVOLA, M.M., REVETTA, R.P. and SANTO DOMINGO, J.W. (2006). Identification of active bacterial communities in a model drinking water biofilm system using 16S rRNA-based clone libraries. *FEMS Microbiology Letters*, **257**: 182–8.

PERMINOW, G., BRACKMANN, S., LYCKANDER, L.G., FRANKE, A., BORTHNE, A. *et al.* (2009). A characterization in childhood inflammatory bowel disease, a new population-based inception cohort from South-Eastern Norway, 2005–07, showing increased incidence in Crohn's disease. *Scandinavian Journal of Gastroenterology*, **44**: 446–56.

RUDI, K., ZIMONJA, M., KVENSHAGEN, B., RUGTVEIT, J., MIDTVEDT, T. *et al.* (2007). Alignment-independent comparisons of human gastrointestinal tract microbial communities in a multidimensional 16S rRNA gene evolutionary space. *Applied and Environmental Microbiology*, **73**: 2727–34.

RUDI, K., BERG, F., GAUSTAD, E., TANNES, T. and VATN, M. (2009a). Ratios between Alpha-, Beta- and Gamma-proteobacteria in tap water determined by the ProteoQuant assay. *Letters in Applied Microbiology*, **50**: 1–6.

RUDI, K., TANNAES, T. and VATN, M. (2009b). Temporal and spatial diversity of the tap water microbiota in a Norwegian hospital. *Applied and Environmental Microbiology*, **75**: 7855–7.

SCHMEISSER, C., STOCKIGT, C., RAASCH, C., WINGENDER, J., TIMMIS, K.N. *et al.* (2003). Metagenome survey of biofilms in drinking-water networks. *Applied and Environmental Microbiology*, **69**: 7298–309.

SIM, W.H., WAGNER, J., CAMERON, D.J., CATTO-SMITH, A.G., BISHOP, R.F. *et al.* (2010). Novel *Burkholderiales* 23S rRNA genes identified in ileal biopsy samples from children: Preliminary evidence that a subtype is associated with perianal Crohn's disease. *Journal of Clinical Microbiology*, **48**: 1939–42.

STARK, R., KONIG, H.H. and LEIDL, R. (2006). Costs of inflammatory bowel disease in Germany. *Pharmacoeconomics*, **24**: 797–814.

SZEWZYK, U., SZEWZYK, R., MANZ, W. and SCHLEIFER, K.H. (2000). Microbiological safety of drinking water. *Annual Review of Microbiology*, **54**: 81–127.

TROSVIK, P., SKANSENG, B., JAKOBSEN, K.S., STENSETH, N.C., NAES, T. *et al.* (2007). Multivariate analysis of complex DNA-sequence electropherograms for high-throughput quantitative analysis of mixed microbial populations. *Applied and Environmental Microbiology*, **73**: 4975–83.

TROSVIK, P., RUDI, K., NAES, T., KOHLER, A., CHAN, K.S. *et al.* (2008). Characterizing mixed microbial population dynamics using time-series analysis. *ISME Journal*, **7**: 707–15.

VAN DEVENTER, S.J.H. (1999). Review article: Targeting TNF alpha as a key cytokine in the inflammatory processes of Crohn's disease – the mechanisms of action of infliximab. *Alimentary Pharmacology and Therapeutics*, **13**: 3–8.

VENTURA, G.M.D., BALLOY, V., RAMPHAL, R., KHUN, H., HUERRE, M. *et al.* (2009). Lack of MyD88 protects the immunodeficient host against fatal lung inflammation triggered by the opportunistic bacteria *Burkholderia cenocepacia*. *Journal of Immunology*, **183**: 670–6.

WILLIAMS, M.M., DOMINGO, J.W., MECKES, M.C., KELTY, C.A. and ROCHON, H.S. (2004). Phylogenetic diversity of drinking water bacteria in a distribution system simulator. *Journal of Applied Microbiology*, **96**: 954–64.

YU, A.P., CABANILLA, L.A., WU, E.Q., MULANI, P.M. and CHAO, J.D. (2008). The costs of Crohn's disease in the United States and other Western countries: A systematic review. *Current Medical Research and Opinion*, **24**: 319–28.

ZIMONJA, M., RUDI, K., TROSVIK, P. and NAES, T. (2008). Multivariate curve resolution of mixed bacterial DNA sequence spectra: identification and quantification of bacteria in undefined mixture samples. *Journal of Chemometrics*, **22**: 309–22.

Part III

Crisis management

16

A proactive attempt to control an outbreak of foodborne disease in humans after detection of *Salmonella* during surveillance and control of pork in Denmark

A. Wingstrand, National Food Institute, Technical University of Denmark, Denmark, T. Jensen, Danish Veterinary and Food Administration, Denmark, G. Sørensen, National Food Institute, Technical University of Denmark, Denmark, S. Gubbels, Statens Serum Institut, Denmark and European Centre for Disease Prevention and Control, Sweden, L. J. Porsbo, National Food Institute, Technical University of Denmark, Denmark, S. Ethelberg, Statens Serum Institut, Denmark, G. O. Bjerager, Danish Veterinary and Food Administration, Denmark and K. Kjeldsen, M. Torpdahl and K. G. Kuhn, Statens Serum Institut, Denmark

Abstract: Based on Kuhn *et al.* (2012), this chapter describes the course and investigation of a large Danish foodborne outbreak in 2010, caused by a rare *Salmonella* Typhimurium phage type with the majority of patients residing in the eastern and southern parts of Denmark. It describes the interview studies conducted and the changes in the *Salmonella* subtypes and human demography during the outbreak. It also suggests interpretations of these changes. Comprehensive Danish register data on herds, slaughterhouses and *Salmonella* surveillance were utilized for the tracing of products and herds and for assessing the risks of herds, slaughterhouses and meat production plants being the source of infection in the outbreak.

Key words: *Salmonella*, outbreak investigation, epidemiology, typing, tracing.

16.1 Introduction to the case

Foodborne outbreaks in Denmark are reported either by consumers to the central or Regional Veterinary and Food Control Authorities (RVFCA) or by physicians via the regional medical officer to Statens Serum Institut (SSI). Furthermore, clusters of human cases may be discovered in the laboratory or through the laboratory-based surveillance systems where, as a guideline, the number of cases indicating a potential outbreak is >5 cases in two subsequent weeks, depending on pathogen, demography, etc. Investigation of cross-regional foodborne outbreaks in Denmark is coordinated at weekly meetings in a Central Outbreak Management Group (COMG) with appointed members from SSI, the Danish Veterinary and Food Administration (DVFA) and the National Food Institute, Technical University of Denmark (DTU Food).

16.1.1 Danish surveillance and control of *Salmonella* in pigs and pork

Denmark has a long history of nationwide monitoring and control of *Salmonella* in the production chain from 'farm-to-fork' (Wegener *et al.*, 2003; Anonymous, 2009; Anonymous, 2010). Surveillance and control of *Salmonella* in Danish pig production have proven efficient in reducing the number of human *Salmonella* cases from pork. A large proportion of the pig herds is infected subclinically at a low level, and outbreaks of clinical infection (salmonellosis) in pigs are rare.

Most pig herds are monitored serologically for *Salmonella* antibodies in blood samples or muscle fluid (Nielsen *et al.* 1998). Slaughter pigs from herds with higher antibody levels (approximately 1% of the herds in 2010) are slaughtered at specific slaughterhouses under special hygienic precautions. Until January 2011, fecal samples from these herds were additionally cultured for *Salmonella*. This practice was part of the *Salmonella* DT104 surveillance and control program, which was discontinued in 2011. Instead, a new control program was implemented with focus on *Salmonella* resistant to antimicrobial agents of critical importance for treatment of human infections. The occurrence of *Salmonella* in pork is monitored by culture of swab samples from pig carcasses collected at the slaughterhouses (Danish Order No 1722 of 22 December 2010).

In 2009, the overall *Salmonella* prevalence in a total of 24 505 carcasses was estimated to be 1.1%. Approximately one-third of the *Salmonella* isolates from pork were *Salmonella* Typhimurium (*S.* Typhimurium). Culture results from the carcass swabs further serve to monitor slaughterhouse contamination. An increased prevalence in a slaughterhouse leads to special follow-up and further interventions by the authorities and the industry to reduce contamination. Additionally, control of *Salmonella* in domestic and imported pork is conducted by the authorities, while export control and internal monitoring (own-check) programs are carried out by the food industry.

16.1.2 Typing of *Salmonella* isolates

In Denmark, all *Salmonella* of human and non-human origin are subject to serotyping. Until January 2011 antibiotic resistance profiling and phage typing were conducted on *S.* Typhimurium isolates as part of the *Salmonella* DT104 control program. This program was discontinued in 2011, and an effort to control *Salmonella* resistant to antimicrobial agents of critical importance for treatment of human infections was introduced instead. Phage typing is still conducted on a selection of isolates for specific purposes. For the resistance profiling, susceptibility to a standard panel of 15 antimicrobial agents is tested. Further, *Salmonella* isolates from human cases are typed by multiple loci variable number of tandem repeats analysis (MLVA) (Larsson *et al.*, 2009). Where relevant, MLVA typing of veterinary and food isolates is also conducted. A shared database between SSI and DTU Food facilitates the comparison of molecular typing results of *Salmonella* isolates from different sources.

16.1.3 How did it all begin?

In late March 2010, the Zoonosis Laboratory at DTU Food reported a high number of *S.* Typhimurium isolates with an unknown phage type and resistance to ampicillin, streptomycin and sulphamethoxazole (ASSu) (Kuhn *et al.*, 2012). The phage type lysis pattern was not described in the scheme from the World Health Organization Collaborative Centre for phage typing of *Salmonella* – Health Protection Agency (HPA), Colindale, UK, which is why the phage type was initially assigned as Reaction Does Not Conform (RDNC).

The isolates originated from the *Salmonella* surveillance of carcasses in a medium-sized pig slaughterhouse (hereafter called Slaughterhouse A), from pork sampled in a cutting plant closely associated with Slaughterhouse A, or from meat production plants. DVFA was able to trace most of the meat from the production plants back to Slaughterhouse A. The early suspicion of Slaughterhouse A and the related cutting plant was further strengthened by a previous outbreak of *S.* Typhimurium U288 from the same cutting plant/slaughterhouse constellation in 2008, resulting in human cases of the outbreak strain in Denmark, Sweden and Norway (Bruun *et al.*, 2009).

In the following week more food isolates of the *S.* Typhimurium RDNC ASSu strain were reported mainly from pork (cuts and mince) originating from Slaughterhouse A. Different contaminated meat batches were withdrawn from the market in Denmark and later from Sweden, but most of the meat had already been consumed at the time of withdrawal.

By the end of March, only two human cases of *S.* Typhimurium with the same RDNC phage type profile had been notified, and the outbreak strain was not isolated from food in the following 2 weeks. Thus, it appeared as if a potential food safety issue from the contamination of Slaughterhouse A and pork products had been avoided. Considering the lag phase between human disease onset and registration of a case, it was decided to monitor the situation thoroughly in the following weeks.

16.1.4 What were the challenges involved?

At this point of time, when only two human cases had been identified, it was crucial, in order to prevent or stop the development of an outbreak, to trace and delimit contaminated food batches, and withdraw risky products from the market. Despite established systems for food traceability, the tracing of pork meat in such a diversified and large market as Denmark may be comprehensive and complex.

Further, there were attempts to identify the source of meat contamination in order to stop further contamination and prevent new outbreaks originating from the same source. Due to the prior knowledge on contamination of pork originating from Slaughterhouse A with *S.* Typhimurium RDNC ASSu, and a previous outbreak originating from the same establishment in 2008, the COMG from early on had primary focus on investigation of the pork production and distribution chain from herds supplying pigs to Slaughterhouse A and further to the consumers, attempting to establish a link between the patients and pork from Slaughterhouse A.

Several pork isolates of this rare *S.* Typhimurium RDNC phage type originated directly from pork and environment in Slaughterhouse A. Therefore it became a key question in the investigation whether the contamination was a result of continuous input from a herd or a persistent contamination of equipment and environment in Slaughterhouse A.

16.2 Significance of the case

In Denmark, a total of 2129 cases of human salmonellosis were registered in 2009 (Anon., 2010). In general one-third of the Danish *Salmonella* cases and approximately half of the registered *Salmonella* outbreaks are caused by *S.* Typhimurium, and the majority of the *S.* Typhimurium cases are domestically acquired. A *Salmonella* source account, based on the comparison of subtypes of *Salmonella* isolates from humans, food, and food animals, attributes approximately a quarter of the cases to different food sources, mainly Danish pork, table eggs and imported meat. The remaining 75% of the human cases, which are not attributed to a specific source, are associated with recent travel abroad or are of unknown source (Hald *et al.*, 2004; Anonymous, 2010).

16.2.1 Regulatory aspects

There is no general EU or national legislative restriction concerning the presence of *Salmonella* in fresh meat, but there is a zero-tolerance policy for *Salmonella* in minced meat, meat preparations and ready-to-eat (RTE) meat products (Commission Regulation (EC) No 2073/2005 of 15 November 2005). Contamination of these products and other products posing a particular risk to human health will lead to withdrawal from the market (Council Regulation (EC) No 178/2002, art. 14).

According to the EU legislation, traceability systems for food are in place in all Danish food producing establishments (Council Regulation (EC) No 178/2002, art. 18 and 19).

All holdings with production animals in Denmark, including slaughterhouses, markets and rendering plants, are identifiable through a unique number in the Central Husbandry Register (CHR number), and all movements of pigs between CHR numbers or for export must be registered in a central pig movement database (Danish Order No 59 of 18 January 2010).

16.2.2 Economic and market aspects

Compared with the area and a population of 5.5 million inhabitants (2009), the animal food production in Denmark is relatively large. In 2009, 18.9 million pigs, 0.5 million cattle and 100 million broilers were slaughtered and more than 60 000 tons of table eggs were produced (Anonymous, 2010). Compared with other pig producing nations, the Danish pig herds are medium-sized, and pigs are slaughtered at few large slaughterhouses and a limited number of smaller slaughter plants. The large and diversified production of pork and RTE pork products takes place in numerous large and small plants, including some Danish-owned plants located in the neighboring countries. The large export of meat and particularly pork significantly contributes to the national finances, and food scares may potentially affect the national economy.

16.3 The outbreak history continues

Later, in April 2010, more than a dozen human patients infected with the rare *S.* Typhimurium RDNC strain had been confirmed, and COMG initiated a thorough outbreak investigation. In the following weeks the outbreak strain was isolated from more samples of mainly pork and fresh pork products, and the number of human cases increased steadily.

16.3.1 Strain similarity

The outbreak strain was defined by the MLVA profile JPX.0007.DK (3-12-9-NA-211). Three isolates with a MLVA profile differing in one locus (two in STTR6 and one in STTR10) were also included (Larsson *et al.*, 2009). As MLVA profile 0007.DK is common among other human and non-human *S.* Typhimurium phage types (e.g. DT120), phage typing was important for the case definition. In June, the RDNC phage type of the outbreak strain was characterized as phage type U323 by the HPA (Colindale, UK).

Subtype variations are frequently observed within a *Salmonella* outbreak, and previous experience and thorough knowledge of the typing methods are required to delimit and link the human and food isolates belonging to an outbreak. In the second half of the Danish outbreak, approximately one-third of the human as well as non-human isolates were resistant to streptomycin and sulphamethoxazole (SSu) or were fully sensitive to the antibiotics. This shift could be interpreted as a shift in the epidemiology of the outbreak (e.g. source or vehicle).

16.3.2 Demography of human cases

Between March and October 2010, infection with the outbreak strain was registered in a total of 172 patients. The outbreak curve had two peaks in April and July. Most infections were domestically acquired and patients mainly resided in the Eastern and Southern parts of the country, with the highest incidence (6.5 cases per 100 000 inhabitants) in the county where Slaughterhouse A and the associated cutting plant were situated.

One patient lived in Greenland, where Danish food products are common. An urgent enquiry through the Epidemic Intelligence Information System of European Center for Disease Prevention and Control (ECDC) identified 12 possible cases in Sweden. Meat from Slaughterhouse A and the cutting plant had been exported to Sweden in the outbreak period, but a link between the imported pork and the Swedish cases could not be established.

Overall, more women than men were infected and the median age was high. In the second half of the outbreak, the proportion of women and the proportion of children between 1 and 5 years of age increased considerably. Based on prior knowledge of Danish food consumption patterns (Sisse Fagt, pers. comm.), the shift in the outbreak demography suggested that a change in outbreak source had taken place from primarily fresh pork (cuts mince and prepared) towards a combination of pork and products of pork with more appeal to women and young children.

16.3.3 Hypothesis-generating interviews and case–control study

Hypothesis-generating patient interviews conducted by SSI in May and June did not point to one specific food item as the source, but most patients had eaten pork or RTE pork products, and several cases reported having tasted raw, minced pork. This in itself is not conclusive, as consumption of pork is very common in Denmark. On the other hand, this finding does not contradict the prior hypothesis that pork was the source of the outbreak.

In the second half of the outbreak period, the case interviews showed that an increasing proportion of patients had eaten an RTE tea-sausage (fermented spreadable sausage originally eaten at tea-time) from a specific company, containing raw pork. The results of a case–control study conducted by SSI on cases from the second half of the outbreak supported that the tea-sausage was a source of infection, particularly of children and young adults. As only approximately half of the cases in the study reported having eaten the tea-sausage, other sources may have contributed as well.

16.3.4 Farms delivering slaughter pigs to Slaughterhouse A

The herds delivering pigs to Slaughterhouse A were ranked according to their likelihood of being the source of contamination. This was done on the basis of data from the CHR register (including the Danish central pig movement database), the bacteriological and serological *Salmonella* herd surveillance data, the deliverance size and pattern and the herd mortality. Deliveries of pigs from the herds were compared with the dates of slaughter from which pork batches were found culture-positive with the outbreak strain. In selected herds the assessment led to full mapping of the trade pattern and visits from the RVFCA to confirm register data and ensure the absence of clinical signs of salmonellosis.

In one herd in particular, the days of delivery frequently matched the culture-positive slaughter days. The herd was a major supplier of pigs to Slaughterhouse A, but no clear indication of high *Salmonella* levels was found in the herd.

16.3.5 A pig farm harboring the outbreak strain

Only one pig farm was known to harbor the outbreak strain. Pigs from this farm were received for slaughter at Slaughterhouse A (only once) several months prior to the first isolation of the outbreak strain at Slaughterhouse A. Thus, this farm was not considered to be a possible continuous source of contamination of Slaughterhouse A. Pigs from the farm were regularly slaughtered at another slaughterhouse (Slaughterhouse B).

The outbreak strain was isolated from pork from this slaughterhouse only once during the outbreak period, whereas routine pork samples from Slaughterhouse B were found positive six times with a *Salmonella* strain varying from the outbreak strain on one MLVA locus. However, this Salmonella subtype was not isolated from human cases.

16.3.6 Isolation of the outbreak strain from food and slaughterhouse production environment

In total, more than 100 isolates of the *S.* Typhimurium U323 outbreak strain were detected on more than 50 occasions, mainly from pork, including minced and prepared pork, and from the slaughterhouse environment. Like the patients, most food isolates clustered in two periods in the spring and summer. Most isolates originated from Slaughterhouse A and more than a dozen different food producers or wholesalers, from which the contaminated products could be traced back to Slaughterhouse A. Cleaning and disinfection procedures in the cutting plant associated with Slaughterhouse A were found satisfactory and the outbreak strain was not isolated from equipment.

16.3.7 Troubleshooting

After the first isolation of the outbreak strain from the slaughterhouse environment in mid-May 2010, Slaughterhouse A became subject to intensified surveillance and repeated audits from the RVFCA and the industry until the end of August, and extended sampling for hygiene control was conducted. Since the outbreak strain was isolated in swabs from equipment at Slaughterhouse A, even after the plant was closed down for thorough cleaning and disinfection at the end of May, the establishment had most likely been persistently contaminated. The *Salmonella* outbreak strain was not isolated from the slaughterhouse environment after repeated cleaning of the slaughterhouse in mid-June. The last slaughter day associated with positive meat samples was early July.

Contaminated meat and meat products were withdrawn from the Danish market ten times during the outbreak. On four occasions, the entire daily output from an establishment was withdrawn due to likely cross-contamination. Several contaminated batches had passed onto the market, and most of the meat had been consumed at the time of withdrawal.

All batches of the tea-sausage from the specific company still within the shelf-life were withdrawn from the market early in September. *Salmonella* was not isolated from samples of the withdrawn tea-sausages. Trace-back from the incriminated batches of tea-sausage revealed that the company producing the sausage had received pork from the cutting plant associated with Slaughterhouse A in the period when contamination of meat was detected, but had not received pork known to be contaminated with the outbreak strain. Contamination of other meat products from the company producing the tea-sausage or contamination of products from other food companies could not be excluded.

16.4 Resolution and outcomes

Repeated isolation of the outbreak strain from equipment at Slaughterhouse A even after thorough cleaning suggested a persistent contamination of the production environment. However, it remains uncertain which slaughter pig herd was the source of the contamination, and whether the outbreak strain was introduced to the slaughterhouse environment only once or repeatedly during the outbreak.

Contaminated pork has undoubtedly led to the contamination of minced and prepared meat and meat products, e.g. tea-sausage containing raw pork. The tea-sausage was pointed out as an important food product, contributing to a significant proportion of the human infections in the second phase of the outbreak (July–August) when more women and children fell ill. Fresh pork, including minced and prepared meat, was most likely a source of infection throughout the entire outbreak period. In spite of this, a statistical association between consumption of fresh pork and disease was not found.

The number of registered cases declined steadily after the peak in August, 1 month after the last slaughter day in Slaughterhouse A, which was associated with a culture-positive meat sample. The human incidence was low even before withdrawal of the tea-sausage in early September, after which only a few people became ill.

16.5 Comments and personal reflections

In the outbreak, contaminated fresh pork, including minced and prepared pork, was traced back to Slaughterhouse A prior to the rise in the number of human cases, and the outbreak group had already focused on pork from this establishment before the first peak of human cases appeared. However, it was frustrating to see the outbreak grow in numbers and length despite this early warning.

Patient interviews and a case–control study identified a specific and not very common RTE product (tea-sausage) as the source of a sub-outbreak. Further, the overwhelming microbiological evidence with near-certainty identified fresh Danish pork as a source of human infections, but the analytical epidemiology did not confirm fresh pork as an outbreak source. It demonstrates the challenge in analytical outbreak epidemiology when investigating the significance of very common and frequent exposures like the consumption of fresh pork in Denmark. This reinforces the need for alternative approaches to point out common exposures as sources in human foodborne outbreaks.

The current *Salmonella* culture, typing and subtyping methods used to obtain a high microbiological discrimination in outbreak investigations are time-consuming. In general, shortening the time from sampling to a level of typing sufficient for outbreak investigation will lead to a more real-time picture of the outbreak status and fewer recall (remembrance) problems, particularly in case interviews. Faster microbiology may further enable withdrawal of contaminated products at an earlier stage and thereby limit the outbreak.

Although traceability systems are in place, tracing of pork meat that has entered the production chain can be a complicated discipline. A more automated registry system for tracing products exists in other countries and may be considered, although in the large, diversified and changing Danish production of pork and pork products establishing such a standardized registration system will be a challenge.

By the time the detection and verification of the *Salmonella* outbreak strain in the products are completed and a withdrawal can be issued, products with a short shelf life (e.g. fresh meat) are likely to have been consumed. Thus, in the actual outbreak, the frequent and long-lasting contamination of pork underscores the importance of stopping the outbreak at the primary source of meat contamination.

16.6 Acknowledgements

The authors are grateful to the personnel at RVFCA-Meat Inspection Units for assistance in the establishments involved. We also thank the personnel at the Epidemiological Surveillance unit at SSI for conducting telephone interviews, the laboratory staff at DTU Food, the clinical microbiological human laboratories at SSI and RVFCAs for the comprehensive culture and subtyping of *Salmonella* isolates. Finally we thank Sisse Fagt, DTU Food for providing data on human consumption of pork in Denmark.

16.7 Critical questions for discussion

1. What are the methodological principles of the *Salmonella* subtyping methods: serotyping, phage typing, MLVA-typing and antimicrobial resistance profiling?

2. What does pre-harvest surveillance of *Salmonella* reflect when based on serology and culture methods, respectively?
3. What are the strengths and limitations of the use of patient interviews in the investigation of foodborne outbreaks?
4. How can the demography of a foodborne outbreak support hypothesis-generation?
5. What is the difference between the study design in a cohort study and a case–control study?
6. What are the physio-chemical characteristics of salami and tea-sausage that may increase or reduce the risk of foodborne infections from these products?
7. What does the Danish *Salmonella* source account estimate?

16.8 References

ANONYMOUS. 2009. *Annual Report on Zoonoses in Denmark 2007*. National Food Institute, Technical University of Denmark.

ANONYMOUS. 2010. *Annual Report on Zoonoses in Denmark 2009*. National Food Institute, Technical University of Denmark.

BRUUN, T., SØRENSEN, G., FORSHELL, L.P., JENSEN, T., NYGARD, K. *et al.* 2009. An outbreak of *Salmonella* Typhimurium infections in Denmark, Norway and Sweden, 2008. *Eurosurv.*, **14**: 1–6.

HALD, T., VOSE, D., WEGENER, H.C. and KOUPEEV, T. 2004. A Bayesian approach to quantify the contribution of animal-food sources to human salmonellosis. *Risk Anal.*, **24**: 255–69.

KUHN, K.G., SØRENSEN, G., TORPDAHL, M., KJELDSEN, M.K., JENSEN, T. *et al.* 2012 A long-lasting outbreak of *Salmonella* Typhimurium U323 associated with several pork products, Denmark 2010. *Epidemiology and Infection*, available on CJO, doi: 10.1017/S0950268812000702.

LARSSON, J., TORPDAHL, M., PETERSEN, R.F., SØRENSEN, G., LINDSTEDT, B.A. *et al.* 2009. Development of a new nomenclature for *Salmonella* Typhimurium multilocus variable number of tandem repeats analysis (MLVA). *Eurosurv.*, **15**: 15.

NIELSEN, B., EKEROTH, L., BAGER, F. and LIND, P. 1998. Use of muscle fluid as a source of antibodies for serologic detection of *Salmonella* infection in slaughter pig herds. *J. Vet. Diagn. Invest.*, **10**: 158–63.

WEGENER, H.C., HALD, T., LO FO WONG, D., MADSEN, M., KORSGAARD, H. *et al.* 2003. *Salmonella* control programmes in Denmark. *Emerg. Infect. Dis.*, **9**: 774–80.

17

Managing and learning from an anthrax outbreak in a Swedish beef cattle herd

R. Knutsson, V. Båverud, M. Elvander and E. Olsson Engvall, National Veterinary Institute (SVA), Sweden, K. Eliasson, National Board of Agriculture (SJV), Sweden and S. Sternberg Lewerin, Swedish University of Agricultural Sciences (SLU), Sweden

Abstract: This chapter describes an unexpected anthrax outbreak in a group of Swedish beef cattle kept indoors during the winter season of 2008. Anthrax is a serious zoonotic disease, caused by the pathogenic agent *Bacillus anthracis*. The way in which the outbreak was managed, the costs involved and the lessons learned are discussed. Research needs, crisis management and preparedness gaps are also identified.

Key words: anthrax, *Bacillus anthracis*, crisis management, decontamination, diagnostics, preparedness, risk assessment.

17.1 Introduction to the case

Anthrax is a zoonotic disease, caused by the pathogenic agent *Bacillus anthracis*, which affects both animals and humans (WHO 2008). In humans the bacterium can cause three forms of the disease:

- cutaneous (skin) anthrax,
- inhalation anthrax,
- gastrointestinal (bowel) anthrax.

B. anthracis must be considered from a food safety perspective, but foodborne anthrax cases are rare. When they do occur, foodborne anthrax outbreaks occur mainly in the developing world, where poorly developed veterinary and food-safety programs may result in the consumption of anthrax-contaminated raw or undercooked meat (Sobel *et al.* 2002). For instance, during 2009–2010, anthrax emerged as a zoonotic disease in Bangladesh with findings of 195 animal cases and

706 human cases being reported (Ahmed *et al.* 2010). Moreover, during autumn 2010, in one of the anthrax outbreaks that took place in Bangladesh, human infections were caused by handling infected cattle or eating contaminated beef.

Ruminants are considered to be the species most sensitive to anthrax, and outbreaks in cattle are common worldwide. During recent years, outbreaks have been reported in several countries, e.g. in Australia, Canada and Italy (Durrheim *et al.* 2009; Himsworth and Argue 2009; Fasanella *et al.* 2010; Mongoh *et al.* 2007). Such outbreaks are rarely associated with human cases in countries where a higher food safety standard and personal protection measures can be afforded.

17.2 The Swedish outbreak in 2008: how did it all begin?

In December 2008, a single anthrax outbreak occurred in a Swedish beef cattle herd in the county of Halland (Sternberg Lewerin *et al.* 2010). The outbreak began with the sudden death of animals, seven animals died within 10 days. The herd consisted of 90 animals kept in a barn on deep straw bedding. The unusual circumstances of this outbreak were the unexpected and atypical cases in a non-endemic area as well as, at first, the lack of typical clinical signs.

This was the first anthrax case in 27 years in Sweden and it was unexpected since it occurred in the winter season, in animals kept indoors and fed only hay. Initially, only respiratory symptoms were recorded, and clinical and pathological features typical for anthrax were not seen on the carcasses in the first cases (Elvander 2009). The carcasses were reportedly normal in appearance, which imposed additional diagnostic and laboratory challenges in terms of necropsy and bacteriological analysis. The uncertainty related to the unexpected and non-typical clinical signs made it difficult to obtain a common view of the situation. In addition, the information given by the owner regarding feeding practices (later revised) did in no way support an introduction of *B. anthracis* spores in the feed. After the confirmation of anthrax, further diagnostic tests included antimicrobial susceptibility testing of *B. anthracis* in a Biosafety Level 3 (BSL-3) laboratory environment, as well as analysis of environmental samples and extended sampling in order to trace the source of infection.

17.3 The case history continues

The herd of mixed breed animals, 45 beef cows with their calves and young stock, was brought indoors in the middle of October 2008. The first animal died on 29 November 2008 and the outbreak continued, with a lack of typical clinical signs, with ten more dead animals until anthrax was diagnosed on 12 December 2008 (Sternberg Lewerin *et al.* 2010). On 10 December three carcasses were sent to a regional laboratory for necropsy. The investigating veterinarians immediately suspected anthrax due to the appearance of the enlarged spleen, non-coagulated blood and massive internal bleeding.

The National Veterinary Institute (SVA) was contacted and blood samples from the three animals were sent to SVA by courier. These blood samples arrived on 11 December and anthrax was confirmed on 12 December by culture and polymerase chain reaction (PCR) methods. In total, 13 animals died and the carcasses were disposed of by incineration, including those sent to autopsy. The remaining animals in the herd received prophylactic antibiotic treatment before being euthanized (Elvander 2009). The reason for euthanasia was that there was nowhere to keep the animals during sanitation and the owner wanted to cease animal production. Seventy-eight animals were euthanized on the farm and the carcasses were transported to an animal by-product facility. The lack of mobile animal waste incinerators at this point in time made the transportation of carcasses necessary.

A broad set of laboratory methods were used during investigation of the outbreak (Sternberg Lewerin *et al.* 2010). They included necropsy and histology according to routine diagnostic procedures, bacteriological investigations based on microscopy, culture methods, real-time PCR and strain characterization using Multiple-Locus Variable number tandem repeat Analysis (MLVA) typing and antimicrobial susceptibility testing. The real-time PCR assay was not yet fully validated, which was why the final diagnosis was based on the culture method.

Another concern during the laboratory work was biosafety issues, especially considering the number of samples that were analysed. In this case, the first samples were submitted for bacteriology before any anthrax suspicion. Therefore, the first samples were handled in a BSL2 laboratory at SVA (Båverud 2009). Laboratory investigations were also performed for confirmatory tests after decontamination of BSL-2 laboratories and the incinerator at SVA. In addition, samples were taken on three occasions in an effort to trace the source of infection: samples at the directly after the outbreak in December 2008, in November 2009 (soil sampling at a suspicious anthrax grave) and in June 2010 for environmental river bank sampling of anthrax spores (see Fig. 17.1).

17.4 Significance of the case

Bacillus anthracis is a Gram-positive bacterium closely related to other *Bacillus* spp. in the *B. cereus* group. The pathogenic *B. anthracis*, causing anthrax in humans and animals, is listed on several high-priority agents lists such as the CDC Bioterrorism list as a category A agent (CDC 2010), the USDA/APHIS Agricultural Select Agent List (USDA/APHIS 2010) and the Australian Group (A) List of Biological Agent for Export Control (Australia Group 2010). A fully virulent *B. anthracis* strain contains the two mega plasmids pXO1 and pXO2. These plasmids encode various virulence factors, such as lethal factor on pXO1 and poly-D-glutamic acid capsule on pXO2 (Quinn *et al.* 1994). The *B. anthracis* endospore-forming rod is a species in the *B. cereus* group, which is composed of the following species: *B. anthracis*, *B. cereus* (sensu stricto), *B. mycoides*, *B. thuringiensis* and *B. weihenstephanensis* (Guinebretiere *et al.* 2002). To genotype *B. anthracis* on a subspecies level, Multiple-Locus Variable number

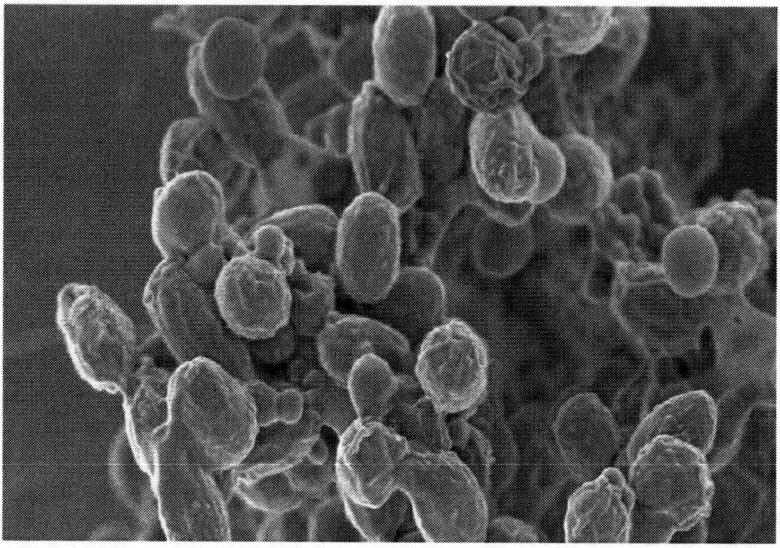

Fig. 17.1 Scanning electron microscopy image of *Bacillus anthracis* 4229 spores. The spore morphology of species within the *Bacillus cereus* group has different characteristics in terms of spore length, spore diameter and the structure of exosporium. However, most *B. anthracis* strains have similar spore morphology. (Source: Joakim Ågren (SVA) and Leif Ljung Uppsala University, Sweden.)

tandem repeat Analysis (MLVA) is the state-of-the-art technology (Keim *et al.* 2000). Systematic microbiological analysis of the spore has recently identified a number of morphological variants based on phenotypic characteristics and the ability to sporulate, i.e. various morphotypes (Rasko *et al.* 2011). The life cycle of the monomorphic species *B. anthracis* is well known and the spores can survive for more than 100 years in the soil (De Vos 1990).

The susceptibility of infection varies among host species, with cattle and sheep being the most susceptible. Anthrax was relatively common in cattle and sheep throughout Sweden, as well as in other European countries, during the first half of the twentieth century. In the period 1901–1957 approximately 5000 animals on 4500 holdings were diagnosed with anthrax in Sweden (Elvander 2009). A major outbreak was reported in 1956–7 in the county of Halland due to imported meat and bone meal (MBM) (Rutqvist and Swahn 1957). According to an old leaflet with instructions on handling anthrax, the infected herd was put under quarantine for 10 days after an infected carcass had either been burned or torched and buried in dry soil with minimum 1.5 m of soil above (Florén 1917). However, with the vast number of anthrax cases it was well known that this was not always done properly, sometimes resulting in burial in wetlands without proper disinfection.

The International Organization for Standardization (ISO) has recently developed guidance for burial of animal carcasses to prevent epidemics (ISO 28901:2011, 2011).

17.4.1 Regulatory aspects

B. anthracis in feed and anthrax in animals is a disease that is included in the Swedish Epizootic Act (SFS 1999:657). This means that it is notifiable upon suspicion and the authorities must investigate, trace the source and take measures to eradicate the infection. Details about the anthrax outbreak are made public. Samples containing *B. anthracis* shall, according to the Swedish Work Environment Authority (AFS 2005:1), be handled in a BSL3 laboratory. According to this legislation, measures for personal protection of staff who may be exposed to the organism should be based on risk assessments. Anthrax is also included in the OIE Animal Health Code (OIE 2011a).

17.4.2 Economic and market aspects

The economic burden of this anthrax case was mainly related to decontamination activities on the farm, and sanitation of the production facility that had processed the contaminated animal by-products from the farm. The animal by-product facility closed on 16 December 2008 and all production was shut down for a month. The reason for the stop in the production was that many possibly infected carcasses had been transported to the facility for destruction. The company was financially compensated with 6 million SEK. An incinerator facility was built on the farm and 60 million SEK was spent on the decontamination of the farm, feed, manure and soil over a period of 1.5 years. This is one of the most expensive single sanitation operations ever in Sweden. The reason for the severe economic impact might be due to an overestimation of the risk. However, the outbreak had no consequences for the consumers' perception of food of animal origin or for the beef market. The interest from national media was modest and no humans became ill. However, the staff involved in the outbreak were treated with antibiotics to prevent anthrax.

17.5 Resolution and outcomes

The outcome from this anthrax case has highlighted a number of issues and some questions remain to be answered. Some of the identified issues are summarized as follows:

(i) The source of the infection has not been identified. None of the environmental samples taken were positive except for two soil samples taken from the courtyard of the farm where the infected carcasses had been stored. However, the most likely source of infection was anthrax-contaminated hay harvested on a dried-up river bank that had not been used before (Sternberg Lewerin *et al.* 2010).
(ii) Challenges of how to perform a cost-benefit analysis and risk assessment during the decontamination phase for making proper judgments to justify the costs.
(iii) Public fear may make some measures difficult to undertake, knowing that proper risk communication is a major part of the handling of any outbreak.

(iv) Gaps in the decontamination policy have been identified and practical recommendations generated from this case will in the future enable improved biological decontamination response.
(v) Questions about the establishment of joint awareness and coordinated decision making by all involved parties during an anthrax outbreak and improved communication strategies are being developed.
(vi) Work safety and biosafety issues have been a concern during the whole outbreak, but improved education, awareness and training will improve the response. For instance, transportation of anthrax-contaminated carcasses should be avoided, if possible.

It is usually difficult to determine the exact location of historical anthrax graves and remaining spores may have travelled several meters and kilometers due to geological events since the burial of the carcasses. However, anthrax should be considered as a differential diagnosis in any case of sudden death in ruminants, especially in association with disturbance (digging, drainage work, draught or flooding) of the soil on pastures or fields where hay or silage was harvested.

An assessment of the risk of spreading the infection should be made at every step, from suspicion to diagnosis and control. If information about old burial sites is available, it must be used for assessment of the risk of exposure to anthrax spores. Proper assessment of the risk of contamination and the need for decontamination of various surfaces is needed for a cost-effective cleaning and disinfection of affected premises as well as deciding upon relevant measures for personal protection of personnel. Risk assessments, whether qualitative or quantitative, must be documented, for transparency and trust.

17.6 Commentary

The cattle on the farm was the only animal species infected but the consequences were rather severe in terms of labor efforts and costs. The case highlights experiences related to atypical clinical signs in non-endemic areas as well as diagnostic needs for environmental and feed samples. Decontamination of the farm has been extensive and it has been one of the most expensive single sanitizing procedures in Swedish history. Overestimation of the risk due to public fear might be one reason for the high cost. Good preparedness for quick risk assessments in various situations when handling such cases is vital, as is good collaboration and communication between different authorities.

Apart from the fact that the clinical and pathological features in the first carcasses were atypical for anthrax, leading to a difficulty in obtaining a joint situational awareness, the case presented several other challenges. These included the fact that appropriate isolation and characterization of *B. anthracis* strains require a broad diagnostic portfolio (Knutsson 2011). Sampling and sample preparation of feed and environmental samples for detection of *B. anthracis* spores also require trained staff and validated laboratory methods. After the acute response phase, another challenging issue was the time-consuming decontamination work on the farm. The large volume and sheer weight of contaminated material made this procedure complicated. In total, approximately 300 tonnes of feed, 150 tonnes of timber and 900 tonnes of straw bedding and manure were burned on site (SOU 2010:106).

This case has identified preparedness gaps and research needs for early detection, diagnostics and decontamination. To be able to trace the source of infection, improved sampling strategies and diagnostic capabilities for isolation and characterization of *B. anthracis* in environmental and feed samples are needed. Rapid field-based tests are also required for earlier detection and improved monitoring, and research activities related to dose–response studies of *B. anthracis*. Geo-historical research is also needed to map out present anthrax graves for risk assessment purposes and improved decontamination methods on site. Crisis management includes many challenges, such as to make an early diagnosis, to make operational and strategic decisions, estimate the risks and to learn from a crisis (Boin *et al.* 2005). Below are some more detailed comments about research needs, crisis management and preparedness.

17.6.1 Early detection

To avoid further exposure of people and animals, early detection is essential. This entails awareness of the disease, by veterinarians as well as animal owners. A field test would aid the early assessment of animal carcasses, especially as transport of samples from potentially infected animals is strictly regulated and therefore expensive and complicated. Even without such a test, it is important that all suspect cases are immediately notified and sampled, to allow rapid diagnosis.

17.6.2 Bacteriological diagnostics

A broad set of complementary laboratory methods is needed for detection and isolation of vegetative cells and spores of *B. anthracis* (WHO 2001). Laboratory work related to *B. anthracis* must be carried out in authorized laboratories with staff trained in biosafety. Commonly used methods include:

- Direct microscopy: square-ended, blue bacilli in short chains surrounded by a pink capsule can be seen in direct microscopy from blood smears from the animal (polychrome methylene blue stain) (Quinn *et al.* 1994).
- Cultivation: blood and organ samples can be cultured on blood agar and incubated in aerobic atmosphere at 37 °C (Quinn *et al.* 1994). Typical colony appearance with hemolysis can be seen after spore cultivation.
- Real-time PCR: a variety of PCR-based methods have been developed targeting different biomarkers such as *pag*A (Baeumner *et al.* 2004), *gyr*A (Hurtle *et al.* 2004), BA5345 (Antwerpen *et al.* 2008), BA5510 (Olsen *et al.* 2007), prohage lambdaBa03 (PL3) (Wielinga *et al.* 2011). The real-time PCR assay applied to this anthrax outbreak targets three genes; i.e. the *rpoB* gene on the chromosome and the virulence genes *lef* and *cap* located on the pXO1 and the pXO2 plasmids (Sternberg Lewerin *et al.* 2010). Since the outbreak in 2008, substantial work has been done on the development of robust real-time PCR-based methods.

17.6.3 Decontamination

The spores of *B. anthracis* are notoriously difficult to inactivate. For field purposes, such as preliminary decontamination on a farm, aldehydes (4% glutaraldehyde or 10% formaldehyde) are usually recommended (OIE 2011b), since most other chemicals are less effective in the presence of organic material. These chemicals are not fully effective at temperatures below 10 °C. For final disinfection of such surfaces (i.e. after thorough cleaning), aldehydes, sodium hypochlorite (5–10%), hydrogen peroxide (3%) or peracetic acid (1%) are recommended (OIE 2011b). Previously, calcium oxide has also been recommended but recent data indicate that the calcium may contribute to the preservation of the spores (Himsworth 2008).

Disinfection of larger indoor areas can be achieved by fumigation, but this is technically more demanding. Some studies have been performed to assess the efficacy of various commercial disinfectants against *Bacillus* spores (Heninger *et al.* 2009; Oie *et al.* 2011) but most of these do not represent field conditions. Some studies attempt to mimic field conditions (Wood *et al.* 2011) but most are still conducted at room temperature.

More research is needed on suitable disinfectants for various situations, especially at lower temperatures and where a high content of organic material is to be expected. Methods to determine whether disinfection has been efficient need also be refined, as the commonly used indicator spores are not suitable for all disinfectants. An important issue is the sporulation and viability of *B. anthracis* spores in relation to environmental temperature and humidity (Minnet 1950). Further research to determine the risk of sporulation at different temperatures and humidity is also needed for cost-effective decontamination efforts.

17.7 Critical questions

1. How may early detection of anthrax cases be achieved, and what are the critical elements of a sensitive clinical surveillance for the disease?
2. What are the most important needs for correct sampling, detection and isolation of vegetative cells and spores in different types of anthrax-contaminated samples?
3. Which aspects that affect the choice of disinfectant in various situations should be considered and what aspects are the most important?

17.8 Acknowledgments

The writing of this chapter has been supported by grants from the Swedish Civil Contingencies Agency (Anslag 2:4 Krisberedskap), Sweden, and the framework of the EU-project AniBioThreat (Grant Agreement: Home/2009/ISEC/AG/191) with the financial support of the Prevention of and Fight against Crime Program of the European Union, European Commission – Directorate General Home Affairs. This publication reflects the views of the authors only, and the European Commission cannot be held responsible for any use which may be made of the information contained therein. We thank Mr Jeffrey Skiby (DTU-Food, Denmark) for critical reading and suggestions.

17.9 References

AHMED, B.N., SULTANA, Y., FATEMA, D., ARA, K., BEGUM, N. *et al.* (2010). 'Anthrax: An emerging zoonotic disease in Bangladesh.' *Bangladesh J Med Microbiol*, **04**(1): 46–50.

ANTWERPEN, M.H., ZIMMERMANN, P., BEWLEY, K., FRANGOULIDIS, D. and MEYER, H. (2008). 'Real-time PCR system targeting a chromosomal marker specific for *Bacillus anthracis*.' *Mol Cell Probes*, **22**(5–6): 313–5.

AUSTRALIA GROUP (2010). 'The Australia Group – List of Biological Agents for Export Control.' Available at http://www.australiagroup.net/en/biological_agents.html.

BAEUMNER, A.J., LEONARD, B., MCELWEE, J. and MONTAGNA, R.A. (2004). 'A rapid biosensor for viable *B. anthracis* spores.' *Anal Bioanal Chem*, **380**(1): 15–23.

BOIN, P., HART, P., STERN, E. and SUNDELIUS, B. (2005). *The Politics of Crisis Management: Public Leadership Under Pressure*. Cambridge University Press, Cambridge, UK.

BÅVERUD, V. (2009). 'Health Surveillance of Staff.' In *The Anthrax Outbreak in Sweden – Lessons Learned*. Nordic Biosafety Network Abstract Compendium, Solna, Sweden. SVA Dnr 2010/112, SUA Report Serie 17 ISSN 1654–7098.

CDC (2010). *CDC Bioterrorism Disease Agent List*. Available at http://www.bt.cdc.gov/agent/agentlist-category.asp.

DE VOS, V. (1990). 'The ecology of anthrax in the Kruger National Park, South Africa.' *Salisbury Medical Bulletin*, (**68S**): 19–23.

DURRHEIM, D.N., FREEMAN, P., ROTH, I. and HORNITZKY, M. (2009). 'Epidemiologic questions from anthrax outbreak, Hunter Valley, Australia.' *Emerg Infect Dis*, **15**(5): 840–2.

ELVANDER, M. (2009). 'The outbreak – an overview from farm to destruction of cows.' In *The Anthrax Outbreak in Sweden – Lessons Learned*. Nordic Biosafety Network Abstract Compendium, Solna, Sweden. SVA Dnr 2010/112, SUA Report Serie 17 ISSN 1654–7098.

FASANELLA, A., GAROFOLO, G., GALANTE, D., QUARANTA, Q., PALAZZO, L. *et al.* (2010). 'Severe anthrax outbreaks in Italy in 2004: considerations on factors involved in the spread of infection.' *New Microbiol*, **33**(1): 83–6.

FLORÉN, A.G. (1917). *Mjältbranden och dess bekämpande, Meddelande nr 108*. Djurskyddets Expedition, Stockholm, Sweden.

GUINEBRETIERE, M.H., BROUSSOLLE, V. and NGUYEN-THE, C. (2002). 'Enterotoxigenic profiles of food-poisoning and food-borne *Bacillus cereus* strains.' *J Clin Microbiol*, **40**(8): 3053–6.

HENINGER, S.J., ANDERSON, C.A., BELTZ, G. and ONDERDONK, A.B. (2009). 'Decontamination of *Bacillus anthracis* spores: Evaluation of various disinfectants.' *Appl Biosaf*, **14**(1): 7–10.

HIMSWORTH, C.G. (2008). 'The danger of lime use in agricultural anthrax disinfection procedures: the potential role of calcium in the preservation of anthrax spores.' *Can Vet J*, **49**(12): 1208–10.

HIMSWORTH, C.G. and ARGUE, C.K. (2009). 'Clinical impressions of anthrax from the 2006 outbreak in Saskatchewan.' *Can Vet J*, **50**(3): 291–4.

HURTLE, W., BODE, E., KULESH, D.A., KAPLAN, R.S., GARRISON, J. *et al.* (2004). 'Detection of the *Bacillus anthracis* gyrA gene by using a minor groove binder probe.' *J Clin Microbiol*, **42**(1): 179–85.

ISO (2011). *ISO 28901:2011: Soil quality — Guidance for Burial of Animal Carcasses to Prevent Epidemics*. The International Organization for Standardization, Geneva, Switzerland.

KEIM, P., PRICE, L.B., KLEVYTSKA, A.M., SMITH, K.L., SCHUPP, J.M. *et al.* (2000). 'Multiple-locus variable-number tandem repeat analysis reveals genetic relationships within *Bacillus anthracis*.' *J Bacteriol*, **182**(10): 2928–36.

KNUTSSON, R. (2011). 'A tracing tool portfolio to detect *Bacillus anthracis*, *Clostridium botulinum* and Noroviruses: bioterrorism is a food safety and security issue.' *Int J Food Microbiol*, **145** Suppl 1: S121–2.

MINETT, F.C. (1950). 'Sporulation and viability of *B. anthracis* in relation to environmental temperature and humidity.' *J Comp Pathol*, **60**:161–76.

MONGOH, M.N., DYER, N.W., STOLTENOW, C.L. and KHAITSA, M.L. (2007). 'Characterization of an outbreak of anthrax in animals in North Dakota: 243 cases (2005)'. In *40th Annual Conference of the American Association of Bovine Practitioners*. Vancouver, British Columbia, Canada, 20–22 September 2007. *Bovine Practitioner*, **41**: 101–9.

MULLER, J.D., WILKS, C.R., O'RILEY, K.J., CONDRON, R.J., BULL, R. *et al.* (2004). 'Specificity of an immunodiagnostic test for anthrax.' *Aust Vet J*, **82**: 220–2.

OIE (2011a). *OIE Code*. Available at http://www.oie.int/index.php?id=169&L=0&htmfile=chapitre_1.8.1.htm.

OIE (2011b). *OIE Code Disinfection*. Available at http://www.oie.int/index.php?id=169&L=0&htmfile=chapitre_1.4.13.htm.

OIE, S., OBAYASHI, A., YAMASAKI, H., FURUKAWA, H., KENRI, T. *et al.* (2011). 'Disinfection methods for spores of *Bacillus atrophaeus, B. anthracis, Clostridium tetani, C. botulinum* and *C. difficile*.' *Biol Pharm Bull*, **34**(8): 1325–9.

OLSEN, J.S., SKOGAN, G., FYKSE, E.M., RAWLINSON, E.L., TOMASO, H. *et al.* (2007). 'Genetic distribution of 295 *Bacillus cereus* group members based on adk-screening in combination with MLST (Multilocus Sequence Typing) used for validating a primer targeting a chromosomal locus in *B. anthracis*.' *J Microbiol Methods*, **71**(3): 265–74.

QUINN, P., CARTER, M., MARKEY, B. and CARTER, G.R (1994). '*Bacillus* species.' In *Clinical Veterinary Microbiology*, pp. 178–83. Wolfe Publishing, London.

RASKO, D.A., WORSHAM, P.L., ABSHIRE, T.G., STANLEY, S.T., BANNAN, J.D. *et al.* (2011). '*Bacillus anthracis* comparative genome analysis in support of the Amerithrax investigation.' *Proc Natl Acad Sci USA*, **108**(12): 5027–32.

RUTQVIST, L. and SWAHN, O. (1957). 'Epizootologiska och bakteriologiska undersökningar vid mjältbrandsepizootien i Sverige 1956–1957.' *Nord Vet Med*, **9**: 641–63.

SOBEL, J., KHAN, A.S. and SWERDLOW, D.L. (2002). 'Threat of a biological terrorist attack on the US food supply: the CDC perspective.' *Lancet* **359**(9309): 874–80.

SOU (2010:106). *Folkhälsa – Djurhälsa. Ny ansvarsfördelning mellan stat och näring. L. Statens Offentliga Utredningar*. Elanders Sverige AB, Stockholm.

STERNBERG LEWERIN, S., ELVANDER, M., WESTERMARK, T., HARTZELL, L.N., NORSTRÖM, A.K. *et al.* (2010). 'Anthrax outbreak in a Swedish beef cattle herd – 1st case in 27 years: Case report.' *Acta Veterinaria Scandinavica*, **52**(1): 7.

USDA/APHIS (2010). *Agricultural Select Agent*. Available at http://aphis.usda.gov/programs/ag_selectagent.

WHO (2001). *Guidelines for the Surveillance and Control of Anthrax in Human and Animals*. 3rd ed. Available at http://www.who.int/csr/resources/publications/anthrax/whoemczdi986text.pdf.

WHO (2008). *Anthrax in Humans and Animals*. 4th ed. World Health Organization, WHO Press, Geneva. Available at http://www.who.int/csr/resources/publications/anthrax_webs.pdf.

WIELINGA, P.R., HAMIDJAJA, R.A., AGREN, J., KNUTSSON, R., SEGERMAN, B. *et al.* (2011). 'A multiplex real-time PCR for identifying and differentiating *B. anthracis* virulent types.' *Int J Food Microbiol*, **145** Suppl 1: S137–44.

WOOD, J.P., CHOI, Y.W., ROGERS, J.V., KELLY, T.J., RIGGS, K.B. *et al.* (2011). 'Efficacy of liquid spray decontaminants for inactivation of *Bacillus anthracis* spores on building and outdoor materials.' *J Appl Microbiol*, **110**(5): 1262–73.

18

Mounting an effective response to an outbreak of viral disease involving street food vendors in Indonesia

D. Orlando, Dennis, M. Suci Hardianti and F. M. C. Sigit Setyabudi, Universitas Gadjah Mada, Indonesia

Abstract: This chapter focuses on the viral agents of hepatitis, especially foodborne transmission of hepatitis A virus (HAV). A recent outbreak (2008) due to the unhygienic practices of food stall personnel and street vendors is discussed. An overview of the actions taken to control the outbreak and prevent another occurrence is presented. The case can be used as an integrated model for the development of recommendations to control and prevent similar foodborne viral disease outbreaks in developing countries.

Key words: hepatitis A virus (HAV), developing countries, ready-to-eat food, street food, food safety education.

18.1 Introduction to the case

In the period April–August 2008, epidemics of hepatitis A (HA) were reported to have increased sharply in Yogyakarta Province (Indonesia) with approximately 500 human clinical cases, including some at Universitas Gadjah Mada (UGM). UGM Public Relations Office reported that a recorded outbreak of HA on campus consisted of 122 students and seven administration staff. Contaminated food was suspected to be the likely cause.

18.1.1 How did it all begin?

Following the worrying development of an HA outbreak on the university campus, the transmission of HAV in food became the subject of a comprehensive collaborative study. The initial steps involved a risk assessment and epidemiological study of the outbreak in Yogyakarta that was processed by the Field Epidemiology Training Program (FETP), Faculty of Medicine-UGM. Subsequently, a study was conducted on producer and consumer perceptions before and after a food safety training program that was conducted by the Faculty of Agricultural Technology and Centre for Food and Nutrition Studies (CFNS) at UGM.

18.1.2 What were the challenges involved?

Indriani (2008) reported that four consumption habits, regardless of age, showed statistically significant correlation with the outbreak rate among the 406 cases that included 203 patients and 203 controls during the FETP investigation. First, consumption of food prepared in food stalls increased the risk of HA infection 7.8-fold compared with consumption of food prepared at home. In addition, the risk of HA infection was 1.9 times higher when consuming food that was served by traders using their bare hands than consuming food served using utensils. An infection risk 2.7 times higher was also found when consuming food at food stalls that washed their utensils but did not use any sanitizer. Finally, the lack of hand washing both by the consumers and the food traders significantly contributed to the infection risk, increasing it by as much as 5.9-fold.

Kurniawan and Fallah (2009), also reporting on the case, stated that there was inappropriate food processing in food stalls; and street vendors around the university campus were often associated with a lack of knowledge of good hygiene practices. The authors also found that 91% of the workers at food stalls had never obtained any hygiene training, even though 88% of them had been in business for more than 6 years. Thus, among all the other risk factors mentioned earlier, poor hygiene practices played the most important role in this outbreak. It was considered that HAV had probably been transmitted to food products by workers due to contamination with faeces during food handling (Cliver, 1985).

However, regardless of the results described above and despite intensive work, the investigation could not obtain adequate evidence to identify the source of virus infections. Laboratory investigation among the traders failed to prove that they were the source of the HA outbreak. Similar results were also reported by Staes (2000) during an investigation of an HA outbreak in the US. Although proof was not obtained, the university authorities decided to implement an outbreak control and prevention plan. The plan focussed on the street food vendors, reducing the likelihood of future cases.

18.2 Significance of the case

18.2.1 Hepatitis A infection

It must be noted that ready-to-eat, processed or cooked foods must be free of HAV contamination, since the minimum dose of HAV required to initiate an infection after ingestion can be very low, perhaps in some cases a single particle (Cliver, 1985). Infection with HAV usually occurs by the faecal-oral route of transmission.

The viruses enter the hepatocyte via an interaction with a cellular receptor, followed by un-coating of the viral particle and release of RNA into the cell. The work of nonstructural proteins in the replication process produces newly assembled HAV particles, which will eventually be secreted by the cell across the apical membrane of the hepatocytes into the biliary canaliculus before then reaching the bile and small intestine (Martin and Lemon, 2006).

Since HAV has no lipid envelope, it is stable when excreted from the infected liver to the bile to enter the gastrointestinal tract. It has been found to survive in experimentally contaminated fresh water, seawater, wastewater, soils, marine sediment, live oysters, and creme-filled cookies. HAV is extremely resistant to degradation by environmental conditions, a property that allows its maintenance and spread within populations.

Several symptoms can manifest differently among infected persons (Table 18.1). The mechanism responsible for hepatocellular injury in hepatitis is suggested to be more attributable to an immunopathologic response to infection of the hepatocyte,

Table 18.1 A summary of the symptoms reported among patients with hepatitis A

Period of infection	Duration	Symptoms
Prodromal phase (early symptoms of infection)	Several days to more than a week	General symptoms: fever, fatigue, malaise, myalgia, arthralgia. Gastrointestinal symptoms: anorexia with disorders of taste and smell, sore throat, nausea, vomiting, diarrhoea, constipation, hepatomegaly, splenomegaly, abdominal discomfort. Symptoms from respiratory tract: cough. Skin symptoms: rash, urticaria, exanthematous skin eruptions.
Acute phase	Three weeks	Gastrointestinal symptoms: anorexia, nausea, vomiting, hepatomegaly, hepatic tenderness, splenomegaly, pale stools. Skin symptoms: yellowish discoloration of the mucous membranes (conjunctivae, sclera), jaundice, pruritus. Other symptoms: dark, golden-brown urine, biochemical and haematological changes.
Chronic phase (does not occur in all cases)	Several weeks to 11 months after the onset of illness	Biochemical and haematological changes.

Source: Cook and Rzeżutka, 2006.

rather than to a direct cytopathic effect of the virus. As explained (WHO, 2000), 'The presence of anti-HAV IgM is detectable about three weeks after exposure, its titre increases over four to six weeks, then declines to nondetectable levels generally within six months of infection. Anti-HAV IgA and IgG are detectable within a few days of the onset of symptoms. IgG antibodies persist for years after infection and provide lifelong immunity. The development of antibody to HAV coincides with a decrease in quantity of viremia and faecal shedding of the virus. Saliva and faeces generally do not contain neutralizing antibodies.'

18.2.2 Economic and market aspects

Approximately 80 000 students are enrolled in UGM, including undergraduate, graduate and postgraduate students. Most of them stay in boarding houses near the campus. Their time is mostly spent doing academic and other activities on the campus. Therefore they often buy ready-to-eat food at the food stalls. Currently, the number of food stalls around the university campus is increasing due to the increasing numbers of students. The food stalls are actually helpful from the consumer perspective as they provide a convenient and affordable way to eat, although not all of them meet the hygiene standards set by the regulations. The HA outbreak in 2008 was very serious since there were both short- and long-term economic losses, as well as potential loss of market. Kusnanto (2011) estimated that greater expenditure for outbreak eradication between May and September 2008 would be needed. The expenditure would be spent on vaccine and genotype analysis of the virus.

18.3 The case history continues

The university authorities initiated an integrated plan as the response to the outbreak. The handling plan became a model to control the spread of HA outbreaks in order to reduce foodborne disease transmission. The model was initiated by the assessment of food stalls and street vendors around the campus area. The stall holders and street vendors were then provided with routine training and assisted with certification in good hygiene practices. Moreover, the street vendors were relocated; a new location was provided in which they could run their businesses.

Since the infected people were mostly students, the university authorities decided to assign the task of conducting routine training and assisting with the certification of the traders in good hygiene to the Faculty of Agricultural Technology and CFNS-UGM. The training aimed at the application of a scheme of hazard analysis and critical control point (HACCP)-like principles, which placed emphasis on all apart from the last two of the seven HACCP principles. This HACCP-like scheme was initially promoted by Forsythe and Hayes (1998) as a way to remove obstacles to the adoption of HACCP by small and medium enterprises (SMEs). Furthermore, the responsibility for treatment and personal medical consultation for HA prevention among students was assigned to Gadjah

Mada Medical Centre. A standardized location for the food traders was also provided by the university, namely Food Court Plaza Campus.

18.3.1 The establishment of Food Court Plaza Campus

Food Court Plaza Campus is a part of the UGM Master Plan for Campus Development (RIPK 2005–2015) which embraces the Educopolis Campus concept. Some of the food stalls and street vendors, which were scattered around the campus area, have been relocated to the Food Court Plaza Campus. It is managed jointly by *Direktorat Pengelolaan dan Pemeliharaan Aset* UGM, PT. Gama Multi Usaha Mandiri (GMUM) and Paguyuban PKL Bulevar. Management is handled by GMUM, which arranges monthly meetings with the traders to maintain harmonized dialogue between them. In addition, various training programs on entrepreneurship and sanitation are also scheduled regularly.

Since February 2009, 52 food stalls and street vendors have been relocated to Food Court Plaza Campus which is equipped with tents, carts, tables, chairs and a hotspot area for the customers (Fig. 18.1). Moreover, the traders also have access to faucets for washing their utensils.

Food Court Plaza Campus can improve traders' conditions and serving performance as the traders are under monitoring and evaluation from the university authorities. Moreover, the Plaza may improve the image of UGM as a comfortable and environmentally sound campus. Kurniawan and Fallah (2009) described Food Court Plaza Campus as a proper response to the challenge of ensuring the availability

Fig. 18.1 Food Court Plaza Campus.

of food at affordable prices for students who have a limited monthly living allowance (less than 100 000 000 rupiah), while maintaining a good standard of food safety.

18.4 Commentary and personal reflections

The present case shows the importance of good hygiene practices to proper food handling by food stall holders and street vendors. Campuses are areas vulnerable to outbreaks of HA or similar outbreaks of foodborne disease with intermediate endemicity patterns due to student mobility and typical age. Responses to similar outbreaks in the future should aim to limit economic and productivity losses. The challenge remains for interested researchers to isolate the source of the HA outbreak. Hopefully, this chapter will provide valuable information about the integrated efforts that drew on the competences of all related disciplines to control viral disease in the campus area.

18.5 Acknowledgements

The authors thank Hari Kusnanto (Field Epidemiology Training Program UGM), Eni Harmayani (Center for Food and Nutrition Studies UGM) and Muhammad Prasetya Kurniawan (Faculty of Agricultural Technology UGM) for providing the research data and discussion during preparation of this manuscript. This manuscript would not have been written without the encouragement of Nuryono (Faculty of Mathematic and Natural Science).

18.6 Critical questions for discussion

1. What efforts other than those mentioned in this chapter are needed in order to identify the source of the outbreak?
2. What study design is needed to answer the aforementioned question?
3. Which kind of (urgent and more comprehensive) actions should be taken, should a similar outbreak happen again on the same campus?
4. Considering how the money could otherwise have been spent, do you think the relocation of food stalls and street vendors was a cost-effective intervention?
5. Was the integrated handling plan among the departments at the university implemented fully?

18.7 References

CLIVER, D.O. 1985. Vehicular transmission of hepatitis A. *Pub. Health Rev.*, **13**: 235–92.

COOK, N. and RZEŻUTKA, A. 2006. Hepatitis viruses. In Y. Motarjemi and M. Adams (eds), *Emerging Foodborne Pathogens*, pp. 283–308. CRC Press, Boston.

FORSYTHE, S.J. and HAYES, P.R. 1998. HACCP and Product Quality. In *Food Hygiene Microbiology and HACCP*, pp. 276–324. Aspen Publisher, Gaithersburg.

INDRIANI, C. 2008. Laporan Penelitian Analitik Faktor Risiko Hepatitis A di Provinsi DIY Tahun 2008. Program Studi Ilmu Kesehatan Masyarakat. Fakultas Kedokteran UGM.

KURNIAWAN, M.P. and FALLAH, M.A.F. 2009. Analisis Persepsi dan Edukasi Sistem Keamanan Pangan Bagi Produsen dan Konsumen Food Court Kampus UGM. Program Studi Teknologi Industri Pertanian. Fakultas Teknologi Pertanian UGM.

KUSNANTO, H. 2011. Personal communication and discussion on Hepatitis A: Wabah dan Beban Sakit di Masyarakat on 13 and 18 May 2011.

MARTIN, A. and LEMON, S.M. 2006. Hepatitis A virus: From discovery to vaccines. *Hepatology*, **43**(2). Suppl. 1: 164–71.

STAES, C.J., SCHLENKER, T.L., RISK, I., CANNON, K.G., HARRIS, H. *et al.* 2000. Source of infection among persons with acute hepatitis A and no identified risk factors during a sustained community-wide outbreak. *Journal of Pediatrics*, **106**(4) e54.

UGM PUBLIC RELATIONS. 2008. *Hepatitis A mewabah di DIY, 129 warga UGM menjadi Korban*. Available at http://www.ugm.ac.id/index.php?page=rilis&artikel=1466. Accessed 20 April 2011.

WHO/CDS/CSR/EDC. 2000. *Hepatitis A*. Available at http://www.who.int/emc. Accessed 16 August 2011.

19

Handling of the media during an incident in Romania involving dioxin contamination of yoghurts

A. I. Nicolau, Dunarea de Jos University of Galati, Romania

Abstract: The case study describes what happened in 2007 when the mass media released the news that two batches of fruit yoghurts produced by the Romanian unit of a global company were suspected of being contaminated with dioxin and two lots of its products were banned by local food safety authorities. The mass media's comments on these events and their effects on consumers and the company image are presented. Details of the measures taken by the company to ensure consumer safety and the strategy devised for communicating with consumers are revealed, along with measures taken by national and European authorities involved in food safety assurance.

Key words: yoghurt, dioxin, guar gum, withdrawal, mass media, food scare.

19.1 Introduction to the case

19.1.1 How did it all begin?

On Wednesday, 22 August 2007 a Romanian newspaper published an article reporting the potential presence of dioxins in two types of fruit yoghurts produced by the Romanian branch of a multinational corporation (hereafter called Company X). The next day, the news was rapidly taken up by other media outlets. During the following days, as the representatives of the food company and the National Sanitary-Veterinary and Food Safety Authority (NSVFSA) did not react to the news, the journalists kept on presenting facts on dioxins and reminded the readers of the case of the Ukrainian President Victor Yuschenko, who in 2005 was allegedly poisoned with a high dose of dioxin. To make the matter worse, other food issues that are negatively perceived by consumers, such as food additives, were commented upon.

Consumers reacted promptly to the news and accompanying comments by not buying the fruit yoghurts supposed to be contaminated with dioxin, even though units of the incriminated batches were no longer on supermarket shelves. Supermarkets' sales declined as a result of consumer reluctance to buy yoghurts and other dairy products produced by both Company X and other companies. The reaction of Company X came 10 days after the beginning of what journalists had named 'the dioxin crisis': a press conference was held by Company X to assure consumers that the foods the company produced had not presented a danger to consumers at any time.

19.1.2 What were the facts beyond the press releases?

On 14 August 2007, NSVFSA was notified via the Rapid Alert System for Food and Feed (RASFF) that an ingredient contaminated with dioxin had reached some Romanian food processors. NSVFSA immediately warned its Food Safety Offices in all 42 counties of Romania and asked the inspectors working there to identify the food processors using the ingredient and take the necessary legal measures to avoid the consumption of foods containing this ingredient. The contaminated ingredient was supposedly a fruit preparation manufactured in the Czech Republic, to which guar gum (E412) was added as a thickener. The guar gum, which contained 12–156 pg of dioxin per gram of fat, was imported from India by a Swiss company and delivered to several customers across the European Union, including the Czech manufacturer (Swissinfo, 2007).

The above fruit preparation was used by Company X in two types of fruit yoghurts. According to the company's quality records the quantity of guar gum in the yoghurt jars was very low (0.02 g/125 g yoghurt), but NSVFSA applied the precautionary principle and asked the company to start the withdrawal procedure for two batches of fruit yogurt suspected of being contaminated with dioxins.

Since the dairy company had a traceability system in place, it was able to rapidly locate the suspect yoghurt jars, remove them from its storage house and send samples to a reference laboratory for analysis for dioxins. By the end of August, the laboratory results proved that the dioxin concentrations in incriminated yoghurts were 13–37-fold lower than the maximum tolerable level for such products (Reuters, 2007). Finally, Company X did not reintroduce the suspect yoghurts to the market. This decision was justified by the fact that the cold chain was interrupted during the withdrawal process.

19.1.3 Challenges for Company X and the National Sanitary-Veterinary and Food Safety Authority (NSVFSA)

At that time, Romania did not have official laboratories to detect dioxins in foodstuffs and the samples were sent to a reference laboratory for dioxins in Hungary. Besides the time taken to send the yoghurt samples to another country, the company also had to wait several days for the results because conventional dioxin tests require lengthy sample preparation. The rapid tests for dioxin were not suitable in this case due to their lower precision. While Company X and NSVFSA were waiting for the laboratory report from Hungary, the mass media kept writing about the presence of dioxin in yoghurts.

19.2 Significance of the case

19.2.1 Significance of the chemical agent

Dioxins and dioxin-like compounds include a range of toxic and environmentally persistent substances. The terms most often refer to 29 congeners of polychlorinated dibenzo-*p*-dioxins (PCDD), polychlorinated dibenzofurans (PCDF) and dioxin-like polychlorinated biphenyls (PCB). Dioxin and furan congeners are formed as unintentional by-products during combustion or burning processes, as well as during some industrial processes such as paper pulp bleaching and the manufacture of chlorinated pesticides (EFSA, 2010). Dioxins accumulate in fatty tissues, from where they exert their toxic effects on the human body. According to the World Health Organization (WHO), 90% of exposure to dioxins is through food, and 80% is from food of animal origin (WHO, 2010).

19.2.2 Regulatory aspects

At the time of the incident, the maximum levels for dioxins and dioxin-like PCBs in food had to comply with EC Regulation 1881/2006 (European Commission 2006). According to this legislation act, the maximum level for the sum of dioxins and furans in milk and dairy products is 3 pg/g fat and the maximum level for dioxin, furan and dioxin-like PCBs is 6 pg/g fat, while foods exceeding these limits have to be excluded from sale. According to Article no. 19 of the General Food Law Regulation (EC) 178/2002, food business operators are obliged to withdraw food that is not in compliance with food safety requirements, and to recall the food if it has reached the consumer.

19.2.3 Economic and market aspects

After the publication of the articles in Romanian newspapers on 22 August 2007 relating to dioxin presence in the yoghurts produced by Company X, the shares of Company X fell by 13 eurocents (18 dollar cents), or 0.2%, to 54.70 euros ($74.22) on the following morning. The slight fall was nonetheless enough to send it to the bottom three performers of the Paris CAC 40 index, which rose 55.99 points, or 1.0%, to 5574.16 (Laurent, 2007). As a result of the dioxin crisis, the dairy company withdrew 44 000 plastic jars, each containing 125 g of yoghurt, which were not reintroduced into the food chain. The company's sales decreased for a 3-month period (August–October). According to *Ziarul Financiar* (*The Financial Newspaper*), sales decreased by 8% compared with those from June and July in the same year. The company incurred financial costs in its efforts to overcome the dioxin crisis, most of them due to a campaign promoting the safety of the products.

In order to regain consumers' trust, the company spent 35% more money than in 2006 on advertising their products in a food safety context (Tudor, 2008).

19.3 The case history continues

Dioxin-contaminated guar gum affected not only a company from Romania, but several others located across Europe, especially major international corporations. In all the countries where these companies are located, the mass media acted in a similar way. In Hungary, for example, the mass media accused food companies and the government of putting at risk not only the lives of Hungarians, but also those of non-Hungarians, who at that time were participating in the Hungarian Grand Prix or Sziget Festival, and suggested that those wishing to visit Hungary in the future should have a comprehensive health insurance plan (Horvath, 2007).

Since the dioxin crisis of 2007, Europe has been through other similar crises. In early 2008, the mass media reported on the levels of dioxins in the buffalo milk used to make mozzarella cheese in Italy, while in 2011 the mass media reported on the levels of dioxins in eggs from German farms, pork products from the Irish Republic and cattle from Northern Irish farms.

The story has been the same every time: the mass media transformed the incidents into food scares and did not comment on the mechanisms that companies and governments were using to protect consumers. In forthcoming years, the number of reported incidents of dioxin contamination is expected to increase as a result of intensive monitoring programs in the European countries. Should we expect the same reaction from the mass media every time a new laboratory report is released?

19.4 Resolution and outcomes

Because in 2007 the guar gum containing dioxin had a European dimension, the EC's Food and Veterinary Office (FVO) suspended all imports of guar gum from India Glycols (the producer of dioxin-contaminated guar gum) and blocked the consignments that remained. The FVO inspectors then carried out an inspection in India from 3–10 October 2007. After visiting the guar gum plant and other related places, the inspection team concluded that there was not sufficient evidence to determine the cause of the contamination incident, and that adequate controls were not in place to prevent further incidences of contamination from taking place in the future.

The following year, 1–12 October 2008, FVO organized a follow-up inspection in India to assess the control measures put in place by the Indian authorities to prevent contamination of guar gum with dioxin and PCP. Unfortunately, the inspectors did not find any improvement in the control system or any significant reduction in the associated risks.

The outcomes of these visits were:

- Development of special legislation related to this contamination incident:
  - Commission Decision 2008/352/EC of 29 April 2008 imposing special conditions governing guar gum originating or consigned from India due to risks of contamination with pentachlorophenol and dioxins posed by these products.
  - Commission Regulation (EU) No 258/2010 of 25 March 2010 repealing Decision 2008/352/EC and imposing special conditions on the imports of guar gum originating in or consigned from India due to risks of contamination by pentachlorophenol and dioxins (applicable from 15 April 2010).
- Development of a strategy at the EU level to reduce population exposure to dioxin.

19.5 Commentary and personal reflections

This case shows the importance of monitoring contaminants in food and food additives, as well as the importance of tools that can be successfully operated to assure food safety (RASFF, traceability). The European Union has to continually improve its legislation and tools involved in food safety assurance. The case also demonstrates that intervention by the mass media is not always welcome. The media informed the Romanian consumers about a potential chemical risk associated with food, but also scared them with the headlines they used. Even if the mass media did not succeed in instilling panic, they succeeded in causing anxiety among consumers. Although there might have been a problem related to the presence of dioxins in Company X's yoghurt, reviewing the facts revealed that the media made the crisis. This was later recognized by the mass media themselves, in headlines such as 'Dioxin from Company X's yoghurt – a mediatic intoxication' (Gândul, 2007) and 'Company X Hysteria' (Popescu, 2007).

Judging from the facts, the dualism of the mass media is obvious. After incriminating Company X for putting consumers at risk, which was not the case, the mass media transformed itself into the company's allies in its efforts to regain consumer trust and to rebuild its image. Company X's efforts to explain that it did not put its profits above consumers' food safety were supported by the mass media. At this stage, headlines changed to 'Romania Company X yoghurt safe for human consumption' (Anonymous, 2007), 'Company X acts over Romanian dioxin fears' (Merrett, 2007), while TV news programmes dedicated significant time to interviews with company representatives, who explained in detail how the yoghurts were safe and healthy. Both newspapers and TV stations also hosted industry advertising messages that aimed to offset the media's previous messages relating to the unsafe nature of the products.

The same pattern of events has been seen in other countries in similar situations (see the headlines and newspapers article covering the historical dioxin crises, which are presented in Table 19.1). The author strongly believes that consumers have to be informed regularly not only of the food safety problems that a Member

Table 19.1 Newspapers' headlines reflecting incidents related to dioxin presence in food or feed

Headline	Newspaper	Incident			
		Year	Country	Dioxin contaminated food or feed	Dioxin source that led to contamination
Food scare is leaving tables bare in Belgium	*The New York Times* (USA) 6 June 1999	1999	Belgium	Eggs and chickens	Feed
Chicken and eggs withdrawn in Belgian health scare	*The Sunday Sun* (UK) 30 May 1999				
Coke acknowledges tainting in Belgium	*The Daily Gazette* (USA) 16 June 1999	1999	Belgium	Soft drinks	Fungicide applied to transport pallets
Coca Cola health scare was 'mass hysteria'	*The Guardian* (UK) 2 July 1999				
Mozzarella with dioxin poisons Italy (La mozzarella à la dioxine empoisonne l'Italie)	*Le Point* (FR) 28 March 2008	2008	Italy	Mozzarella	Garbage
Alert for dioxin in mozzarella (Alerte à la dioxine dans la mozzarella)	*Le Post* (FR) 20 March 2008				
Mozzarella with dioxin: Italy will recall the suspected products (Mozzarella à la dioxine: l'Italie va retirer les produits suspects)	*24 Heures* (CH) 28 March 2008				
EU dioxin alert	*Budapest Business Journal* (HU) 9 August 2007	2007	Switzerland and EU countries	Yoghurt, dairy products, fruit product	Swiss thickening agent based on contaminated guar gum originating from India

(*Continued*)

Table 19.1 Continued

Headline	Newspaper	Incident			
		Year	Country	Dioxin contaminated food or feed	Dioxin source that led to contamination
Dioxin returns in Belgium (La dioxine de retour en Belgique)	*La Dernière Heure* (BE) 22 August 2007				
Dioxin in Irish pork: alert (De la dioxine dans le porc irlandais: état d'alerte)	*Le Matin* (CH)	2008	Ireland and other EU countries	Pork meat	Feed
Pork with dioxin in Ireland. UE: alarm even in Italy (Maiale alla diossina in Irlanda. L'Ue: allarme anche in Italia)	*Corriere della Sera* (I) 7 December 2008				
Minimum risk in Italy (Ma in Italia rischi minimi)	*Corriere della Sera* (I) 8 December 2008				
Eggs: food scandal in Germany (Œufs: scandale alimentaire en Allemagne)	*Le Figaro* (FR) 5 January 2011	2011	Germany and other European countries	Eggs and meat	Feed
Dioxin: the criminal track (Dioxine: la piste criminelle)	*Le Matin* (CH) 7 January 2011				
The criminal machinations of the feed industry	*Der Spiegel* (DE) 1 October 2011				
Hundred dioxin pigs slaughtered in Germany (Horderden/ofgeslocht in Duitsland dioxinevarkens)	*De Telegraaf* (NL) 7 January 2011				

German dioxin scare spreads to meat	*De Telegraaf* (NL) 9 January 2011
No dioxin found in German chickens and eggs in Belgium, according to AFSCA* (Pas de dioxine trouvée dans les poulets et oeufs allemands en Belgique, selon l'AFSCA)	*La Derniere Heure* (BE) 10 January 2011

* Federal Agency for the Safety of the Food Chain (Belgium).

State encounters, but also of ongoing actions by organizations working to ensure consumer safety. Information has to be released by specialists, not by journalists who sometimes mislead consumers and periodically frighten them with false stories.

It is essential for government authorities and food companies to effectively communicate and provide evidence that they are reducing, mitigating, or minimizing a particular risk. Because it is difficult for government authorities and food companies to communicate directly with consumers, they have to learn to communicate via the mass media. This communication channel will save resources and help in transmitting accurate information to consumers. The timing of communication is also important. Both governmental authorities and food companies have to be one step ahead of the mass media in order to prevent speculation.

19.6 Critical questions for discussion

1. How do dioxins exert toxic effects on humans?
2. What are the other typical products that have been involved in dioxin contamination?
3. What is the RASFF and how does it function?
4. What is the precautionary principle in European legislation?
5. What is 'traceability' and how was a traceability system used in the case described above?
6. Did Company X act correctly when it decided to withdraw all of the yoghurts, or would it have been better to have a public recall? What is the difference between a product withdrawal and a product recall?
7. After deciding to break the silence, what other measures could have been taken by Company X to manage the dioxin crisis?
8. What is your opinion on the mass media's involvement in this case and in other food-related issues?
9. Who is in the position to deliver a food safety message most effectively: government officials, the mass media or company representatives?
10. Compare the way in which governments, companies and the mass media acted in the guar gum dioxin crisis of 2007 with their actions in other food incidents involving dioxins that have subsequently taken place.

19.7 References

ANONYMOUS (2007). *Romania Company X yoghurt safe for human consumption. Money*, 30 August 2007. Available at http://www.money.ro/iaurtul-danone-este-sigur-pentru-consum_18244.html.

EFSA (2010). Results of the monitoring of dioxin levels in food and feed. *EFSA Journal*, **8**(3): 1385.

EUROPEAN COMMISSION (2006). Commission Regulation (EC) No 1881/2006 of 19 December 2006, setting maximum levels for certain contaminants in foodstuffs. *Official Journal of the European Union*, L 364/5–364/24.

GÂNDUL (2007). *Dioxina din iaurtul Danone – o intoxicație mediatică* (Dioxin from Danone's yoghurt – a media intoxication). Available at http://www.gandul.info/news/dioxina-iaurtul-danone-intoxicatie-mediatica-895545. (Accessed 5 November 2011.)

HORVATH, J. (2007). Hungary's tainted food industry. *Ohmy News International*. Available at http://english.ohmynews.com/articleview/article_view.asp?no=377820&rel_no=1. (Accessed 6 May 2011.)

LAURENT, L. (2007). *Danone's Dioxin Scare*. Available at http://www.forbes.com/2007/08/23/danone-dairy-cadbury-markets-equity-cx_ll_0823markets07.html. (Accessed 6 May 2011.)

MERRETT, N. (2007). Danone acts over Romanian dioxin fears. *Dairy Reporter*, 23 August 2007. Available at http://www.dairyreporter.com/Regulation-Safety/Danone-acts-over-Romanian-dioxin-fears.

POPESCU, A. (2007). *Isteria Danone* (Danone Hysteria). 23 August 2007. Available at http://www.saptamana.com/link.php?id=719205. (Accessed 5 November 2011.)

REUTERS (2007). *Romania Danone yogurt safe for human consumption*. Available at http://www.reuters.com/article/2007/08/29/idUSL29747438. (Accessed 5 May 2011.)

SWISSINFO (2007). *Dioxin found in Swiss food thickener*. Available at http://www.swissinfo.ch/eng/Home/Archive/Dioxin_found_in_Swiss_food_thickener.html?cid=6030650. (Accessed 5 May 2011.)

TUDOR, D. (2008). Danone reveals impact of 2007 dioxin scandal. *Ziarul Financiar*, 30 January 2008. Available at http://www.zf.ro/zf-english/danone-reveals-impact-of-2007-dioxin-scandal-3052146. (Accessed 1 July 2011.)

WHO (2010). *Dioxins and Their Effects on Human Health*. WHO Media centre, Fact sheet N°225. Available at http://www.who.int/mediacentre/factsheets/fs225/en/. (Accessed 5 May 2011.)

Part IV

Farm-level interventions

20

Is it possible to reduce foodborne *Campylobacter* infections in humans through vaccination of animals?

J. Hoorfar, Technical University of Denmark, Denmark

Abstract: Vaccination has been used successfully over the years to eradicate many serious diseases, but what about human foodborne pathogens, such as *Campylobacter*? Most human cases of *Campylobacter* infection are associated with consumption of poultry products. Vaccination of poultry to prevent early colonization or to reduce the *Campylobacter* colonization level may be a viable intervention strategy in the future; however, no commercial *Campylobacter* vaccine is currently available. This case considers the rationale for such a strategy, the forms it could take and the challenges it would involve.

Key words: zoonoses, infection, food, intervention, immunity.

20.1 Introduction to the case

Raw chicken is one of the most common carriers of *Campylobacter*. Humans are often exposed to the bacteria when cooked meat is placed on unwashed cutting boards previously used for trimming raw chicken, or when chicken is not cooked adequately. The chicken gut, in which high numbers of *Campylobacter* can be found, is usually the source of contamination of chicken carcasses during the slaughter process.

The recent scientific opinion by the European Food Safety Agency (EFSA Panel 2010) states that the source of up to 80% of human campylobacteriosis is poultry; however, only 40% is poultry meat-related. This indicates that there are other transmission routes from primary production to humans besides food consumption. This underscores the fact that prevention of *Campylobacter* colonization in primary production can contribute to the reduction of campylobacteriosis in humans.

In principle, there is no need to protect poultry against *Campylobacter*, since the organism is commensal in poultry. Therefore, a treatment capable of diminishing the numbers of bacteria in chicken gut and thus the infective pressure would be an efficient control measure to reduce the risk of human infections due to contaminated meat.

Currently there is no commercial vaccine available to control *Campylobacter* infections in poultry (Lin 2009). In Denmark, a project (www.camvac.dk) was initiated to develop a comprehensive and harmonized assessment of different vaccine candidates. This is the story of the idea behind this rather unusual approach to control the most prevalent foodborne pathogen in the Western world.

20.2 Significance of the case

On-farm control of *Campylobacter* would have the greatest impact of all control measures currently under consideration because the intestine of living poultry, cattle and hogs is the only amplification point for *Campylobacter* throughout the food chain (Wagenaar *et al.* 2008). The colonization level can be up to 10^{10} cfu/g faeces (Sahin *et al.* 2002). Three general strategies for on-farm control of *C. jejuni* are:

- Reduction of environmental exposure through biosecurity measures,
- An increase in host resistance to reduce gut carriage by, e.g., vaccination,
- The use of antimicrobial alternatives, such as bacteriophage, etc., to reduce colonization.

Rosenquist *et al.* (2003) suggest that a 2-log reduction in the *C. jejuni* populations on poultry carcasses could bring about a 30-fold reduction in human campylobacteriosis cases. They further conclude that, to obtain a similar reduction in the incidence of campylobacteriosis, the flock prevalence should be reduced approximately 30 times (e.g. from 60% to 2%) or kitchen hygiene improved approximately 30 times (e.g. from 21% of consumers not washing the cutting board to 0.7%). The infectious dose of *Campylobacter* is estimated to be 500 cfu (Doores 1999).

Several risk assessment models (Christensen *et al.* 2001, Rosenquist *et al.* 2003, Nauta *et al.* 2005, Uyttendaele *et al.* 2006, Havelaar *et al.* 2007, Nauta *et al.* 2007, Lindqvist and Lindblad 2008) have come to similar conclusions as to the implementation of cost-effective control strategies:

1. All models found a negligible effect of logistic slaughter (= processing positive flocks after the negative flocks),
2. The most effective intervention measures aim at reducing the *Campylobacter* concentration (cfu/g), rather than reducing the prevalence,
3. At the stage when the consumer handles the food, cross-contamination was generally considered to be more relevant for human infection than undercooking,

4. Those animals/flocks that were most heavily contaminated, in other words the tails of the distributions describing the variability in *Campylobacter* concentrations between meat products and meals, determine the risks, not the mean values of those distributions.

Intervention should therefore aim at reducing these heavily infected animals. While biosecurity measures are practical and have helped to some degree, hygiene barriers can be breached. All other potential control measures are still under development.

20.2.1 Clinical, regulatory and market aspects

Campylobacter is currently the food bacterium that infects the most people in Denmark and in many other countries. In 2008, almost 3500 people were registered as sick with *Campylobacter* in Denmark. However, it is estimated that the actual figure is probably 10–20 times higher because many do not seek medical help (Ministry of Agriculture, Food and Fisheries, Denmark. http://www.uk.foedevarestyrelsen.dk/Food_Safety/Campylobacter/forside.htm, accessed 2 July 2011). Most people are infected during the summer, as our control of food hygiene often fails when we barbeque in the outdoors, and much greater numbers of *Campylobacter* are found in poultry in the summer. The sneaky thing about *Campylobacter* is that it takes only a few bacteria to fall ill. Just a few drops of juice from raw meat are enough to cause stomach pain and intense diarrhoea.

In a European study by EFSA, *Campylobacter* was found in 76 out of the 396 Danish broiler flocks tested, which is an incidence of 19.2% (EFSA 2010). A few other EU countries had lower incidences than Denmark: Estonia (2%), Finland (4.1%), and Sweden (12.4%). At the opposite end of the scale, *Campylobacter*

Table 20.1 Number of farms in broiler production in Denmark in 2010

	No. of holdings	No. of houses/flocks	Livestock (capacity)
Rearing period (grandparent)	4	14	90 000
Adult period (grandparent)	5	11	80 000
Rearing period (parent)	15	90	130 000
Adult period (parent)	44	145	710 000
Hatcheries	5		
Broilers	242	589	

Source: Danish Veterinary and Food Administration and Danish Agriculture and Food Council. For other statistics visit www.lf.dk.

was found in 100% of the broiler flocks in Luxembourg, 97% of those from Malta, 87.7% of those in Spain and 82.9% of those in Portugal.

Poultry statistics are shown in Table 20.1. In Denmark, a voluntary intervention strategy implemented in 2003 aims to reduce the incidence of *Campylobacter*-positive results for broiler flocks. This strategy encompasses directives for monitoring the infection status of broilers at pre-harvest stages and at harvest while poultry houses and slaughterhouses take specific hygienic measures, and it provides an economic incentive to farmers delivering *Campylobacter*-negative flocks. A significant decrease in the prevalence of *Campylobacter*-positive broiler flocks, from 38% in 2003 to 29.9% in 2006, was attained. This decrease may be attributed to the intervention program (Olsen *et al.* 2009).

20.3 Why vaccinate poultry and not humans?

It is clear that a potential vaccine for human use would be the best strategy. This would, however, require an effective vaccine and a standard vaccination program similar to vaccination against measles and mumps. In poultry, despite several control measures including intensive biosecurity measures, it has not been possible to further reduce or eliminate the colonization of *Campylobacter* in chickens significantly. In addition, there are poultry production methods, such as free-range or organic farms, where biosecurity measures are not possible. An alternative approach is vaccination, which could create new technologies for on-farm control and provide chicken producers with an innovative and practical tool, especially in organic production and free-range chickens. Similar cost-effective vaccinations have been developed against avian influenza and coccidiosis that can be used as a model (Walker *et al.* 2010; Sharman *et al.* 2010).

With regard to the chicken immune response to vaccinations, administering *Salmonella* orally to newly hatched chickens results in extensive gut colonization and a strong adaptive immune stimulus that exerts a variety of rapid protective effects (Van Immerseel *et al.* 2005). These include specific colonization-inhibition (competitive exclusion) in which the protective bacteria exert a profound resistance to establishment and colonization by other related bacteria.

20.4 What vaccination strategy to use?

Effective vaccination strategies directed against *Campylobacter* in broiler chickens (yet to be developed) must provide protection from exposure, presumably from the first day of life to the point of slaughter; eliminate colonization by the time of slaughter; or reduce the *Campylobacter* numbers in the caeca. Given the 5–6-week average life span of broiler chickens, there is a very short period in which to induce effective antibody responses in an immunologically immature animal. Most vaccines tested are inactivated whole-cell preparations, administered systemically or orally, sometimes in combination with adjuvants (Cawthraw *et al.* 1994; Khoury and Meinersmann 1995; Rice *et al.* 1997). The three main challenges in development of a vaccine in poultry are the identification of cross-protective antigens, the induction of a rapid and effective immune response, and the development of novel adjuvants to further stimulate immunity against *Campylobacter* antigens (reviewed by Zoete *et al.* 2007).

An alternative, or complementary, strategy is passive immunization. Orally administered anti-*Campylobacter* antibodies have both therapeutic and prophylactic properties in chickens (Stern *et al.* 1990). One approach may be to immunize parent flocks to produce passively protected chicks. Detectable circulating maternal IgG antibodies showing bactericidal effects are present in young chicks (Cawthraw *et al.* 1994; Sahin *et al.* 2001). Three-day-old chicks from *C. jejuni*-colonized hens exhibit a 2–4-day delay in colonization compared with chicks derived from non-colonized hens, by both homologous and heterologous strains (Sahin *et al.* 2003). The improvement and extension of the protective capacity of any such maternally-derived immune protection requires further investigation. The choice of vaccine candidates and delivery systems will dictate the outcome of any immunization. Parental immunization, though feasible for breeder flocks, would be too costly for broilers.

Oral delivery systems would be less expensive and easier to administer. Killed vaccines have little, if any, protective capacity even when delivered with mucosal adjuvants (Rice *et al.* 1997; Cawthraw *et al.* 1994), unless the vaccine and the adjuvants are linked together (Khoury and Meinersmann 1995). Nevertheless, *in ovo* (in egg) vaccination may overcome some of these problems and can certainly induce immune responses, though the protective capacity of such responses has yet to be determined (Noor *et al.* 1995). Concerning whole-cell vaccines, much effort has been put into the development of flagella subunit vaccines, a subunit of the flagella being the immunodominant antigen recognized during infection (Martin *et al.* 1989; Wenman *et al.* 1985).

20.5 Resolutions and outcome

A *Campylobacter* vaccination strategy should be easy to implement, cost-effective (<10 US cents per chicken), and significantly reduce the colonization of breeder and broiler chicken guts with *C. jejuni*. To do this, we need to investigate different vaccines and delivery systems in comparable and harmonized experimental setups. In addition, one should assess the potential of current vaccine candidates on maternal immunity from breeder to broiler flocks. It is also important to investigate the bioethical aspect, consumer perception and producer acceptance of different vaccination strategies.

The cost-efficiency of a vaccine strategy will depend on a number of factors other than the vaccine efficiency. Issues such as flock size, climate and biosecurity have an important influence on the colonization pattern of *Campylobacter* in a flock. Therefore, the effects of these factors have to be considered when deciding whether or not to vaccinate.

20.6 Critical questions for discussion

1. Do you know of any vaccine that has been able to prevent foodborne pathogens from entering the animal production chain?
2. What is the difference between colonization and infection?
3. Is it possible to eradicate *Campylobacter* from animal herds?
4. In poultry production, if just the breeder flocks were vaccinated against *Campylobacter*, would this protect the offspring from colonization with the pathogen?
5. What is the difference between humoral and cellular immune responses, in terms of a *Campylobacter* vaccination strategy?
6. What is the best vaccination route for the induction of mucosal immunity?
7. Can you mention three examples of biosecurity measures at the herd level?
8. Why is it not possible to use biosecurity measures in organic or free-range poultry production systems?
9. What is the most cost-effective control strategy to date against *Campylobacter* in chicken production?

20.7 Acknowledgements

The work was supported in part by grant no. 09-067131 from the Danish Agency for Science, Technology and Innovation.

20.8 References

CAWTHRAW S., AYLING R., NUIJTEN P., WASSENAAR T., and NEWELL D.G. 1994. Isotype, specificity, and kinetics of systemic and mucosal antibodies to *C. jejuni* antigens, including flagellin, during experimental oral infections of chickens. *Avian Dis*, **38**: 341–9.

CHRISTENSEN B., SOMMER H., NIELSEN N., and ROSENQUIST H. 2001. *Risk Assessment of* Campylobacter jejuni *in Chicken Products*. The Danish Veterinary and Food Administration, Copenhagen, Denmark.

DOORES S. 1999. *Food Safety: Current Status and Future Needs*, pp. 8–9. ASM Press, Washington, DC.

EFSA PANEL ON BIOLOGICAL HAZARDS (BIOHAZ). 2010. Scientific opinion on quantification of the risk posed by broiler meat to human campylobacteriosis in the EU. *EFSA Journal*, **1**: 1437.

HAVELAAR A.H., MANGEN M.J.J., DE KOEIJER A.A., BOGAARDT M.-J., EVERS E.G. *et al.* 2007. Effectiveness and efficiency of controlling *Campylobacter* on broiler chicken meat. *Risk Analysis*, **27**: 831–44.

KHOURY C.A., and MEINERSMANN R.J. 1995. A genetic hybrid of the *C. jejuni* flaA gene with LT-B of *E. coli* and assessment of the efficacy of the hybrid protein as an oral chicken vaccine. *Avian Dis*, **39**: 812–20.

LINDQVIST R., and LINDBLAD M. 2008. Quantitative risk assessment of thermophilic *Campylobacter* spp. and cross-contamination during handling of raw broiler chickens evaluating strategies at the producer level to reduce human campylobacteriosis in Sweden. *Int J Food Microbiol*, **121**: 41–52.

MARTIN P.M., MATHIOT J., IPERO J., KIRIMAT M., GEORGES A.J. *et al.* 1989. Immune response to *Campylobacter jejuni* and *Campylobacter coli* in a cohort of children from birth to 2 years of age. *Infect Immun*, **57**: 2542–6.

NAUTA M.J., JACOBS-REITSMA W., EVERS E.G., VAN PELT W., and HAVELAAR A.H. 2005. *Risk Assessment of* Campylobacter *in the Netherlands via Broiler Meat and Other routes*. RIVM report no. 250911 006, Bilthoven, The Netherlands.

NAUTA M.J., JACOBS-REITSMA W., and HAVELAAR A.H. 2007. A risk assessment model for *Campylobacter* in broiler meat. *Risk Analysis*, **27**: 845–61.

NOOR S.M., HUSBAND A.J., and WIDDERS P.R. 1995. *In ovo* oral vaccination with *C. jejuni* establishes early development of intestinal immunity in chickens. *Br Poult Sci*, **36**: 563–573.

OLSEN K.N., LUND M., SKOV J., CHRISTENSEN L.S., and HOORFAR J. (2009). Detection of *Campylobacter* bacteria in air samples for continuous, real-time monitoring of *Campylobacter* colonization in broiler flocks. *Appl Environ Microbiol*, **75**: 2074–8.

RICE B.E., ROLLINS D.M., MALLINSON E.T., CARR L., and JOSEPH S.W. 1997. *Campylobacter jejuni* in broiler chickens: Colonization and humoral immunity following oral vaccination and experimental infection. *Vaccine*, **15**: 1922–32.

ROSENQUIST H., NIELSEN N.L., SOMMER H.M., NØRRUNG B., and CHRISTENSEN B.B. 2003. Quantitative risk assessment of human campylobacteriosis associated with thermophilic *Campylobacter* species in chickens. *Int J Food Microbiol*, **83**: 87–103.

SAHIN O., ZHANG Q., MEITZLER J.C., HARR B.S., MORISHITA T.Y. *et al.* 2001. Prevalence, antigenic specificity, and bactericidal activity of poultry anti-*Campylobacter* maternal antibodies. *Appl Environ Microbiol*, **67**: 3951–7.

SAHIN O., MORISHITA T.Y., and ZHANG Q. 2002. *Campylobacter* colonization in poultry: Sources of infection and modes of transmission. *Animal Health Res Rev*, **3**: 95–105.

SAHIN O., LUO N., HUANG S., and ZHANG Q. 2003. Effect of *Campylobacter*-specific maternal antibodies on *C. jejuni* colonization in young chickens. *Appl Environ Microbiol*, **69**: 5372–9.

SHARMAN P.A., SMITH N.C., WALLACH M.G., and KATRIB M. 2010. Chasing the golden egg: Vaccination against poultry coccidiosis. *Parasite Immunol*, **32**: 590–8.

STERN N.J., MEINERSMANN R.J., and DICKERSON H.W. 1990. Influence of antibody treatment of *C. jejuni* on the dose required to colonize chicks. *Avian Dis*, **34**: 595–601.

UYTTENDAELE, M., BAERT, K., GHAFIR, Y., DAUBE, G., DE ZUTTER, L. *et al.* 2006. Quantitative risk assessment of *Campylobacter* spp. in poultry based meat preparations as one of the factors to support the development of risk-based microbiological criteria in Belgium. *Int J Food Microbiol*, **111**: 149–63.

VAN IMMERSEEL F., METHNER U., RYCHLIK I., NAGY B., VELGE P. *et al.* 2005. Vaccination and early protection against non-host-specific *Salmonella* serotypes in poultry: Exploitation of innate immunity and microbial activity. *Epidemiol Infect*, **133**: 959–78.

WAGENAAR J.A., JACOBS-REITSMA W., HOFSHAGEN M., and NEWELL D.G. 2008. Poultry colonization with *Campylobacter* and its control at the primary production level. In I. Nachamkin, C. M. Szymanski and M. J. Blaser (eds). *Campylobacter*, 3rd ed., pp. 667–8. ASM Press, Washington, DC.

WALKER P.G., CAUCHEMEZ S., MÉTRAS R., DUNG DO H., PFEIFFER D. *et al.* 2010. A Bayesian approach to quantifying the effects of mass poultry vaccination upon the spatial and temporal dynamics of H5N1 in Northern Vietnam. *PLoS Comput Biol*, **6**: e1000683.

WENMAN W.M., CHAI J., LOUIE T.J., GOUDREAU C., LIOR H. *et al.* 1985. Antigenic analysis of *Campylobacter* flagellar protein and other proteins. *J Clin Microbiol*, **21**: 108–12.
ZOETE M.R., VAN PUTTEN J.P.M., and WAGENAAR J.A. 2007. Vaccination of chickens against *Campylobacter*. *Vaccine*, **25**: 5548–57.

21

The free-range meat paradox: the conflict between free-range poultry production systems and biocontainment against zoonotic microorganisms

L. S. Christensen, Islands Organic Ltd, Denmark

Abstract: This case study takes as its starting point the paradox of free-range poultry production – the fact that this production system (which is favourably viewed by a certain segment of consumers) does not allow biocontainment against zoonotic microorganisms, so may increase the risk of transmission of foodborne disease. It demonstrates how basic research-driven studies into the colonization dynamics of *Salmonella* and *Campylobacter* in free-range geese have revealed approaches that could potentially be used to mitigate the risk of foodborne transmission of these pathogens. The case also outlines the positive benefits of these studies for the company producing the geese and possible directions for future studies.

Key words: real-time monitoring, colonization dynamics, zoonotic microorganisms.

21.1 Introduction to the case

Due to concerns about animal welfare and meat quality there is increased consumer demand for free-range poultry. However, biological containment (i.e. the pursuit of a production strategy that is the opposite of free-range) is in many countries the predominant strategy for the control of zoonotic microorganisms such as *Salmonella* and *Campylobacter* in the primary production of poultry and other animals. This paradox is, to a large extent, ignored.

Salmonella and *Campylobacter* are predominant causative agents of gastroenteritis in humans in the industrialized world (Todd, 1997). Among the major sources of the pathogens are undercooked poultry meat, and, for *Salmonella*, in addition eggs (Neimann *et al.*, 2003; Anonymous, 2010). Obviously, the introduction of these microorganisms to free-range production systems cannot be prevented. Investigations have suggested that free-range broiler chickens are almost invariably colonized with *Campylobacter* (Heuer *et al.*, 2001) while the prevalence of *Salmonella* is more variable (Bailey and Cosby, 2005). Little is known about the colonization dynamics of *Salmonella* and *Campylobacter* in host species and transmission between species, in particular their transmission between the environment and domestic animals. The possibility of implementing risk-mitigating interventions under free-range production systems likewise remains to be explored.

Free-range poultry provide a unique opportunity to study the epidemiological dynamics of *Salmonella* and *Campylobacter* in poultry populations and their interaction with the environment under conditions that are as close to natural as possible. The basic knowledge generated might serve as the platform on which novel approaches to mitigate the risks of these zoonotic infections can be developed. These studies also might serve as a model to mirror the colonization dynamics of cryptic microorganisms in the vast reservoir of emerging pathogens.

21.1.1 How did it all begin?

A scientist bought a farm on Sejerø, a small Danish island located to the north-west of Zealand (see Fig. 21.1) and established organic farming there, albeit on a part-time basis. Among the most significant conditions with regard to farming on this island are the relatively large pasture areas, suitable mainly for extensive meat production, and the lack of predators such as foxes. Consequently, the island had a long tradition of small farms producing poultry and small ruminants. Until a few years ago, domestic geese used to walk in the streets in the villages, and sheep are still occasional and uninvited visitors to conventional and intensively-managed neighbouring farms. The part-time farm developed into an enterprise producing free-range geese: Islands Organic Ltd.

The principles of production at Islands Organic Ltd include the following:

- Goslings are housed in heated stables for the first 3 weeks of life and gradually allowed access to open pasture (Fig. 21.2).
- From 6 weeks of age the geese mainly graze large areas of fenced pasture and access is given to 'harvest' on the root fields containing wheat or barley (Fig. 21.3).
- Supplementary food is given when needed and for a period of 1 month prior to slaughter.
- Achieving a sustainable balance between the number of geese and the amount of feed produced on the farm and/or the island is an important objective.
- The geese are preferably slaughtered on location using mobile slaughter facilities.
- The production is certified as organic.
- The production complies with the principles of the Danish Animal Welfare Society, and the products are recommended by this society.

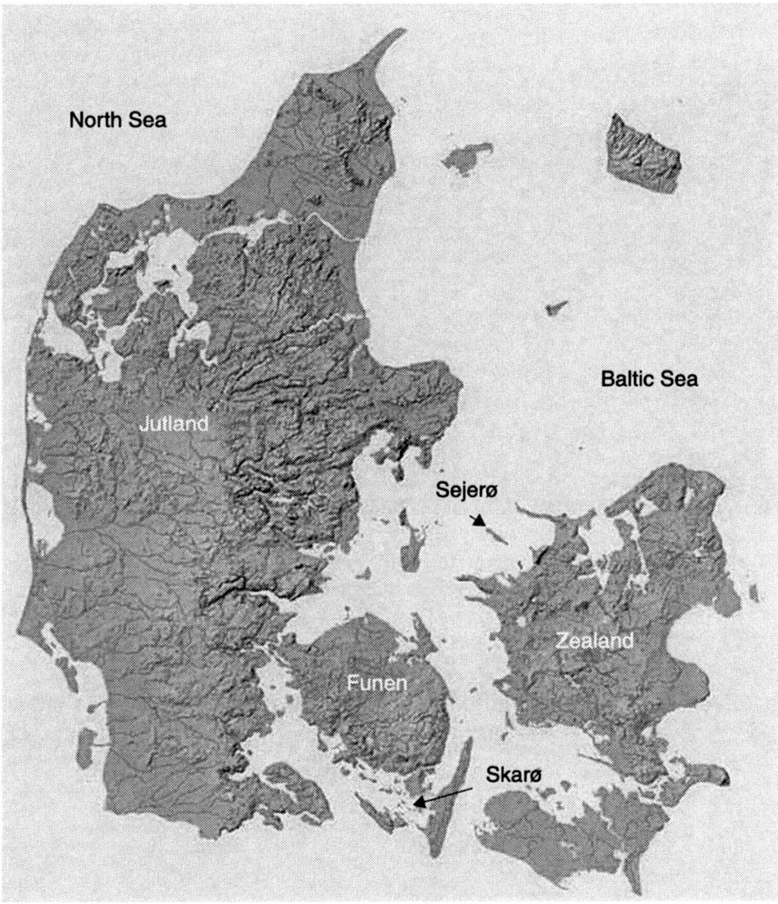

Fig. 21.1 Map of Denmark showing the location of the fox-free islands with free-range domestic geese.

- An ancient breed of grey geese found on Sejerø is being reared to explore whether the birds have advantageous adaptive characteristics in terms of taste and disease resistance.

The fact that the birds graze on large areas of pasture 'spiced' by sea mist gives their meat good flavour. The exercise this involves results in a high body mass index with meat that is perfectly textured by lipids, yet with a significantly lower amount of depository fat in the body compared with conventionally produced geese. Islands Organic Ltd is a member company of the brand Island Specialties®, which is described in more detail in Chapter 38 (Christensen *et al.* 2012).

The enterprise grew quickly to supply major retail companies in Denmark, and export requests were received from countries as distant as Japan. In an attempt

Fig. 21.2 Among other quality parameters for the production of free-range geese on small islands is the sea view.

Fig. 21.3 The lack of foxes allows the geese to find the last grains on a wheat field.

to accommodate these requests, network production was developed involving sub-contractors from other small islands raising geese in compliance with the principles defined. This resulted in a production system involving the splitting and merging as well as translocation of flocks, in some cases to remote islands (Fig. 21.4). Due to the absence of mammalian predators, these islands have a high density of wild birds and mammals and they serve as rest locations for large

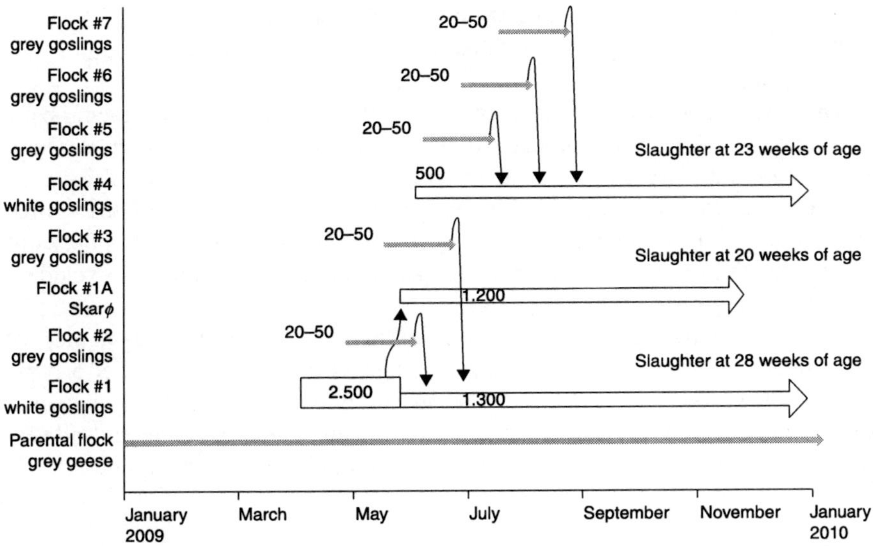

Fig. 21.4 History of flock splitting, merging and relocation during the production season of 2009. Grey bars represent small flocks of ancient breed of grey goslings being merged with the larger flocks of white goslings (white bars) purchased day-old. Flock #1 was split at 6 weeks of age and half of the flock was relocated to the distant island of Skarφ.

numbers of migratory birds. Thus, this production system provides optimal conditions for investigating issues such as:

- The multiplicity of introductions from environment and geographical clustering of environmental microorganisms,
- The impact of within-flock and between-flock pathogen transmission versus environmental exposure,
- The multiplicity and dynamics of co-circulating strains, and
- The impact of protection (homologous and non-homologous immunity) in individuals and flocks developed by colonization.

21.2 Significance of the case

21.2.1 Regulatory aspects

Since it is not possible to avoid the exposure of free-range (mostly organic) poultry to *Salmonella* and *Campylobacter*, and since there are no approved approaches to control the prevalence of these microorganisms in free-range poultry, regulatory aspects and codes of practice in the poultry industry vary to a large extent between countries. Vaccination against *Salmonella* using live vaccines is allowed in some countries but not in Scandinavia. In Denmark, broiler flocks have to be analysed for both *Salmonella* and *Campylobacter* and certified free of

Salmonella if sold as fresh meat (Ministry of Food, Agriculture and Fisheries, 2009) but monitoring in geese for either of the two bacteria is neither required nor practised in general.

The studies conducted at the flock level reveal the challenges of tracing back the origin of an infection in humans to its origin in primary production. For both *Salmonella* and *Campylobacter*, identifying the same strain in clinical cases as well as in a flock of poultry often will be a strong indication of the source responsible for the outbreak. However, given the diversity often seen in flocks of poultry, the failure to detect the strain in question in a flock can hardly exclude it from being the source of transmission.

A basic science-driven interest in, firstly, the colonization dynamics of *Salmonella* and *Campylobacter* at a host flock level and, secondly, applied aspects in terms of approaches of control, at the current time in particular for *Salmonella*, has the potential to bring us to a situation where new standards can be set and new regulatory measures put in place to increase food safety of organic poultry.

21.2.2 Economic and market aspects

The company's basic science-driven interest in the colonization dynamics of *Salmonella* and *Campylobacter* at a host flock level and its practical approaches to increase the safety of its products have become prominent marketing parameters, alongside the more traditional parameter of the quality of the meat produced.

A few countries, such as Sweden, have a policy of zero-tolerance toward *Salmonella* in ducks and geese, and as Sweden, like all other countries, has no means of controlling *Salmonella* colonization in these birds, Islands Organic Ltd has experienced an increase in exports to Sweden.

21.3 The case continues

Due to this farmer's particular interest in the mechanisms of the genesis of species and emergence of infections, the temptation could not be resisted to use the unique production system on his farm as a target for longitudinal monitoring, i.e. detection and typing studies, in particular of the colonization dynamics of *Salmonella* and *Campylobacter* (Christensen *et al.*, 2011a, 2011c; Josefsen, 2011). The ways in which the knowledge generated might be used to improve food safety on the farm became clearer during the course of the studies.

21.4 Resolution and outcomes

It was revealed that the dynamics of *Salmonella* colonization in flocks were highly divergent. One scenario was the rapid colonization of all individuals in the flock with one particular strain of *Salmonella* Enteritidis (phage type 29) followed by a subsequent decline in colonization with one incidence of transient reappearance

(Fig. 21.5). Throughout the lifetime of this flock, no other serovar or strain of *Salmonella* was observed despite the fact that multiple serovars were present in the flock's environment.

The other scenario was more complex, with no significant dissemination of a particular strain or serovar of *Salmonella* in the flock, but rather the transient appearance of multiple serovars and strains throughout its lifetime.

One of the outcomes of these studies (by outcomes we mean ways in which the knowledge generated can be applied to improve food safety) has been that it is now possible to conduct a reliable risk assessment of contamination with *Salmonella* at the time of slaughter. Another is that, for most scenarios of colonization, the time of slaughter can be chosen to assure that there is no contamination, or only low-risk or low-level contamination with *Salmonella*. Yet another outcome is that the highly diverging colonization dynamics of *Salmonella* might reveal putative approaches to impact the dynamics of colonization in individual flocks.

The colonization dynamics of *Campylobacter* were more uniform than those of *Salmonella* and yet also more complex. Once introduced to the flock, *Campylobacter* quickly colonized all the birds, and they remained colonized until

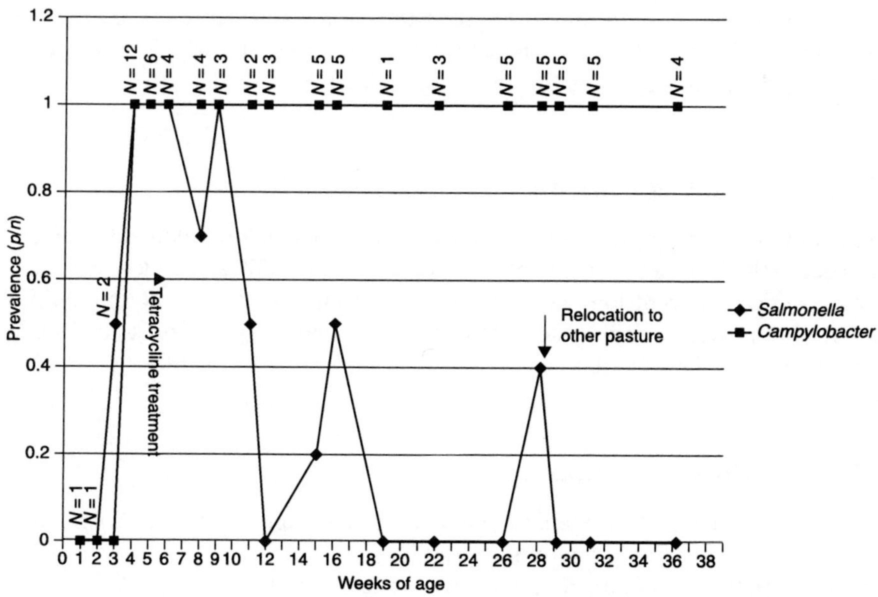

Fig. 21.5 Sampling regimen and the prevalence of *Salmonella* and *Campylobacter* during lifetime of Flock #1. Number of droppings (*N*) are indicated at each sampling point. Prevalence (*p*/*n*): number of positive droppings per number of droppings. The timeline shows the history of the flocks of geese splitting, merging and being relocated during the production season of 2009. Flock #1 was split at 6 weeks of age and half of the flock was relocated to the distant island of Skarø. This graph relates to only the flock on Sejerø.

slaughter (Fig. 21.5). Genotyping studies revealed multiple introductions reflecting the environmental exposure and management of the particular flock. In addition, they showed a fluctuation over time revealing a most prominent impact of flock prevalence of genotypes on the dynamics in individuals (Christensen *et al.*, 2011c). Compartmentalization of the colonization dynamics in sub-populations reflecting social structures in the flock, however, was also seen (L. S. Christensen, unpublished data). The ways in which this knowledge can be applied to improve food safety are not quite obvious but studies are continuing, in particular focusing on quantification of *Campylobacter* throughout the lifetime of the flocks as well as during slaughter and subsequent processing of meat.

21.5 Commentary

The geese are only produced on a small scale, so these studies could not be conducted at the expense of the production company. Instead, the studies were supported by grants for research and development (see below) and this will remain the situation for the foreseeable future. The studies, along with similar studies on the presence of *Campylobacter* in conventional broilers (Olsen *et al.*, 2009) and for foot-and-mouth disease virus in the exhaled breath of cattle (Christensen *et al.*, 2011b), serve among other purposes to demonstrate the potential of real-time infection monitoring as a tool to improve infection status and food safety. This technique will be particularly valuable when the costs of analysis have fallen further and monitoring is demonstrably cost-effective.

21.6 Acknowledgements

The studies referred to were supported by the European Union-funded Integrated Project BIOTRACER (contract FOOD-2006-CT-036272) under the 6th Framework. The contribution from colleagues and co-authors of the studies referred to is greatly appreciated. Photographs are courtesy of Christian Harbou and Inge Marie Ringgaard.

21.7 Critical questions

1. What are the major differences in colonization dynamics between *Salmonella* and *Campylobacter* in free-range geese?
2. How can the risks of *Salmonella* in free-range geese be mitigated?
3. How might interventions affect *Salmonella* colonization dynamics in geese? Which interventions would comply with your national regulations?
4. Why were the dynamics of colonization by *Campylobacter* apparently not affected by interventions in the same way as the dynamics of colonization by *Salmonella*?

5. Please explain the likely costs and benefits of making infection status at a farm level available to customers and to the public. Would a cost–benefit analysis support such an action or not?

21.8 References

ANONYMOUS (2010). *The Community Summary Report on Trends and Sources of Zoonoses, Zoonotic Agents and Food-Borne Outbreaks in the European Union in 2008*. European Centre for Disease Control and Prevention (ECDC) and European Food Safety Authority (EFSA). Available at http://www.efsa.europa.eu/en/scdoc/1496.htm.

BAILEY, J.S., and COSBY, D.E. (2005). *Salmonella* prevalence in free-range and certified organic chickens. *Journal of Food Protection*, **68**: 2451–3.

CHRISTENSEN, L.S., BONNICHSEN, L., MØLBAK, L., JOSEFSEN, M.H., CHRISTENSEN, J. *et al.* (2011a). Colonization dynamics of *Campylobacter* strains in flocks of free-range geese. *Veterinary Microbiology*. Submitted.

CHRISTENSEN, L.S., BREHM, K.E., SKOV, J., HARLOW, K.W., CHRISTENSEN, J. *et al.* (2011b). Detection of foot-and-mouth disease virus (FMDV) in exhaust air of cattle: perspectives for a real-time epidemiological contingency. *Journal of Virological Methods*, **77**: 44–8.

CHRISTENSEN, L.S., JOSEFSEN, M.H., PEDERSEN, K., CHRISTENSEN, J., BONNICHSEN, L. *et al.* (2011c). Real-time monitoring of *Salmonella enterica* in free ranging geese. *Applied and Environmental Microbiology*, **77**: 3160–2.

CHRISTENSEN, L.S., HOORFAR, J., SØRENSEN ØHAVET APS, J., and BISGAARD, S. (2012). Documenting the '*terroir*' aspects of award-winning Danish conserves: A model for the development of authentic food products. In J. Hoorfar (ed.), *Case Studies in Food Safety and Authenticity*. Woodhead Publishing, Cambridge, UK.

HEUER, O.E., PEDERSEN, K., ANDERSEN, J.S., and MADSEN, M. (2001). Prevalence and antimicrobial susceptibility of thermophilic *Campylobacter* in organic and conventional broiler flocks. *Letters in Applied Microbiology*, **33**: 269–74.

JOSEFSEN, M.H., BONNICHSEN, L., LARSSON, J., NIELSEN, E.M., FRICKER, M. *et al.* (2011). Evaluation of *flaA* short variable region sequencing, multilocus sequence typing and Fourier transform infrared spectroscopy for discrimination between *Campylobacter jejuni* strains. *Food Analytical Methods*. Available at http://www.springerlink.com/content/45323631411443601/.

MINISTRY OF FOOD, AGRICULTURE AND FISHERIES (2009). Order no 1462 of 16/12/2009. Bekendtgørelse om salmonellose hos fjerkræ samt *Salmonella* og *Campylobacter* i slagtefjerkræ m.m. Denmark.

NEIMANN, J., ENGBERG, J., MØLBAK, K., and WEGENER, H.C. (2003). A case-control study of risk factors for sporadic *Campylobacter* infections in Denmark. *Epidemiology and Infection*, **130**: 353–66.

OLSEN, K.N., LUND, M., SKOV, J., CHRISTENSEN, L.S., and HOORFAR, J. (2009). Detection of *Campylobacter* bacteria in air samples for continuous real-time monitoring of *Campylobacter* colonization in broiler flocks. *Applied and Environmental Microbiology*, **75**: 2074–8.

TODD, E.C.D. (1997). Epidemiology of foodborne diseases: A worldwide review. *World Health Status Quarterly*, **50**: 30–50.

22

How does Sweden control *Salmonella* before it enters the food chain?

P. Häggblom, National Veterinary Institute, Sweden

Abstract: In the spring of 2003 *Salmonella* was detected in a fattening pig herd in south-central Sweden. Because of the identified serotype there was an early suspicion of feed contamination, although monitoring did not reveal any *Salmonella*-positive samples in the feed mill. The finding in the pig herd resulted in a very large and costly outbreak investigation that went on for several months. The case demonstrates aspects of the Swedish programme for *Salmonella* control in primary production.

Key words: *Salmonella*, feed contamination, Swedish programme for *Salmonella* control in primary production.

22.1 Introduction to the case

In May 2003, *Salmonella* Cubana was detected following routine sampling of a fattening pig herd in south-central Sweden. Because of the serotype detected in faecal samples from the pigs, there was an early suspicion of feed contamination and the producer of the pelleted feed was contacted. According to the feed operator, the data from the *Salmonella* monitoring at the feed mill did not indicate contamination in any of the three production lines. However, after official inspection and sampling of the feed mill, several positive samples were identified on the pig feed production line.

The evidence of feed mill contamination, in accordance with the legal framework, initiated a very large investigation by the authorities, the operators of the feed mill in question and the operators of farms that had received potentially contaminated feed during the critical period, primarily in the province of Östergötland and also in neighbouring provinces.

Official veterinarians were contracted to carry out investigations of potentially infected pig farms. Due to the incomplete information from the company about the customers that had bought the feed, it took considerable time to find out that a total of 77 pig farms and three dairy farms had received potentially infected feed from the feed mill during the critical period.

The situation described above should be considered together with the fact that the prevalence of *Salmonella* in food producing animals in Sweden has been demonstrated to be very low (EFSA 2010). This favourable situation can be attributed to the programme for *Salmonella* control in primary production that started in 1961 to ensure safe animal food products for human consumption. A *Salmonella* control programme that monitors the critical control points in the processing line in feed mills has been in place since 1991. All findings of *Salmonella* in feed, animals or food are notifiable by law and actions are taken when positive samples are identified to prevent further contamination or infection. The present outbreak in fattening pigs was the largest reported in the country where feed was the source of infection.

22.2 Significance of the case

In the EU the pathogen responsible for the largest number of reported foodborne outbreaks is *Salmonella*. The most important food vehicles are eggs followed by pig meat (EFSA 2010). Pigs are infected orally by *Salmonella* that can originate from a contaminated farm environment, infected rodents, infected pigs, or contaminated feed. The aim of the Swedish control programme is to prevent infections in primary production of food producing animals, thus reducing human exposure.

22.2.1 Regulatory aspects

In the legal process it was concluded that the feed company could not be held economically responsible for the outbreak according to the Swedish zoonosis regulation (Anonymous 1999b). The reason was that the regulation could not be used as a basis for right of recourse against the company for the money that was paid to the farmers.

22.2.2 Economic and market aspects

The official cost of the outbreak was around 150 million Swedish kroner (15 million Euros). Considerable market effects on the feed and pig production of the region were observed, affecting the feed supply, feed withdrawal, slaughter and migration of animals, etc.

Sampling, cleaning and disinfection plans for the large number of infected pig farms, including feed storage facilities and animal facilities, had to be developed within a short period of time. The laboratory capacity where the large number of samples could be analysed was another limiting factor in the investigation. Animal welfare became an important aspect during the outbreak, primarily due to overcrowded stables because of the individual sampling of pigs before slaughter.

The daily media coverage of the outbreak could be followed in newspapers, television, different radio channels and on the internet for several weeks. The major consumer issue during the course of the outbreak was the risk of acquiring human infections from the consumption of Swedish pig meat. After the outbreak it was concluded that no human cases of salmonellosis could be linked to the outbreak.

22.3 The case history continues

On 14 May 2003 *Salmonella* Cubana was demonstrated in faecal samples from routine monitoring of a fattening pig herd. The pig farm had received heat-treated pig feed (concentrate) from a nearby feed mill at the beginning of May. On 27 May the sampling of the farm was expanded to other groups of animals and also the stored commercial feed. On 3 June, 15 samples from the processing equipment (scrapings and dust samples) and eight feed samples were collected at the feed mill. Five *Salmonella* positive samples were identified from one processing line taken after the heat treatment (pelleting). *Salmonella* was not detected in any of the stored feed samples that were taken out when the pig feed was delivered to the farm.

On 10 June the processing line for pig feed was stopped and the feed mill was prohibited by the authorities from selling pig feed. The following day, more than 100 samples were collected from the contaminated processing line, 19 of which tested positive for *Salmonella.* Positive samples were detected in the pellet cooler and the aspiration system (Fig. 22.1). In addition, samples of delivered feed were investigated and 24 out of the 26 samples collected 5–8 May were positive for *Salmonella.*

It was estimated that the feed mill had intermittently delivered contaminated pig feed from the end of April until 10 June. For that reason, and parallel to the work in the feed mill, major efforts were made to identify all potentially infected herds. The investigation revealed that *S.* Cubana was detected in 49 of the 77 pig farms that had received feed during the critical period (Österberg *et al.* 2006).

22.3.1 Results from *Salmonella* investigations in the feed mill

In addition to the sampling of the pig feed line, an extensive sampling of the entire feed mill was carried out; however, *Salmonella* was only detected in the premises close to the pig feed production line and not on the lines for layer or cattle feed.

The mill's five coolers were situated in a defined and restricted area; however, wet coatings containing *Salmonella* were only detected in the cooler of the contaminated production line.

The air inlet for the coolers was situated on the roof of the feed mill and connected with a dust filter (Fig. 22.2). Several samples from the filter and the air

Fig. 22.1 Exhaust from a pellet cooler, where the opening allows for inspection, cleaning and sampling. Because of the large amount of water that is removed from the pelleted feed during the drying and cooling process, moisture condensation in the exhaust may lead to a build-up of wet coatings that give rise to bacterial growth. In the described case, *S.* Cubana was isolated in this bacteriologically fragile area of the feed mill.

Fig. 22.2 Pellet cooler (counter current) where air for cooling and drying the moist and warm pellets (behind the glass window) is drawn from within the feed mill through the bed of material. Airborne microorganisms present in the feed mill premises are a possible source of recontamination and for that reason air drawn into the mill should be taken from uncontaminated areas. Air filtration and dust control can prevent recontamination.

inlet were found to be positive for *S.* Cubana. *Salmonella* was also detected in dust samples in the aspiration system where air from the contaminated pellet cooler was released onto the roof of the feed mill.

Pig feed samples of finished feed from February, March and April 2003 were retrospectively investigated for *Salmonella* in attempts to establish the time point at which the production line had become contaminated. Out of 101 samples collected from April 5 were positive for *Salmonella*. Out of 72 samples investigated from February and March, one sample was positive in March and no samples from February.

22.3.2 Technical equipment in the feed mill

The feed mill was built as early as 1962, and certain areas had not been modified since then. Between 1999 and 2000 the feed mill was reconstructed and new equipment was installed, e.g. coolers for pelleted layer feed. The processing line for expanded pig feed was also reconstructed using old equipment (e.g. a modified cooler). The two production lines for pelleted feed and one expander line were separated after heat treatment, except for the storage bins for finished feed.

On the contaminated production line, pig feed concentrates and soybean meal were processed in an expander at 76 °C. After start-up, the mash was recirculated four to five times for up to 10 minutes before the correct temperature was reached and production could continue. This process was used intermittently where the temperature gradually approached ambient temperature. Evidence of previous technical problems in this production line was revealed during the investigation.

22.3.3 Cleaning and disinfection of the feed mill

The processing equipment, including the expander, was to a large extent dismounted or removed from the premises before cleaning and disinfection. Conveyers and storage bins were opened and feed material and dust was removed physically or by vacuum cleaning. All surfaces were inspected to be clean to ensure that no remaining organic matter was present before disinfection. Disinfectant experts were contracted and formaldehyde was used as the disinfectant for the contact spraying and fogging of the premises. After cleaning and disinfection a careful sampling of the premises with more than 200 swab samples was performed to ensure absence of *Salmonella*. *Salmonella* positive samples were not detected in any of those samples.

22.3.4 Source of infection in the feed mill

Despite several investigations, the source of infection could not be found. The indication was that the most likely source was a contaminated raw material for compound feed (Fig. 22.3).

Fig. 22.3 Unloading of feed ingredients into intake pit of a feed mill. Dust formation is a factor inherent in feed manufacturing. *Salmonella* in the dust may be a source of contamination of feed mill premises as well as finished feed. Measures taken to control and reduce dust in feed mills are therefore important. Efficient ventilation systems should be installed to control the dust. However, dust should preferably be removed by vacuum cleaning and not by compressed air. Accumulated dust from the premises and sweepings should be destroyed. *Salmonella*-contaminated dust was a likely source of cooler contamination in this case study.

22.4 Resolution and outcomes

Several government investigations were initiated after the outbreak, e.g. review of the *Salmonella* control programme for feed. However, it was not suggested that the present control strategy for feed should be changed. A summary of the outbreak was presented for the Swedish parliament. The feed mill was restarted about 2 months after production had stopped.

22.5 Commentary

More than 45 000 feed and faeces samples were investigated for *Salmonella* during the outbreak using the NMKL 71 method (Anonymous 1999a). Several private and government laboratories were involved in the work.

Pulsed field gel electrophoresis typing of collected *S.* Cubana strains using three cleavage enzymes showed that a single clone was involved in the outbreak. However, a previous isolate of *S.* Cubana from the feed mill isolated in 1997 was not identical to the outbreak strain.

In the investigation of the feed mill, it was concluded that a significant multiplication of *S.* Cubana had occurred in feed ingredients on the production line. This conclusion was based on the fact that *S.* Cubana was not detected in the premises in raw materials, and also that a clone identical to the outbreak clone was present in, e.g., the cooling equipment of the feed mill.

Salmonella Cubana was also detected on the flat roof of the feed mill (an area occasionally covered with water), close to the exhaust from the pellet cooler. The most probable explanation for this was that the heat-treated finished feed was the source, because bacterial growth could have taken place in the wet coatings inside the pellet cooler and thus contaminated the feed. Other coolers situated in the same room did not contain wet coatings and *Salmonella* was not detected. Most probably the cooling air carried *Salmonella* bacteria into the cooler where growth took place in the wet coatings. Other explanations may be that the recirculation of pig feed introduced contaminated feed into the cooler.

The time period when *Salmonella* was detected in the finished feed indicated that the situation was continuously getting worse, starting in March 2003 and leading later to the development of a massive level of contamination in the pig feed. Increased outdoor temperatures contributed to a more rapid growth of *Salmonella* in the mill. The reason why the contamination of the cooler was not detected earlier in the weekly monitoring was because sampling was not carried out according to instructions.

22.6 Acknowledgements

I wish to thank Julia Österberg for valuable comments on the manuscript. I also would like to thank the staff at the feed mill and the Swedish Board of Agriculture for valuable discussions.

22.7 Critical questions

1. What does the phrase '*Salmonella* is notifiable' mean?
2. How is *Salmonella* in feed a vector for meat contamination?
3. Why in this case was the contaminated feed not detected in the weekly *Salmonella* monitoring of the feed mill?
4. How big are the risks of cross- or re-contamination of a heat-treated (*Salmonella* negative) feed in a contaminated environment?
5. What were the best ways to rapidly identify farms that had received the potentially infected feed?
6. What could the feed mill operator have done to speed up the tracking forward of the contamination to the farm level?
7. Were the sampling instructions followed at the feed mill, and what is the role of sampling?
8. Is it important to educate feed mill employees about hygiene and sampling?

22.8 References

ANONYMOUS (1999a). *Salmonella Detection in Foods, Method no. 71*, 5th ed. Nordic Committee on Food Analysis (NMKL), Oslo, Norway.

ANONYMOUS (1999b). *Swedish Zoonoses Act*, SFS 1999:658. Available at www.rixlex.se (in Swedish).

ANONYMOUS (2010). The community summary report on trends and sources of zoonoses, zoonotic agents and food-borne outbreaks in the European Union in 2008. *EFSA J*, **8**: 1496.

ÖSTERBERG, J., VAGSHOLM, I., BOQVIST, S. and LEWERIN, S.S. (2006). Feed-borne outbreak of *Salmonella* Cubana in Swedish pig farms: Risk factors and factors affecting the restriction period in infected farms. *Acta Vet Scand*, **47**: 13–21.

23

Bovine spongiform encephalitis: the case of an unexpected and rapidly spreading epidemic in cattle with serious consequences

N. Skovgaard, University of Copenhagen, Denmark

Abstract: A case involving a misfolded protein that caused a serious epidemic in cattle and, much later also in humans, is detailed. The case is followed throughout the whole food chain and discusses how changes in feed technology may have had an impact on the prevalence of the infectious agent and on human exposure. The transmission in animals and humans after oral administration is explained related, for example, to the symptoms. The impact of the use of mechanically recovered meat (MRM) in meat products is also outlined. Regulatory and economic aspects are further elaborated, and finally the epidemiology is surveyed, and some general comments on the whole case are volunteered.

Key words: bovine spongiform encephalitis (BSE), dry rendering, variant Creutzfeldt-Jakob disease (vCJD), meat and bone meal, mechanically recovered meat (MRM), technology.

23.1 Introduction to the case

In 1985 the first case of a disease in cattle, which later became known colloquially as 'mad cow disease', occurred in the UK. The name illustrates the characteristic symptoms of the disease. After an incubation period of 4–5 years, the animal shows symptoms such as hypersensitivity to sound and light, increased saliva production, nervous movements of the ears, aggressiveness towards other cattle and humans, aversion to passage through doors and turning around, quivering of the muscles and difficulties when trying to stand up. The scientific name for the disease is bovine spongiform encephalitis (BSE).

The first case was followed by a rapidly expanding and geographically widespread epidemic in cattle that by 1990 had reached an annual incidence of 30 000 cases per year in the UK. At its peak in 1992, more than 1000 cases were diagnosed per week. This was in itself a devastating situation for farmers and society. As time passed, however, the situation turned out to be much worse and much more serious.

In 1996, 10 years after the first recognized case of BSE, several cases of a new variant of Creutzfeldt-Jakob disease (CJD) in humans that had not been reported previously appeared in the UK in teenagers and young adults. This was designated variant Creutzfeldt-Jakob disease (vCJD), which could be distinguished from CJD based upon clinical, biochemical and rodent transmission studies. The suspicion that vCJD might have resulted from infections with the BSE agent was subsequently confirmed experimentally, and an increasing number of new cases were identified during the following years.

The rise and fall of annual cases closely resembled that of BSE, but at a thousand-fold lower level. Some scientists considered doomsday to be near, that the diagnosed cases were only the tip of the iceberg, and that the number of cases could be expected to rise to 10 000 or more a year. This, luckily, turned out not to be the case.

By midsummer 2011, a total of 225 diagnosed human cases had been recorded, including cases in other countries to which BSE had been 'imported'. It was revealed early in the outbreak that the BSE agent was a prion, and that meat and bone meal played a decisive role in its dissemination. A ruminant feed ban was subsequently issued in 1989.

23.2 Significance of the case

The implications of the epidemic in relation to both public health and animal health were in the beginning not well understood, either by the scientists or by the authorities. The epidemic was magnified by the continued export of live cattle and of meat and bone-meal feed from the UK to virtually every country in Europe and a number of non-European countries. Cattle were exported all over the world for breeding purposes. Meat and bone meal exports to Europe from the UK used trade channels far outside the reach of the authorities. The problem was that meat and bone meal is an offal product, in which there was little public interest.

In countries that imported large numbers of sub-clinically infected cattle and large amounts of meat and bone meal (rendered from slaughtered carcasses) from the UK, BSE was able to spread through the recycling of the carcasses of both imported and newly infected domestic cattle. These secondary outbreaks were, luckily, substantially less severe than the outbreak in the UK, which is estimated to have affected up to five million cattle.

The economic significance of the case and the changes in regulation that were made in response to it are discussed in Section 23.5.

23.3 Background to the case: the nature of BSE and vCJD

BSE, CJD and vCJD belong to a group of diseases called transmissible spongiform encephalopathies (TSEs) or prion diseases. Prion is an abbreviation of *in*fectious *pr*otein, and the diseases are caused by a change in the function of prions, which are found in nerve cells, from the normal form, designated PrP^{c}(C = cell), to a misfolded form, designated PrP^{sc} (sc = scrapie). The molecular changes that occur when the normal PrP^{c} changes to its pathogenic form, PrP^{sc}, have been fairly well investigated (see Further reading section below).

TSEs have been identified in a number of species and include scrapie in sheep, chronic wasting disease (CWD) in wild deer and elk, transmissible spongiform encephalopathies in ranched raised mink, sporadic Creutzfeldt-Jakob disease in humans and variant Creutzfeldt-Jakob disease in humans, as well as Kuru in humans. The latter disease was perpetuated by ritual cannibalism practices in Papua New Guinea, whereby brain tissue of dead persons was consumed by the other members of the family.

Variant CJD clinically presents as a psychiatric disturbance, or sensory symptoms, or both. The clinical course that develops includes ataxia and dementia, with a course of approximately 1 year. Patients with vCJD are typically young, between 15 and 30 years of age. All available data indicate that BSE is caused by the PrP^{sc} agent that causes scrapie in sheep. The scrapie prion has subsequently passed on from cattle to humans, causing vCJD (Fig. 23.1).

Disease transmission to several non-livestock animal species has occurred, e.g. to pet cats. Pigs are not susceptible to oral BSE infection, and chickens are resistant to infection by any route of administration. Dogs are apparently also resistant to TSE infection.

Scrapie in sheep
↓
Dry rendering using insufficient heat treatment
↓
Survival of the prion
↓
Infectious meat and bone meal fed to cattle
↓
BSE in cattle
↓
'Enrichment' to10^{8-9}/g in the spinal cord, columnae and brain tissue
↓
Bones with prions used to produce 'mechanically recovered meat' (MRM)
↓
MRM added to hamburger meat, patties and similar products at 5–10%
↓
Total doses of billions eventually consumed by teens and young people
↓
Species barrier jumped
↓
Cases of vCJD

Fig. 23.1 Line of transmission of the scrapie prion, PrP^{sc}, from animal to man.

23.3.1 The pathogen

The normal human prion is a glycoprotein of 253 amino acids, with a molecular weight of 33–35 kDa. Although it is in humans encoded in nervous cells by the *Prnp* gene, its function is so far unknown. It is characteristic that the infectious prion is devoid of any informational nucleic acid, but interestingly is still able to replicate! The replication takes place very slowly but, because the misfolded protein is not degraded, concentrations gradually build up.

The infectious prion is further resistant to protease activity, which may be of importance in the pathogenesis. The molecules form insoluble aggregates visible under the microscope as 'plaques' in sections of the central nervous tissue, which is thought to lead to the death of neurons, leading to the development of holes in the brain with a sponge appearance, hence the name 'spongiform'. Almost all infectivity is confined to the central nervous system, and in the brain has been shown to have a titre of 10^6 mean lethal doses, LD_{50}, per gram of tissue.

To understand the dissemination of the prion disease in the food chain, it is important to notice that prion infectivity is remarkably heat resistant and some infectivity can survive autoclaving for prolonged periods. Different strains may further show different heat resistance. In one experiment, one strain held at 126 °C for 2 hours was reduced from 10^7 LD_{50}/g to approximately 10^4/g, while another strain was reduced to less than 100/g.

Prion infectivity is also very resistant to inactivation by chemicals. Treatments that are used to inactivate prion infectivity include exposure to 2.5–5% sodium hypochlorite solution or 1 M NaOH for 1–2 hours.

23.4 The case continues

23.4.1 Production of the culprit: meat and bone meal

Meat and bone meal was shown to be a tremendously efficient vector for BSE and can be considered to be the main culprit in the epizootic spread of the disease. The production of meat and bone meal will therefore be outlined briefly. The main raw materials are animals, or animal material, condemned during meat inspection and fallen or culled animals. These have, according to legislation, to be heat-treated to kill microbes. The raw material is sent to certain approved factories where it is received in a so-called 'unclean section', which is strictly separated from the rest of the factory, the 'clean section'. The raw material is stored in silos, from where it is passed on to a crusher, then reduced to particles with a maximum size of 50 mm to facilitate the subsequent drying and heat treatment.

The most commonly used method of treatment is the so-called dry-rendering process, in which water is removed from the crushed material by drying before the fat is separated from the mass. Sterilization has to be carried out according to the EU by-product regulation at 133 °C for 20 min, which can be performed before (pre-sterilization) or after (post-sterilization) the drying process. Drying is performed at atmospheric pressure, and the temperature slowly rises from 100 °C to approximately 120 °C.

The fat fraction is separated in a screw press where some generation of heat takes place, followed by a cleaning of the fat by centrifugation. The defatted pressed cakes are cooled and milled to meat and bone meal, one of the most valuable protein-rich feeds for animals.

The major question is why BSE began in the mid-1980s, and why in the UK. Earlier, pressure cooking was applied in a number of European countries, but, because there were no requirements for dividing the raw materials into smaller lots for processing at that time, the sterilization effect might, for that reason, quite often have been questionable. Pressure cooking was not applied anywhere in the UK.

Around 1970 many batch dry-rendering plants in the UK were replaced by continuous-rendering systems. There were quite different physical–chemical conditions during the process in the new rendering systems, including shorter holding times and lower cooking temperatures, because of the vacuum conditions. The drying was still performed in warm fat, but the heating effect must have been substantially reduced, because there was a vacuum in the system. A number of traditional dry-rendering plants incorporated, in parallel with this development, vacuum systems that lowered the cooking temperature, and thereby increased the capacity of these plants. The vacuum installations had the effect that both the cooking temperature and the drying time were reduced.

It has often been maintained that the change in fat extraction method, from using petrol extraction to the use of screw pressure, altered the efficacy of the rendering systems in reducing BSE agents, but this is doubtful and has been impossible to prove. The new systems and the traditional batch systems, which were modified to included vacuum systems, had, because of the changed physical–chemical conditions, a significantly reduced destructive effect on BSE agents. Therefore, if the level of infectivity entering the rendering process was near the threshold of transmissibility, even small changes could have made the difference between survival and destruction.

23.4.2 Mechanically recovered meat (MRM), another culprit

Mechanically recovered meat (MRM) is a technique developed years back in Japan to save fish meat left on the bones of fish after filleting. The name is rather misleading in so far as the main part of MRM is not meat in the usual meaning, but rather crushed bone including the bone marrow. MRM consists of a pressure-treated paste that, in relation to vCJD, may include vertebral columns containing spinal cord fragments and para-spinal ganglia. The product may, if originating from cattle in the incubation period, contain high doses per gram of infectious PrP^{sc}.

This cheap raw material was used in meat products as a substitute for real meat, in proportions as high as 20–30%, but usually in the range of 5–10%. Meat products such as hamburgers, sausages, etc., popular dishes with teenagers, might very well have harboured what corresponded to total consumed infectious doses of PrP^{sc} in the order of several billion. This man-made challenge to food safety was surely the most likely cause of human infection. The species barrier between sheep and cattle, and cattle and humans, of the PrP^{sc} prion is not very high, which could have accentuated the risk, but predicting this was outside the realms of possibility to predict: or was it?

23.4.3 Transmission of the prions within the human and animal body

The route of entry of prions is, under natural conditions, the oral route. Prions pass into the intestinal tract unaffected by protease enzymes. All evidence suggests that Payer's Patches in humans represent a portal of entry for orally administered prions from the intestinal tract to the central nervous system (CNS), and the transepithelial migration is supposed to be mediated by M cells. Prions have an affinity for lymphoid tissues and, after having gained access to the body, colonize lymphoid organs such as the spleen, for which reason the spleen in some slaughter animals is categorized as high-risk material (HRM). Evidence suggests that prions are transferred from the lymphoid system to the CNS along peripheral nerves belonging to the autonomic nervous system, and transfer is likely to depend on expression of PrP^{c} by these nerves.

23.5 Resolution and outcomes

23.5.1 Regulatory aspects

The BSE outbreak in the UK, and the subsequent emergence of vCJD in humans, initiated the issue of a large set of regulations within the EU. Regulations regarding BSE/TSE for disposal of so-called specified risk material (SRM), in particular specified bovine offal (SBO), can be found in the TSE regulation document 'Regulation (EC) No. 999/2001 of the European Parliament, and of the Council of May 2001 laying down rules for the prevention, control and eradication of certain transmissible spongiform encephalopathies'. This document has been supported by the by-product regulation, 1069/21 October 2009, which substitutes a number of other regulations, and in which very detailed rules to prevent BSE and vCJD are laid down.

An important general principle in the preventive legislation is that cannibalism is forbidden: in other words, that meat and bone meal from a species must not be fed to the same species. The regulations define risk material from domestic animals in three groups, in which the so-called high-risk material (HRM) is that of special concern in the epidemiology of TSEs. HRM in cattle is defined as the intestines, the tonsils, the skull including the brain and eyes, the spinal cord and part of the columnae, with further details dependent upon the age of the animal. The HRM is heat-treated in factories specially licensed to handle it. HRM meat and bone meal must not be used for any feeding purposes for domestic animals used for consumption, and it is further incinerated in cement works, added to the red-hot melted material. Meat and bone meal from the other two risk categories are only allowed to be used for companion animals and not for animals for consumption. The EU is, however, at the time of writing considering lifting the meat and bone meal ban and providing for its use in feed for swine and poultry to diminish the demand for the import of protein.

The decline in BSE cases in the UK that began in 1993 has been attributed to the ban, issued in 1988, on the inclusion of mammalian meat and bone meal supplements in feed, followed by the Special Bovine Offal ban prohibiting the use of specified bovine offal in human food. It was, however, not until 1996 that the significant role of MRM in the epidemic was realized, and a ban on its use was issued.

An essential element of the preventive measures taken was, and still is, a very comprehensive surveillance program on the prevalence of PrP^{sc} in cattle. At the beginning this comprised all cattle over 24 months of age, but the age limit has gradually been raised as the number of BSE cases markedly dropped, and has recently been changed to 72 months. Samples are drawn from the Obex area of the brain, the area between the cerebellum and the spinal cord. The PrP^{sc} agent has no antigenic abilities, but the complex method applied nevertheless rests partly on an enzyme-linked immunosorbent assay (ELISA) reaction. The legal interventions enforced, however strict they appear when looking back on the epidemic, have had the intended result. By the end of 2009 and in 2010 a total of only 71 and 28 BSE outbreaks respectively were recorded. This forms the basis for an easing of legislation on the use of meat and bone meal as feed for animals, which to some extent is in progress.

23.5.2 Economic and market aspects

The economic impact of the BSE epidemic has been of a very high magnitude, but to put forward exact figures is outside the scope of this chapter. It would be interesting if a rough calculation were to be undertaken. Some elements of the total costs will, however, be mentioned below.

It is assumed that some five million cattle were affected in the UK, and the cost of this is not that difficult to estimate. The epidemic initiated very expensive research programs all over the world. Activities in international and regional organizations such as WHO, OIE, FAO, EU, etc., to prepare reports, discuss at meetings how to handle the disease, and prepare legislation, etc., have been enormous. The cost of a human life appears to be impossible to calculate, but this is nevertheless possible and rules for this have been published in the literature and have also been applied, e.g., during outbreaks of listeriosis.

Detailed rules to secure traceability of both meat and condemned materials, which are very costly to apply and control, have been developed. The ban on the export of beef from the UK was of great economic importance, and so was the downgrading of certain parts of slaughtered carcasses, which have to be condemned. The control of the fulfilment of the enforced legislation in connection with the epidemic is, indeed, very manpower-consuming, not only in the slaughterhouses, but also at later stages in the food chain, and in international trade. As if the above were not enough, it has to be added that a very tight and expensive surveillance program of all cattle over a certain age has been installed, as detailed in Section 23.5.1.

23.6 Commentary

The emergence of the BSE epidemic in the late 1980s, and some 10 years later the new variant of CJD in humans, vCJD, is the most spectacular example in recent times of how changes in food and feed technology may have an impact on food safety and society. Small changes along the food chain may accumulate, sum, and finally result in disaster; and this is what happened. A pertinent and also difficult question to answer is: could it have been prevented?

The main component in the feed and food chain, of a technological nature, which led to the outbreak in animals and humans, appears to be the following: a new dry-rendering technology was adopted around 1970, resulting in a markedly reduced heat-treatment effect, which may have no longer been sufficient to destroy the BSE agent. The big question is, however, was the extreme heat resistance of the prion known at that time? If not, then the dry-rendering industry may be partly or wholly acquitted of being responsible.

The ineffective rendering process initiated a recycling of the infectious agent in cattle with disastrous effect. The subsequent use of MRM made from animals that were in the subclinical period of BSE turned out to be a deadly component of certain meat products, and there was an outbreak of vCJD. The most important question is: could the passage of the scrapie agent, PrP^{sc}, from sheep, via cattle, to humans have been predicted?

23.7 Acknowledgment

The author was at the time of outbreak a member of the EU Veterinary Committee on Public Health and has prepared this chapter based on his personal hands-on experience with controlling the BSE outbreak. The author is thankful to Mr Henrik Holst-Petersen, former Managing Director, DAKA, Denmark, for valuable advice and review of the chapter.

23.8 Critical questions for discussion

1. Explain the molecular mechanisms of the misfolding of the normal prion protein PrP^{c}.
2. Describe the likely replication of the prions.
3. Discuss the transmission of prions.
4. Could the BSE epidemic have been prevented?
5. What do you consider was the most important factor in the emergence of the BSE epidemic?
6. What is, in your opinion, the lesson to be learned from the case?

23.9 Further reading

AIKEN, J. (2001). Prion diseases. In R. G. Labbe, and S. Carcia (eds). *Guide to Foodborne Pathogens*, pp. 347–60. John Wiley & Sons Ltd, Chichester.

EUROPEAN FOOD SAFETY AUTHORITY (EFSA) (2012). *Bovine Spongiform Encephalopathy (BSE)*. Available at http://www.efsa.europa.eu/en/topics/topic/bovinespongiformencephalopathy.htm.

CARTER, J., and SAUNDERS, V. (2007). Prions. In *Virology – Principles and Applications*, pp. 327–34. John Wiley & Sons Ltd, Chichester.

GLENN, C.T. (ed.) (2004). *Prion and Prion diseases. Current Perspectives*. Horizon Bioscience, UK.

THE NATIONAL CREUTZFELDT-JAKOB DISEASE RESEARCH & SURVEILLANCE UNIT (NCJDRSU). Available at http://www.cjd.ed.ac.uk.

OIE (WORLD ORGANISATION FOR ANIMAL HEALTH). Available at http://www.oie.int/en/search/.

WORLD HEALTH ORGANIZATION (WHO) (2012). *Bovine Spongiform Encephalopathy*. Available at http://www.who.int/topics/encephalopathy_bovine_spongiform/en/.

24

The continuing debate over increasing consumption of raw (unpasteurised) milk: is it safe?

K. N. Jordan and K. Hunt, Teagasc Food Research Centre, Ireland and J. Hoorfar, Technical University of Denmark, Denmark

Abstract: The debate on the consumption of raw (unpasteurised) milk continues. On one side there is the risk of illness associated with pathogenic bacteria that may be in the milk, a risk that can be mitigated by pasteurisation of the milk prior to consumption. On the other side, there are those who believe that pasteurisation of milk damages its nutrient qualities, and that hygienically produced milk will not contain pathogens. But who is right? The discussion is illustrated by a case study on the contamination of raw bulk tank milk by *Listeria monocytogenes*.

Key words: raw milk, unpasteurised milk, pathogens, quality control.

24.1 Introduction to the case

For many centuries mankind has consumed raw (unpasteurised) milk. The practice may have started during the transition from hunter-gatherer to settled communities with domesticated animals. The food chain was short, with producers consuming their home-produced milk or giving it to immediate neighbours. The time from production to consumption was no more than a day. However, the food chain is not like that anymore. The requirement for longer shelf-life and wider distribution means that the time from production to consumption is now several days. So is raw milk consumption still a healthy, safe option?

24.1.1 Pasteurisation – killing milk-borne pathogens

In the seventeenth century the discipline of microbiology began. With the discovery of the microscope there was the realisation of an entire 'micro' community, of which mankind was unaware. There was also a realisation that some members of this micro community could be harmful. In 1862, Louis Pasteur and Claude Bernard carried out the first pasteurisation experiments aimed at reducing bacteria in wine and preventing it from spoiling. It was not long before this process was used to reduce bacterial growth in milk and extend its shelf-life.

Currently, the minimum legal requirement for high temperature short time (HTST) pasteurisation is 72 °C for 15 s. Generally, HTST pasteurisation is a continuous-flow system where the milk (or other liquid) is passed through heat exchangers to achieve the required temperature and then rapidly cooled. Alternatively, in batch pasteurisation heating to 63 °C for 30 min can be used. Industrially, HTST is the most commonly used form of pasteurisation for milk. The time/temperature combination of pasteurisation has been specifically chosen because it inactivates known pathogens, thus reducing the risk of transferring disease. But could such heat treatment of milk have a detrimental effect on the beneficial qualities of the milk?

The issue has become topical due to the increasing interest of consumers in raw food, which is gaining momentum, particularly in the Western world. This is thought to be a counter-trend to the intensive modern food processing technologies and a nostalgic wish to revert to old-time natural food.

24.1.2 Effect of pasteurisation on milk quality

Milk is a nutritious food containing protein, fat, lactose, enzymes, etc., that contributes significantly to the diet. Prior to excretion from the cow, milk is practically sterile. It is contaminated by contact with the udder of the cow, the milking equipment, faeces, etc. Disease, particularly sub-clinical mastitis, can also contribute to the bacterial count of milk. Some of the bacteria can be harmful, but some can also be beneficial.

Pasteurisation kills any potential beneficial bacteria as well as harmful bacteria. However, the total bacterial count of well-produced unpasteurised milk is 10 000–20 000 colony forming units (cfu)/ml. At this concentration the amounts are too low to have any significant beneficial impact. For a beneficial impact the number of bacteria needs to exceed 1 000 000 cfu/ml (Shah, 2000). In order to be considered probiotic (thus healthy), the organism must be of human origin (Teitelbaum and Walker, 2002). Bacteria in milk originate from contact with the cow, the milking equipment and contamination from the environment. Bifidobacteria, common probiotic bacteria, are found in the human and animal gastrointestinal tract. They will not be present in milk unless they come from faecal contamination, in which case pathogenic bacteria could also be present.

Other proposed potentially beneficial constituents of milk are bacterial inhibitors such as lactoferrin and lactoperoxidase. However, the concentration of lactoferrin required for activity is 1–8 g/l, which is considerably higher than the 0.1 g/l found in milk (Naidu, 2000). Lactoperoxidase can, in some cases, inhibit the growth of bacteria. However, the chemical thiocyanate must be added to activate the lactoperoxidase. This practice is only permitted in tropical (hot) countries where refrigeration of milk is not possible during transport to processing facilities for pasteurisation (for example UHT milk in Brazil: see Chapter 33, Cardoso *et al.*, 2012). It is not an alternative to pasteurisation, and such milk (or products made from it) is not permitted for sale on the international market.

Lactase (which is necessary for lactose degradation and is absent in cases of lactose intolerance) is also thought by some to be a beneficial constituent of raw milk. There is no indigenous lactase in milk, only that produced by bacteria in the milk, which (ironically) in hygienically produced milk are present in low numbers. Therefore, unpasteurised milk does not contain sufficient lactase to counteract the lactose content of milk (raw or pasteurised), and is not suitable for patients with lactose intolerance.

The protein quality of pasteurised milk is the same as that of raw milk (Andersson and Oste, 1995). While pasteurisation may alter some proteins in the milk, it does not necessarily destroy them and may make some amino acids more available, although there is no evidence for this. There is some emerging anecdotal evidence that some allergies may be prevented by consumption of raw milk. However, to date this has not been scientifically proven, and if it is then a proper risk–benefit analysis will need to be undertaken.

24.1.3 Growth of pathogens in milk

Pathogens can grow in milk. Milk is a nutritious medium containing all the nutrients necessary for bacterial growth. The potential antimicrobial systems in milk, such as the lactoperoxidase system and lactoferrin, are insufficient to prevent pathogen growth. In addition, there is no evidence that the microflora of milk have an effective inhibitory activity towards pathogens. With the shelf-life expectations of modern consumers, the opportunities for growth of pathogens that may be in milk are increased.

24.1.4 The case of *Listeria monocytogenes* in raw milk

The following case provides a real-life example of pathogen contamination of raw milk. In a recent analysis of raw milk we detected *L. monocytogenes* in the bulk tank milk at a dairy farm (Hunt *et al.*, manuscript in preparation). In previous work, approximately 6% of raw bulk milk samples were positive for *L. monocytogenes* (K. N. Jordan and K. Hunt, unpublished), so this result did not raise concern. However, when the same bulk milk was positive several weeks later, concern arose. The serotype (1/2a) and pulsed-field gel electrophoresis (PFGE) patterns of both isolates were indistinguishable, indicating that the same contaminating strain was present in both samples (see Chapter 10, Jordan *et al.*, 2012). The milking parlour at the farm was visually extremely clean, and discussions with the farmer revealed that good hygienic practices were followed during milking. Swabs from the milking equipment were analysed, but were negative, while the bulk milk remained positive. Milk filters (filtering the milk before entry to the bulk tank) were also positive, but were negative after cleaning (filters were changed after milking). All of this pointed to direct contamination of the milk from a cow.

24.2 Significance of the case

24.2.1 Regulatory aspects

In the EU, the sale of raw milk for consumption is allowed. However, National governments are free to apply their own regulations (EU Regulation (EC) No. 853/2004) (European Commission 2004). These have been applied to varying degrees, including banning the sale of raw milk entirely, banning the sale of raw milk beyond a specified distance from the site of production, and allowing the sale of raw milk. This regulation has no impact on the use of raw milk for dairy product manufacture, which is allowed. In the US, there are similar varying degrees of regulation among states on the sale of raw milk for consumption.

24.2.2 Milk-borne pathogens

Although rarely reported, cases of sub-clinical mastitis caused by *L. monocytogenes* have been recorded in the past (Fedio *et al.*, 1990; Gitter *et al.*, 1980; Sanaa *et al.*, 1993; Bourry 1995). The absence of reports in recent years is due to lack of analysis as much as lack of occurrence. In Ireland, the most recent extensive survey for milk-borne pathogens was in 1992 (Rea *et al.*, 1992). At that time, analytical methods were lacking specificity and sensitivity, probably resulting in an underestimation of pathogen contamination.

In addition, analysis for organisms such as *Mycobacterium bovis* (which is the causative agent of tuberculosis) was not undertaken. Therefore, there is a lack of knowledge on the occurrence of pathogens in milk. This case shows the necessity to analyse raw milk regularly for pathogens if the milk is intended for consumption without pasteurisation.

24.3 The case continues

The farmer milked 180 cows. The challenge was to determine which cow (if any) was excreting *L. monocytogenes*. The cows were divided into groups of 20, the milk from each group was pooled and 10 ml (in 90 ml of selective broth) analysed by enrichment using the AFNOR validated method (Oxoid, 2008) of overnight enrichment in ONE broth at 30 °C followed by plating on chromogenic Brilliance agar, and incubated for 48 h at 37 °C. Positive samples are identified by the presence of green colonies surrounded by a zone of precipitation. The milk from all nine groups was negative, while the bulk milk remained positive. This exercise was repeated and on this occasion 60 ml of the pooled milk from 20 cows was analysed by enrichment in 540 ml of broth. One group was positive for *L. monocytogenes*.

Following this culture confirmation, the milk from each of these 20 cows was analysed individually. Milk from one cow was positive. An isolate from this sample had a PFGE profile that was indistinguishable from the original bulk milk samples. After identification of the 'guilty' cow, the first direct count analysis (direct plating onto Brilliance agar) showed that all quarters (a cow's udder has four teats, commonly referred to as quarters) were <10 cfu/ml for *L. monocytogenes*. Subsequent analysis of all quarters showed that the number of *L. monocytogenes* in the milk of one quarter was approximately 80 cfu/ml while the remaining three quarters were negative, even by culture enrichment. Again, an isolate from the quarter and a PFGE profile that was indistinguishable from the first were isolated from the bulk milk.

The somatic cell count (SCC) – numbers above 400 000 indicating infection – of the cow was 75 000. Analysis of the milk from the four quarters showed that the quarter excreting *L. monocytogenes* into the milk had an SCC of 207 000 while the remaining three quarters had counts of 4000, 9000 and 79 000, respectively. Although higher than the others, the count of 207 000 does not indicate infection, and there were no visible signs of infection.

The milk from this cow was omitted from the bulk milk tank, and the bulk milk tank was subsequently negative for *L. monocytogenes* when sampled for up to 2 weeks after omission of the milk from the infected cow.

24.4 Commentary and personal reflections

This case demonstrates the difficulties encountered in analysing raw milk for pathogens. Features of this case, such as sporadic shedding, low pathogen numbers and lack of visible signs of infection, show that single analysis of raw milk samples is insufficient for pathogen detection. In raw milk intended for consumption without pasteurisation, very regular testing for all pathogens (possibly using a cost-effective multiplex test) is necessary to ensure a reduction in risk. Visual inspection of the cows and of the milking parlour/hygienic conditions of milking is not adequate to control contamination of milk with pathogens.

In our view, any potential unproven benefits of raw milk consumption are far outweighed by associated scientifically proven risks. Pasteurisation is a safe method to mitigate these risks, causing limited changes to the milk. Any slight alterations to the milk during pasteurisation are perfectly tolerable when balanced against the alternative.

From a consumer point of view, the increasing use of ultra-high-temperature (UHT) or other intensive treatments may have promoted the counter-trend to raw milk, because UHT milk has a 'burned' taste, which is not normal for milk. Milk producers should develop new processing technologies that could replace pasteurisation in providing a safe product with a long shelf-life, but without the taste side-effect. However, if consumers continue the trend towards drinking raw milk, then milk-borne outbreaks will inevitably increase, and the entire dairy industry could face a reputational challenge.

24.5 Acknowledgement

The assistance of the farmer who owned the farm and cows referred to in this work is greatly appreciated.

24.6 Critical questions for discussion

1. Discuss the possible reasons for the negative result from the first analysis of the pooled milk from the eight groups of 20 cows.
2. What are the main strategies for sample compositing (pooling), and which one is the most sensitive? (See Jarvis, 2007.)
3. Why is *Listeria* contamination so important?
4. Why is consumption of raw milk such a threat to public health?
5. What other microbial contaminants of milk can be a threat to human health?
6. What other organisms can cause mastitis in cows?
7. What are the particular dangers associated with sub-clinical mastitis?
8. How do you treat a cow with sub-clinical mastitis caused by *L. monocytogenes*?
9. Keeping in mind the alterations to milk during product manufacture, discuss the issues surrounding the use of raw milk for dairy product production.

24.7 Further reading

ANONYMOUS. *Scientific References*. Available at http://www.realrawmilkfacts.com/scientific-references#Pathogens. Accessed 19 October 2011.

RYSER E.T. and MARTH E.H. (2007). Listeria, *Listeriosis and Food Safety*, 3rd ed. CRC Press, Boca Raton, Florida.

TENOVER F.C., ARBEIT R.D., GOERING R.V., MICKELSEN P.A, MURRAY B.E. *et al.* (1995). Interpreting chromosomal DNA restriction patterns produced by Pulse-Field-Gel-Electrophoresis: criteria for bacterial strain typing. *J Clin Microbiol*, **33**: 2233–9.

TODAR K. *Listeria monocytogenes* and listeriosis. *Online Textbook of Bacteriology*. Available at http://www.textbookofbacteriology.net/Listeria.html. Accessed 13 October 2011.

US FOOD AND DRUG ADMINISTRATION (2011). *Raw Milk Misconceptions and the Danger of Raw Milk Consumption.* Available at: http://www.fda.gov/Food/FoodSafety/Product-SpecificInformation/MilkSafety/ConsumerInformationAboutMilkSafety/ucm247991.htm. Accessed 18 October 2011.

24.8 References

ANDERSSON, I. and OSTE, R. (1995). Nutritional quality of heat processed liquid milk. In P. F. Fox (ed.), *Heat-induced Changes in Milk*, pp. 279–307. International Dairy Federation, Brussels, Belgium.

BOURRY, A., POUTREL, B. and ROCOURT, J. (1995). Bovine mastitis caused by *Listeria monocytogenes*: Characteristics of natural and experimental infections. *J Medical Microbiol*, **43**: 125–32.

EUROPEAN COMMISSION (2004). European Regulation (EC) No 853/2004 of the European Parliament and of the Council of 29 April 2004 laying down specific hygiene rules for food of animal origin (OJ L 226, 25.6.2004, p. 22).

FEDIO, W.M., SCHOONDERWOERD, M., SHUTE, R.H. and JACKSON, H. (1990). A case of bovine mastitis caused by *Listeria monocytogenes*. *Can Vet J*, **31**: 773–5.

GITTER, M., BRADLEY, R. and BLAMPIED, P.H. (1980). *Listeria monocytogenes* infection in bovine mastitis. *Vet Records*, **107**: 390–3.

HUNT *et al.* Submitted to *The Irish Veterinary Journal*.

JARVIS, B. (2007). On the compositing of samples for qualitative microbiological testing. *Lett Appl Microbiol*, **45**: 592–8.

JORDAN, K.N., HUNT, K., and FOX, E.M. (2012). Troubleshooting the environmental source of contamination in a small food manufacturing plant. In J. Hoorfar (ed.), *Case Studies in Food Safety and Authenticity*. Woodhead Publishing, Cambridge, UK.

NAIDU, A.S. (2000). Lactoferrin. In A. S. Naidu (ed.), *Natural Food Antimicrobial Systems*, pp. 17–102. CRC Press, Boca Raton, Florida.

OXOID (2008). *Rapid Culture Method*. Available at http://jornades.uab.cat/workshopmrama/sites/jornades.uab.cat.workshopmrama/files/Oxoid_Listeria_Precis.pdf. Accessed 20 October 2011.

REA, M.C., COGAN, T.M., and TOBIN, S. (1992). Incidence of pathogenic bacteria in raw milk in Ireland. *J Appl Bacteriol*, **73**: 331–6.

SANAA, M. (1993). *Epidémiologie de la contamination du lait à la ferme par Listeria monocytogenes*. Thèse de doctorat de l'Université Paris XI.

SHAH, N.P. (2000). Probiotic bacteria: Selective enumeration and survival in dairy foods. *J Dairy Sci*, **83**: 894–907.

TEITELBAUM, J.E. and WALKER, W.A. (2002). Nutritional impact of pre- and probiotics as protective gastrointestinal organisms. *Annu Rev Nutr*, **22**: 107–38.

Part V

Safe food production

25

Challenges in food safety certification of a small enterprise producing a speciality product in Cyprus

E. H. Drosinos, S. G. Charalambidou and P. N. Skandamis, Agricultural University of Athens, Greece

Abstract: This is the history of an attempt to put together a food safety management system (FSMS) for a small enterprise in Cyprus. The enterprise produces *zalatina*, a ready-to-eat pork jelly, made of raw materials capable of supporting survival and/or growth of pathogens. According to existing EU legislation, it is compulsory for the food industry to demonstrate and verify the operation of a system that covers its legal obligations with regard to food safety management. The most widely established system for food safety assurance is hazard analysis and critical control points (HACCP) and this is at the core of the FSMS implemented in this case.

Key words: ISO 22000, hazard analysis and critical control points (HACCP), pork jelly, *zalatina, pikti.*

25.1 Introduction to the case

Perishable foods can support growth of pathogenic and non-pathogenic microorganisms. Microbial activity in foods often compromises their sensory characteristics and this is perceived by the consumers as spoilage. In the worst case scenario, pathogen survival and/or growth may lead to consumers suffering from foodborne disease. Therefore, there is a need for safety and quality control systems, which will ensure that products are of good quality and safe for consumers.

According to existing legislation, it is compulsory for the food industry to demonstrate and verify the operation of a system that covers its legal obligations with regard to food safety management. The most widely established system for food safety assurance is hazard analysis and critical control points (HACCP). HACCP requires a scientific evaluation of the potential hazards entering a specific processing line as well as their prevention, elimination or reduction to acceptable levels through proper control measures (which are monitored and verified). Then its application and the corrective interventions (foreseen in case of deviation of critical safety parameters from their respective limits) ensure continuous monitoring and assurance of product safety, in compliance with consumer and legislative demands.

The HACCP system is currently the core of the integrated food safety management system (FSMS) proposed by the ISO (ISO 2005, ISO/ITC 2007). Apart from the seven established HACCP principles (CCFH 1993), the FSMS also encompasses other supporting prerequisite programs (PRPs), which are divided into two groups: operational and basic. The basic PRPs form the fundamental hygiene infrastructure of the industry and include the hygienic design of premises and equipment, personnel hygiene, water quality assurance, as well as the standards for the supply of edible or non-edible raw materials and procedures for cleaning and sanitation, and personnel training. These PRPs are needed to control specific hazards, which are not targeted by the critical control points. The (remaining) PRPs, which are strongly associated with CCPs and aim to control a potentially severe risk, are called operational PRPs.

25.1.1 The speciality product and its manufacture

Zalatina is a jelly, the basic ingredient of which is pork. It is ready to eat without a further cooking or re-heating step. It has a sour smell and accepted range of pH 4.0–6.0. It is manufactured in plastic containers suitable for food and stored in freezers at temperatures of 0–6 °C until it is released to the market. *Zalatina* is perishable and it is likely that microorganisms such as *Salmonella* spp., *Listeria monocytogenes and Escherichia coli* O157:H7 (Table 25.1) are present and capable of surviving in this product. To ensure that these microorganisms are within acceptable limits, specified under the Regulation (EC) 2073/2005 (European Union 2005), raw materials and final products are microbiologically analysed (Table 25.2).

The enterprise that produces *zalatina* is situated in a village in Cyprus. It is a small enterprise, which employs fewer than ten employees. The enterprise is located in a building of 99 m^2 and produces 200–500 packages per day. The high season is the Christmas period. The processing steps for *zalatina* production are illustrated in the flow diagram (Fig. 25.1) where one can see the stages, the points where raw materials and the rest of the ingredients enter, the point where the final product has been produced and the point where waste material is discarded. The ingredients used for the production of *zalatina* are meat from pork heads, spices, salt, vinegar, gelatin, lemon and bitter orange juice.

25.1.2 Market data and tradition

The history of *zalatina* goes back many years, when each rural family had to raise its own domestic pigs, which were destined to meet the family's seasonal meat

Table 25.1 Microorganisms that can appear in *zalatina*

	Salmonella spp.	*Listeria monocytogenes*	*Escherichia coli*
Importance	Usually pathogenic. Associated with poor hygiene or inappropriate process	Sluggish growth at freezing temperatures. Mortality can be 30% of those infected	Index of bad hygiene or inappropriate process. Some strains produce toxins that are temperature-resistant and temperature-sensitive
Growth temperature range (°C)	5.1–47	0–45	3–46
pH of growth	4–9	4.4–9.6	4.4–9.5
Minimum a_w for growth	0.95	0.92	0.95
Highest permissible salt level (%)	8	10	8
Oxygen requirements	Facultative anaerobe	Facultative anaerobe	Facultative anaerobe

Table 25.2 Microbiological criteria for *zalatina* according to Regulation (EC) 2073/2005

Microorganisms	Sampling plan*		Limits	
	N	*c*	*m*	*M*
Listeria monocytogenes	5	0	Absent in 25 g	
Salmonella spp.	5	0	Absent in 25 g	
Escherichia coli	5	2	500 cfu/g	5000 cfu/g

* *N* = number of units comprising the sample; *c* = number of sample units giving values between *m* and *M*.

requirements. The pig was slaughtered before Christmas and all parts of the animal were used for making a diverse range of meat products. The leftovers from the head and the legs were the main ingredients for *zalatina*. Only a few traditional families in Cyprus villages still abide by this custom as the standardized product is available. In Cyprus, there are nine enterprises that produce *zalatina* and only one of them produces this product exclusively. Given the need to increase export activities, the enterprise in question had to implement a food safety management system assuring product safety and minimizing recalls and product waste due to spoilage or pathogen contamination. It is also interesting to note that *zalatina*, a

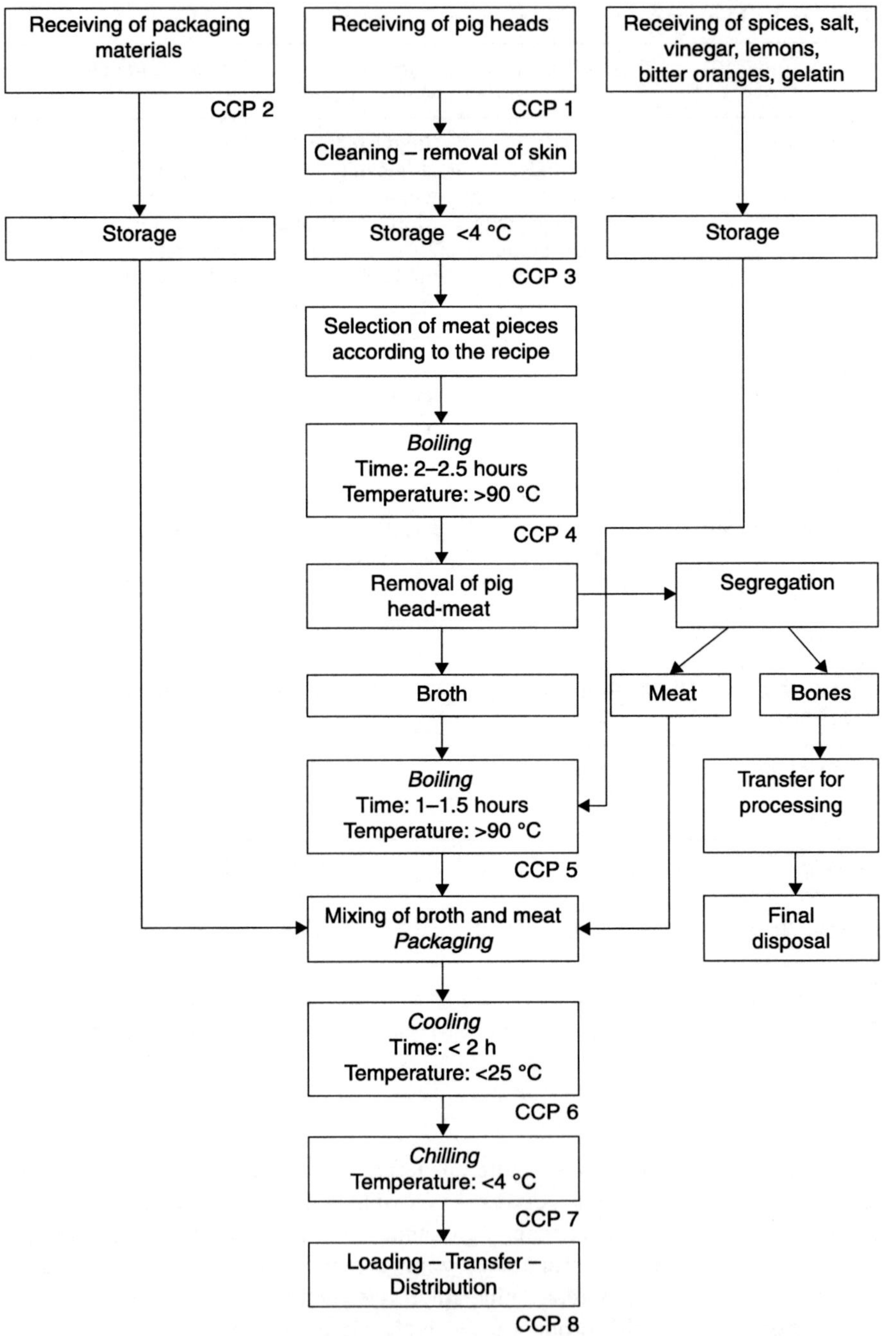

Fig. 25.1 Flow diagram for *zalatina* production. CCP, critical control point.

traditional product, may belong to the category of agricultural products and foodstuffs considered traditional speciality guaranteed (TSG). If the product were to obtain TSG status, this could help to boost exports. A similar product named *pikti* is produced in Greece.

25.1.3 Essential prerequisite programs (PRPs)

For proper operation of an FSMS in an enterprise, the essential (basic) PRPs, which form part of the aforementioned good manufacturing practice (GMP) and good hygiene practice (GHP), must be established in advance. More specifically, these programs include:

- Legislative requirements regarding the premises, its construction, the equipment and its maintenance: there must be proper planning so that the entrance and exit of people, vehicles, raw materials and products are thoroughly checked.
- All the equipment and machinery must be maintained and checked according to the manufacturers' directions.
- The flow of production of the final products must be linear and their path should not be crossed by the paths of other activities, to minimize cross-contamination from the dirty to the clean areas.
- Control of suppliers: the suppliers must apply the programs of good manufacturing practice (GMP) and good hygiene practice (GHP).
- All the instruments used for recording temperature (thermometer, timer, and pH meter) should be standardized and calibrated.
- Cleaning – disinfection: the facilities and equipment must be disinfected after work every day.
- Personal hygiene: all staff and visitors to the enterprise must follow the personal hygiene requirements and they must wear proper clothes before entering production areas.
- Chemical control: detergents, disinfectants, insecticides and other chemical substances that are not food ingredients must be stored separately and must be used accordingly.
- Receiving, storage and distribution: all raw materials and final products must be stored in suitable places under the right conditions (temperature, humidity, ventilation), to ensure safe storage and suitability for human consumption.
- Traceability and withdrawal of products: a batch number must be given to all raw materials and final products. For the fast and total withdrawal of unsuitable or dangerous products a special system needs to be developed and its effectiveness needs to be tested frequently.
- Labelling: there must be instructions for product labelling which adhere to the provisions of legislation.
- The ingredients and the recipe of the final product must be recorded.

25.2 Significance of the case

Procedures based on HACCP are compulsory for food businesses in the EU. The General Food Law (Regulation (EC) 178/2002) (European Union 2002), the Food Hygiene Regulation (Regulation (EC) 852/2004) (European Union 2004) and the regulation for microbiological criteria (Regulation (EC) 2073/2005) (European Union 2005) form the main legislative framework within which the food industry operates in Europe. Cypriot legislation is structured in the same way (i.e., Regulations 150(I) 2003, 776/2003, 362/2003). ISO 22000:2005 deals with the hygiene and safety of the product, but also addresses requirements and procedures for the management of the processes based on the procedures listed in ISO 9001:2008. Table 25.3 indicates

Table 25.3 Procedures of FSMS for *zalatina* production

Procedure	Purpose
Waste management	Management of wastes until the time of their removal from the premises of the enterprise
Examination for presence of foreign materials and pollution	Activities that must be adopted for reducing the possibility of presence of outer elements in the final products
Examination of raw materials and other ingredients for the production of *zalatina*	Activities for the preparation of the ingredients that will be used
Hygienic production of *zalatina*	Following hygiene rules for minimizing the possibilities of infection
Examination of products presented for sale	Handling products at the point of retail sale
Personnel management	Job description, collection of clothing, applications for training
Examination of documents and data	To ensure the existence and the correct operation of a system, issue, change, coding, withdrawal of documents
Treatment of products that do not comply with the above	Activities for acknowledging and treatment of these products
Announcement of and withdrawal of products	Informing interested parties how to spot problematic products
Tracing	'One step back – one step forward'
Management review	Applied during the preparation of records at meetings
Internal audit	Examination of the application of the food safety management system
Continual improvement and updating of the system	Evaluation of improvement and performance of the system
Emergency preparedness and response	Activities that are applied in possible cases of emergency and accidents

procedures of FSMS for *zalatina* production. It is not a legal requirement to achieve certification to ISO 22000:2005; the standard can simply be implemented for the benefits it provides.

25.3 The case continues

25.3.1 HACCP (hazard analysis and critical control points) plan

According to the principles of HACCP, the description of the product and its intended use and the flow diagram, the possible and potential hazards, were acknowledged and evaluated in terms of the safety of the product. The hazards were divided into three groups: microbiological, chemical and physical. For each hazard, control measures were specified. The critical control points in various stages of the manufacturing procedure were defined and the critical limits that must be followed were specified (Table 25.4).

25.3.2 How did we verify the HACCP plan?

Verification was performed by internal audit of the system and microbiological analysis of raw material and final product. The following were performed with samples from pig heads and *zalatina*:

- Determination of *Salmonella* according to International Standard ISO 6579:2002 (ISO 2002),
- Determination of *Listeria monocytogenes* according to International Standard ISO 11290-1:1996 (ISO 1996),
- Determination of *Escherichia coli* according to International Standard ISO 16649-2:2001 (ISO 2001),
- Enumeration of total microbial flora according to International Standard ISO 4833:2003 (ISO 2003).

The results of the microbiological analysis of the pork meat samples (Table 25.5) were negative. The level of the total microbial flora in all samples was less than 5×10^5 cfu/g, thereby within the limits specified by the regulation of microbiological criteria for foods (Reg. (EC) 2073/2005). The results of the microbiological testing of the final product (Table 25.6) were negative as for the presence of these specific microorganisms; however, enumeration of total microbial flora was not carried out. For the verification of the system, an examination of the whole system was carried out (Table 25.7) and this included:

- Examination of the temperature of the various stages of production, of raw materials and of the final product,
- Examination of the boiling time of pig heads and broth,
- Examination of the cooling time after the mixture of broth and meat,
- Examination of the pH of the pig heads and *zalatina*.

Table 25.4 Hazard analysis control measures critical control points – critical limits

Process step	Potential hazards	Control measures (controlled parameters)	CCP	Limits and tolerance	Observation (frequency of control)
1. Receipt, storage of raw materials	M, C	1. Macroscopic and organoleptic examination of meat, pig heads and other raw materials (colour, smell, appearance)	CCP 1	1. Normal physiological and organoleptic characteristics, absence of contamination	1. Examination at each receipt
		2. Examination of pH of pig heads		2. pH 5.4–6.1	2. Examination at each receipt
		3. Examination of the temperature of the transportation vehicles of meat and pig heads		3. Temperature <4 °C	3. Measure and record of temperature
		4. Examination of the microbiological condition of meat and pig heads		4. Total microbial flora $<10^5$ cfu/g *Enterobacteriaceae*: $<10^3$ cfu/g *Staphylococcus*: $<10^2$ cfu/g *Salmonella, E. coli* 0157:H7 'Absence in a 25 g sample'	4. Occasionally and when it is needed according to the results of the macroscopic examination, temperature and pH measurement
		5. Examination of temperature of received pig heads		5. Temperature at the core of pig heads <7 °C	5. Examination at each receipt
2. Receipt, storage of packaging materials	M, C	1. The package is transparent, microbiologically and chemically suitable for foods	CCP 2	1. According to the microbial criteria. Total microbial flora: $<50/cm^2$	1. Visual examination and microbiological tests. In case of contamination, disposal of packaging materials
3. Removal of skin – Cleaning – Selection of meat and raw materials according to recipe	M	1. Fast carrying out of all activities 2. Using knives and cold drinking water		1. Temperature of meat <7 °C 2. According to rules	1. Monitoring of temperature

4. Preserving pork meat and pig heads that need freezing	M	1. Following the program of cleaning/ disinfection of all areas and equipment 2. Temperature of refrigerator 3. Temperature of pig heads and meat	CCP 3	1. Visually clean areas of activity and equipment 2. Temperature of refrigerator <4 °C 3. Temperature of pig heads and meat <7 °C	1. Daily examination 2. Constant examination 3. Constant examination
5. Boiling of pig heads and meat	M	1. Clean equipment 2. Examination of temperature and time of boiling	CCP 4	1. Visually clean 2. Temperature of boiling >90 °C, time 2–2.5 h	1. Daily examination 2. Every batch
6. Separation of meat and bones		1. Examination of cleanliness of equipment and all areas			1. Every batch
7. Boiling of broth after separation of meat and bones	M	1. Examination of temperature and time of boiling the broth	CCP 5	1. Temperature of boiling >90 °C, time 1–1.5 h	1. Every batch
8. Mixing of broth and meat, packaging, cooling	M	1. Care to avoid contamination 2. Examination of room temperature 3. Examination of temperature and cooling time	CCP 6	1. Absence of foreign materials 2. Room temperature <12 °C 3. Reducing temperature of *zalatina* <25 °C, <2 h	1. Every batch 2. Every batch 3. Every batch
9. Chilling of the final product	M	1. Constant temperature of refrigerator	CCP 7	1. Temperature of chilling <4 °C	1. Constant examination
10. Loading and distributing of *zalatina*	M	1. Examination of temperature of final product 2. Examination of microbial flora 3. Recording temperature of room or vehicles of transfer	CCP 8	1. Temperature of *zalatina* <6 °C 2. Absence of pathogens 3. Temperature of freezer <4 °C	1. Examination at every loading 2. Microbiological examination every 3 months 3. A thermograph to record temperature continuously is recommended

M: microbiological, C: chemical, P: physical.

Table 25.5 Results of microbiological analysis in the samples of pork meat

Microbiological analysis	Methodology	Results
Detection of *Salmonella*	ISO 6579:2002	Absent
Detection of *Listeria monocytogenes*	ISO 11290-1:1996	Absent
Enumeration of total microbial flora	ISO 4833:2003	$<5 \times 10^5$ cfu/g
Detection of *Escherichia coli*	ISO 16649-2:2001	Absent

Table 25.6 Results of microbiological analysis in the samples of *zalatina*

Microbiological analyses	Methodology	Results
Detection of *Salmonella*	ISO 6579:2002	Absent
Detection of *Listeria monocytogenes*	ISO 11290-1:1996	Absent
Enumeration of total microbial flora	ISO 4833:2003	Not determined
Detection of *Escherichia coli*	ISO 16649-2:2001	Absent

Table 25.7 Audit of the manufacture of *zalatina*

Process step	Temperature (°C)	Time
Preserving pig heads and meat	2	48 h
Pig heads (pH = 5.8)	0.5	
Boiling pig heads	>100	2.5 h
Boiling of broth	>100	1 h
Mixing of broth and meat:		2 h
Room temperature	12	
Temperature of final product	25	
Chill storage	4	6 months
Delivery of *zalatina* (pH = 4)	5	

25.4 Outcomes and commentary

The FSMS was developed in such a way that the enterprise could easily be certified by a third party. The owner of the enterprise did not wish to be certified, but rather aimed at applying the system for a certain period of time in order to gain experience in food hygiene and safety. Therefore, there was no need for documentation and/or application of procedures, which would not be officially modified by the authorities. For example, the following procedures along with the corresponding documents were excluded:

- Response in situations of emergency,
- Personnel management,
- Continual improvement and updating of the system,
- Internal audit, and
- Management review.

As for CCP 1 of the HACCP plan, which is used during receiving, and storage of raw materials, measuring and recording the value of pH in pig heads and meat could be omitted and monitoring actions could be restricted to temperature measurements. Furthermore, CCP 5 concerning the boiling of the broth could be considered as a control point, provided that the boiling of pig heads was carried out at the right temperature and for the correct amount of time (boiling temperature >90 °C for 2–2.5 h).

The FSMS was verified partly according to the results of the physico-chemical and microbiological analysis. From these results (Tables 25.5, 25.6 and 25.7) it is obvious that the system is effectively applied and that the management and personnel of the enterprise comply with international standards and requirements. The results of the microbiological analysis of the final product were negative (Table 25.6), which may be attributable to the fact that the enterprise uses vinegar for pH reduction to around 4.00.

25.5 Acknowledgments

The authors would like to express their gratitude to the owner of the enterprise for providing the appropriate information for the conduct of this case study. In addition, we would like to thank the consultancies and educational services for their support.

25.6 References

CODEX COMMITTEE ON FOOD HYGIENE (CCFH) 2003. *Hazard Analysis and Critical Control Points (HACCP) System and Guidelines for its Application. Annex to CAC/RCP 1-1969.* Revision 4. Codex Alimentarius Commission, Rome.

EUROPEAN UNION 2002. Regulation (EC) No 178/2002 of the European Parliament and of the Council of 28 January 2002 laying down the general principles and requirements of food law, establishing the European Food Safety Authority and laying procedures in matters of food safety. *Official Journal of the European Communities.* **L31**: 1.

EUROPEAN UNION 2004. Regulation (EC) No 852/2004 of the European Parliament and of the Council of 29 April 2004 on the hygiene of foodstuffs. *Official Journal of the European Union.* **L139**: 3.

EUROPEAN UNION 2005. Regulation (EC) No 2073/2005 of 15 November 2005 on microbiological criteria for foodstuffs. *Official Journal of the European Union.* **L338**: 1.

INTERNATIONAL ORGANIZATION FOR STANDARDIZATION (ISO) 1996. ISO 11290-1:1996. *Microbiology of Food and Animal Feeding Stuffs – Horizontal Method for the Detection and Enumeration of* Listeria monocytogenes *– Part 1: Detection Method.* ISO, Geneva.

INTERNATIONAL ORGANIZATION FOR STANDARDIZATION (ISO) 2001. ISO 16649-2:2001. *Microbiology of Food and Animal Feeding Stuffs. – Horizontal Method for the Enumeration of beta-glucuronidase-positive* Escherichia coli. *– Part 2: Colony-count Technique at 44 Degrees C using 5-bromo-4-chloro-3-indolyl beta-D-glucuronide*. ISO, Geneva.

INTERNATIONAL ORGANIZATION FOR STANDARDIZATION (ISO) 2002. *ISO 6579:2002. Microbiology of Food and Animal Feeding Stuffs – Horizontal Method for the Detection of* Salmonella *spp*. ISO, Geneva.

INTERNATIONAL ORGANIZATION FOR STANDARDIZATION (ISO) 2003. *ISO 4833:2003. Microbiology of Food and Animal Feeding Stuffs – Horizontal Method for the Enumeration of Microorganisms – Colony-count Technique at 30 degrees C*. ISO, Geneva.

INTERNATIONAL ORGANIZATION FOR STANDARDIZATION (ISO) 2005. *ISO 22000:2005 Food Safety Management Systems – Requirements for any Organization in the Food Chain*. ISO, Geneva.

INTERNATIONAL ORGANIZATION FOR STANDARDIZATION AND INTERNATIONAL TRADE CENTRE (ISO/ITC) 2007. *ISO 22000 Food Safety Management Systems: An Easy-to-use Checklist for Small Business. Are you ready?* ISO Copyright Office and International Trade Centre UNCTAD/WTO, Geneva, Switzerland.

26

The dilemma of regulatory testing of dairy foods in Australia: minimising risk to the public while avoiding unnecessary regulatory burden

D. W. Eddy and I. N. Haynes, Dairy Food Safety Victoria, Australia

Abstract: For food safety regulators, ensuring that that the public are protected from unsafe food products is a given expectation. But what constitutes being 'unsafe'? The presence of a microbial pathogen alone does not determine that the product is unsafe. The level of a microbial pathogen and the characteristics of the dairy food in which it is contained will determine whether the microbe will die, survive or be able to grow and reach numbers that could constitute an infectious dose at the point of consumption. Susceptibility of the consumer also plays a significant role in determining whether the product could be determined unsafe. Understanding the circumstances of each case and determining the relative associated risk is an approach that the regulator can take to ensure that risk to the public is minimised, and to avoid unnecessary regulatory burden on industry.

Key words: dairy products, risk assessment, strain typing.

26.1 Introduction to the cases

Australia's dairy industry produces about 9 billion litres of milk each year, of which around 65% comes from the southern state of Victoria (Dairy Australia, 2011). The dairy industry businesses in this state (farms, milk carriers, manufacturers and distributors) are required to be licensed by the statutory authority, Dairy Food Safety Victoria (DFSV). This authority is responsible for ensuring that standards which safeguard public health are maintained in the Victorian dairy industry, whilst ensuring that it performs its functions and exercises its powers efficiently and effectively.

Within the authority, an Operations team deals with the aspects of licensing the dairy businesses and verifying licensees' compliance with approved hazard analysis and critical control point (HACCP)-based food safety programs. Uncharacteristic, however, of similar regulatory organisations, DFSV also has a separate Technical team, which provides assistance to the industry through standards interpretation and guidance as well as technical information on industry queries and problems. Over recent years, DFSV has endeavoured to move toward the adoption of a partnership role with the industry. This has been facilitated by the Technical team through providing a broadening of industry knowledge and understanding wherever possible. The compliance and enforcement activities of the Operations team remain a separate stream within DFSV.

Regulators will at times come across circumstances where the results of the microbiological testing indicate a failure of the defined Standards for a particular food type. Sometimes, the action to be taken, as required by the regulator, may not be a good reflection of the true level of risk involved. In such circumstances, the manufacturers, especially the smaller artisanal cheese makers, can lose a significant proportion of their income through the costs incurred in having to dispose of the implicated product.

26.2 Significance of the cases

The cases presented in this chapter address some significant questions. Can a boundary be drawn, as distinct from the prescribed Standards, to effectively determine what constitutes an unsafe product? Is it possible to have a pathogenic microbe present and yet still have a safe product? Can strict adherence to prescriptive standards, in circumstances where the actual outcome sought is protection of public health, be moderated so as not to cause small manufacturers excessive financial hardship whilst ensuring public safety? It was considered that, with the advent of cost-effective and reliable molecular identification techniques, it should be possible for regulators to be reassured that any deviations from the prescriptive nature of the regulations are balanced by scientific validation of the effectiveness of food safety programs. This approach also aligns with the concepts used in microbial risk management strategies, whereby food safety objectives (FSO) set at the point of product consumption are more informative than public health standards.

26.2.1 Regulatory aspects

Of particular interest to our dairy industry are the pathogens *Listeria monocytogenes* and *Escherichia coli*. In Australia, the only stipulated requirement for *L. monocytogenes* in cheeses is for soft cheese (>39% moisture and >5.0 pH), where the organism must be absent in 25 g of the product (Food Standards Australia New Zealand, 2011a). A user guide document suggests, however, that *L. monocytogenes* should also be absent in 25 g (or 25 ml) of pasteurised milk and cream, dried milk powder, ice cream and ice cream mixes, and infant formulae

(Food Standards Australia New Zealand, 2011b). In the state of Victoria, an industry Code of Practice for Dairy Food Safety mandates that all pathogen levels specified in this Guide must be met (Dairy Food Safety Victoria, 2002). It also requires that manufacturers of all dairy products comply with the Australian Manual for Control of *Listeria* in the Dairy Industry (Australian Dairy Authorities' Standards Committee, 1999), which involves advising the regulator of detections, and initiating clearance and corrective action procedures.

The Food Standards Code Standard 1.6.1 clause 5 also requires that all cheese products be tested for *E. coli*, which may be present (at a level above 10, but less than 100/g) in no more than one of the mandatory five samples tested. Most *E. coli* organisms appear to exist as harmless commensals, and of no threat to human health (Donnenberg and Whittam, 2001). However, we contend that the emergence of particularly pathogenic strains of *E. coli* causing major outbreaks of illness throughout the food industry in recent years clearly indicate that the need to test for *E. coli* is still warranted.

26.2.2 Utilising technology to identify bacterial strains

Hazards exist in all aspects of life. The level of those hazards in the specific circumstances will determine what degree of risk is present. Determining the boundary between the point where the public is exposed to undue risk and the point where the level of risk is so low that the chance should be permitted remains problematic.

The victorian dairy industry has progressed from a quality control system to a quality assurance system over the last 25 years. However, when verification testing of a quality assurance process indicates that contamination has occurred, the prescriptive basis of a quality control system seems to take over. The nature of quality assurance means that each unit process and step within the production system has controls applied to it in order to prevent any problems from occurring. If end-product testing indicates that contamination is present, it is highly unlikely that every step of the process has failed. It is most likely that only one step has an issue that allowed the contamination to occur. The resulting contamination is most likely to be confined to a small range of unwanted organisms. If tests can be undertaken to demonstrate that they are limited in diversity and not pathogenic, then consideration should be given to including a risk analysis of this in the overall assessment for accepting or rejecting the product.

26.3 The case histories continue

26.3.1 *L. monocytogenes* in cheese for the pizza market

In 2009, DFSV commenced working with the official public health microbiological laboratory, the Microbiological Diagnostic Unit (MDU) at the University of Melbourne, to apply genetic typing to all strains of *L. monocytogenes* isolated from dairy product samples. This same laboratory performs the typing of isolates detected in human cases of illness. The tests being applied included pulsed-field gel electrophoresis (PFGE), ribotyping, binary typing, molecular serotyping and multi-locus sequence typing (MLST). The combination of these methods is providing a basis for determining if more rapid methods can match the PFGE method for reliable differentiation between strains.

Around the same time, a manufacturer licensed with DFSV was experiencing intermittent detections (present in 25 g) of *L. monocytogenes* in shredded semi-hard to hard cheese varieties that were destined for the pizza market. Efforts to quantify the level of contamination repeatedly came back with a result of 'not detected' (<10/g). Given that this product was to undergo further heat treatment in a pizza oven at around 180 °C for some minutes and the compositional parameters (pH and A_w) of the product, risk analysis by DFSV in this case indicated that the likelihood of survival of any *L. monocytogenes* present would be remote. The possibility of the bacteria surviving at levels which could constitute an infectious dose if the product was consumed would be unlikely.

The health department, ultimately responsible for taking the regulatory decision with respect to this food product, responded to this situation strictly in the only way possible according to the prescribed legislative requirements, and required that the product not be released for sale. This resulted in many tonnes of product being destroyed.

The specific *L. monocytogenes* strain isolated in this case by the testing laboratory was submitted for molecular typing by the range of tests being applied at that time. The subsequent result indicated that this particular dairy isolated strain had not been linked with human illness (personal communication). Since that time in 2009, other closely related strain isolations from this same manufacturing facility have occurred. All occurrences have been responded to, but the particular strain(s) involved have not yet been linked to cases of human illness.

In almost 3 years, since the molecular fingerprinting project was initiated, there have been approximately 150 samples of dairy product, mainly further processed (cut, shredded or grated) hard cheese, that have tested positive for the presence of *L. monocytogenes*. Interestingly, in these types of product, *L. monocytogenes* is not expected to grow due to pH and water activity hurdles, and will most probably die out slowly. The effectiveness of the HACCP-based food safety programs implemented by manufacturers has indicated that the overall production and processing systems in place are able to produce safe product. Whilst the possibility of *L. monocytogenes* contaminating a product is low, the capability of it being detected in qualitative testing (presence in 25 g of product) remains relatively good. The results of industry contaminations, however, indicate that the possibility of being able to quantify the level of *L. monocytogenes* contamination from a qualitative positive sample is very low given that the vast majority of quantitative testing yields a negative result.

26.3.2 The application of qualitative risk assessment

Whether these low-level contaminations constitute sufficient risk still needs to be determined. None of the molecular fingerprints from these 'dairy' isolates match any strain of *L. monocytogenes* that has been responsible for human illness cases over that period of time. This suggests that, whilst a product may contain very low levels of pathogenic microbes, this is not necessarily an indication of the likelihood of the product to cause illness in the consumer.

In order to support this understanding, DFSV has recently undertaken to develop an approach to enable the application of a qualitative risk assessment to each case where a dairy product test result indicates contamination with a pathogen that contravenes the prescriptive requirements of the Australian Food Standards Code. The specific circumstances of the detection, the product composition and batch details, units produced and likely use of the product are all considered in this assessment. In each case, the results of this qualitative risk assessment, and the recommended actions, have then been passed on to the state Department of Health, which is ultimately responsible for declaring whether the product may be released for sale or must be withdrawn or recalled if already on the market. This process has resulted in closer communications and an increasing level of trust developing between the dairy authority and the health department, which has proven to be a very good means for knowledge sharing.

26.3.3 *E. coli* in cheese

In another incident, a small cheese-maker received results of *E. coli* testing that indicated that the product was over the limit set by the product standard. In this instance, the product was required to be destroyed and not offered for sale. Subsequent testing that our Authority commissioned with a university indicated that only one *E. coli* strain was identical throughout the product and that it was not a pathogenic type. In this instance, knowledge of this additional information would have offered the potential for a manufacturer to understand that corrective action needs to be undertaken, but that the product does not constitute a health risk to the consumer. The potential for additional information to be used to confirm or refute a claim as to whether a product is unsafe should be more fully investigated.

26.4 Resolutions and outcomes

Food safety standards are written to provide levels for microbial pathogens that producers and processors need to confirm products comply with in order to ensure the products are safe to consume. Our work indicates that a prescriptive approach can sometimes place an unfair burden on a manufacturer. Our focus is always to ensure public health and safety, but, where the application of reliable new technology enables greater understanding of the actual risk, we should consider being more flexible in our overall approach.

All dairy businesses in Australia are required to have HACCP-based food safety programs (FSP) in place. These FSP ensure that safety and quality parameters that are required for the protection of consumers are managed within the production and processing environment. Regulators protect the consumer by ensuring that the approved food safety system will result in safe product and that no unsafe product reaches the market. There is also a role for the regulator in assisting in the education of manufacturers to reduce the likelihood of adverse food safety incidents.

It is clearly recognised that the detection of potential pathogens in dairy products indicates that there has been a breakdown in the food safety system of the manufacturer concerned. Immediate remedial action is required to establish the facts, identify the reason for the contamination, and implement corrective actions to prevent a possible recurrence. When regulators can apply a flexible approach using appropriate tools and work cooperatively with the manufacturer, problems are often resolved more efficiently than would otherwise have been the case. Such an approach also increases overall knowledge which can be applied more broadly. This helps management focus on the causes, rather than just eliminating the immediate problem.

26.5 Commentary

Whilst standards exist to protect public health and safety, they tend to be general in scope and not always applicable to each and every product in the category for which they are written. They should be flexible enough in their application to recognise the benefit of additional information and the specific circumstances that may influence whether a food product is indeed unsafe.

Large manufacturers may have the resources and time to conduct further testing to show that, under a given set of circumstances, a product that may contravene the standard does not represent a risk to the consumer. The application of additional testing and research may greatly assist the industry and the regulator to better understand where to draw the 'line in the sand' between safe and unsafe product. But small to medium food businesses are highly unlikely to have this resource capability. Whilst there is clearly a need to ensure that products do not have pathogenic *E. coli* present, it is somewhat detrimental to industry if the presence of non-pathogenic *E. coli* demands removal of that product. The role of the regulator is to apply controls where risks exist, not necessarily to apply a simple approach that makes regulatory decisions easier to implement.

Modern genetic testing methods are becoming more reliable, rapid and cost-effective; making greater use of these tools in regulatory decision-making would appear to be warranted.

Regulators and standards authorities should consider whether access to additional testing might provide greater clarity in their decision-making process.

26.6 Critical questions for discussion

1. Should set standard criteria apply equally to all products without consideration of product variation or degrees of safety to individuals?

2. What types of microbiological or genetic testing could be applied to provide greater definition to regulators between which products are safe or unsafe to the consumer when investigating qualitative product contaminations?
3. Should any additional testing process be able to benefit the business if results indicate that contamination is not from a pathogenic strain?
4. Should a manufacturer be able to claim losses against a regulator for the ordered destruction of product that may ultimately prove not to have been unsafe?
5. The question has been raised as to whether *E. coli*, as an indicator organism, should be removed from standards or regulations. Does having to test for *E. coli* within industry any longer serve a purpose?

26.7 References

AUSTRALIAN DAIRY AUTHORITIES' STANDARDS COMMITTEE (1999). *Australian Manual for the Control of* Listeria *in the Dairy Industry (*Listeria *Manual)*. Available at http://www.dairysafe.vic.gov.au/.

DAIRY AUSTRALIA (2011). *Latest Production and Sales Statistics*. Available at http://www.dairyaustralia.com.au/Statistics-and-markets/Production-and-sales/Latest-Statistics.aspx.

DAIRY FOOD SAFETY VICTORIA (2002). *Code of Practice for Dairy Food Safety. Section 5.3: Standards*. Available at www.dairysafe.vic.gov.au/pdf/DFSV_CodeOfPractice2002.pdf.

DONNENBERG M.S., and WHITTAM T.S. (2001). Pathogenesis and evolution of virulence in enteropathogenic and enterohemorrhagic *Escherichia coli*. *Journal of Clinical Investigation*, **107**(5), 539–48.

FOOD STANDARDS AUSTRALIA NEW ZEALAND (2011a). *Food Standards Code (2011) Standard 1.6.1, Clause 5 – Microbiological Limits for Food*. Available at www.foodstandards.gov.au/thecode/foodstandardscode.cfm.

FOOD STANDARDS AUSTRALIA NEW ZEALAND (2011b). *User Guides to the New Food Standards Code (2011) User Guide to Standard 1.6.1 – Microbiological Limits for Food*. Available at www.foodstandards.gov.au/thecode/assistanceforindustry/userguides/index.cfm.

27

Learning from outbreaks of *Escherichia coli* O157:H7 caused by low pH foods

P. Feng, Food and Drug Administration (FDA), USA

Abstract: Most bacteria cannot grow or survive in low pH environments; hence, increasing the acidity of foods, either through fermentation such as in making sausage, cheese and other dairy products or by the addition of weak acids, such as in pickling, has been used as a method of food preservation since ancient times. Accordingly, foods that have low pH have often been regarded as being low-risk in terms of food safety. This chapter examines the safety of acidic foods and the effectiveness of low pH in controlling enterohemorrhagic *Escherichia coli* (EHEC) serotype O157:H7, a pathogen that has caused foodborne illnesses worldwide.

Key words: acidity, foodborne illness, *Escherichia coli* O157:H7.

27.1 Introduction

Enterohemorrhagic *Escherichia coli* (EHEC) has emerged as a predominant cause of hemorrhagic colitis (HC), which can progress to the life-threatening complication known as hemolytic uremic syndrome (HUS). With a few exceptions, EHEC pathogenicity is associated with the production of several virulence factors, including verotoxins or Shiga toxins (Stx) and the intimin protein, which enables the pathogen to attach to intestinal epithelial cells. Serotype O157:H7 is the predominant pathogen in the EHEC group that is often associated with human infections worldwide.

Serotype O157:H7 was first implicated in a foodborne outbreak in 1982 linked to the consumption of contaminated hamburgers at a fast-food restaurant chain in the US. Subsequent outbreaks, including the large outbreak in 1993 that infected more than 700 persons, have mostly been associated with the consumption of bovine products, giving the impression that it was a 'burger pathogen'. However, in the almost 30 years since being recognized as a human pathogen, serotype

O157:H7 infections have implicated a large variety of vehicles, such as water, salad vegetables, sprouts, spinach, cookie dough, and, seemingly unlikely vehicles, acidic foods (Feng, 1995).

Most bacteria grow optimally at a neutral pH of 7.0. Hence, by tradition, foods that have high acidity (a pH of less than 4.6) have been regarded as low-risk in terms of food safety. However, an outbreak of O157:H7 in 1991 that affected 23 persons, with four cases of HUS, was traced to the consumption of fresh-pressed apple cider, which had a pH value of 3.7–3.9 (Besser *et al.*, 1993). Although the source of the O157:H7 was not found, it was suspected that the implicated cider, made at a farm in Massachusetts, used unwashed, 'dropped' apples, which may have come into contact with cattle manure. Locally produced apple juice was also suspected to have caused a much earlier outbreak in 1980 in Toronto (Canada), which resulted in 14 HUS cases in children. Patient stools tested negative for known enteric pathogens and viruses, but were not tested for EHEC, as this pathogen was relatively unknown at that time (Steele *et al.*, 1982).

When cider was epidemiologically linked to the 1991 outbreak, many food safety experts deemed juices as unlikely vehicles due to their high acidic content. However, within a few years, the ability of O157:H7 to cause infections via acidic vehicles was substantiated when other acidic foods were implicated in outbreaks worldwide. In the fall of 1991, O157:H7 caused an outbreak in England, where 16 children were infected and five cases progressed to HUS (Morgan *et al.*, 1993). The source of the outbreak was epidemiologically linked to the consumption of yogurt, which usually has a pH of 4.0–4.1. The implicated product was made locally using pasteurized milk; hence, it was speculated that the milk was inadequately pasteurized or, perhaps, that it was contaminated post-pasteurization.

In 1993, a series of restaurant outbreaks of O157:H7 in Oregon and Washington was confirmed to have infected at least 48 persons, although it was estimated to have affected ~300 (Jackson *et al.*, 2000). The source of the outbreaks was not identified but investigations implicated items served in the salad bar, especially mayonnaise and mayonnaise-based dressing and sauces (pH of 3.6–4.4). How the mayonnaise was contaminated with O157:H7 was not determined; however, bulk mayonnaise appeared to have been improperly handled and was cross-contaminated with raw meat. Thus, although beef consumption was not linked epidemiologically to these outbreaks, it was suspected that raw meat was the source of O157:H7 and that the outbreaks were caused by lapses in proper food handling.

The spectrum of acidic food vehicles associated with O157:H7 infections was extended in 1994 when dry-cured salami was implicated in an outbreak in Washington and California that affected 23 persons with two cases of HUS (CDC, 1995). Dry-cured salami is not cooked but it is considered a ready-to-eat (RTE) food because it is made via fermentation that drops the product pH followed by dry-curing; both regarded as bacteriostatic processes. However, the profile of the O157:H7 isolates cultured from the implicated salami samples matched those obtained from the patients, thereby confirming the link of salami with the outbreak.

Similarly, cheese is another RTE food because the initial fermentation of milk drops the pH and the subsequent aging process helps to reduce bacteria levels.

However, in 1998, fresh cheese curds made at a Wisconsin dairy caused an outbreak of O157:H7 that affected 55 people, all having symptoms of bloody diarrhea (CDC, 2000). The cheese curds were made using pasteurized milk as required by law, but O157:H7 was isolated from an open package of curds that was implicated in the outbreak. It was suspected that cheese curds were inadvertently made in vats that had previously been used to make batches of raw milk cheddar cheese and became contaminated with raw milk that had O157:H7.

27.2 Significance of the cases

'Potentially hazardous food' is a term often used by food safety agencies to classify foods that require time–temperature control to keep them microbiologically safe for human consumption. Among the factors used to define potentially hazardous food are moisture or water activity (a_w) levels greater than 0.85, as bacteria usually require at least a_w of 0.91 for growth and neutral to slightly acidic conditions, typically between pH 4.6–7.5, a range that is suitable for bacterial growth. As such, pH 4.6 has been used to delineate between 'low-acid food' (pH >4.6) and high-acid food (pH <4.6) and often perceived as a critical limit for bacterial growth or no growth. However, the rash of O157:H7 outbreaks in the 1990s that implicated acidic foods suggested that this pathogen may have some unexpected acid resistance properties.

The apple cider outbreak of 1991 prompted many to study growth characteristics of O157:H7 in acidic conditions. The results of these studies showed that O157:H7 can tolerate high acidity and some strains can even persist in culture media with pH as low as 2.0 (Miller and Kaspar, 1994). In foods, several studies showed that O157:H7 inoculated into cider became undetectable after 7 days at 25 °C, but at refrigerator temperatures (8 °C), slight growth was observed and the organism was still detectable after 20 days (Besser *et al.*, 1993) and may persist up to 31 days (Zhao *et al.*, 1993). Addition of preservatives had mixed results in that 0.1% potassium sorbate had no effect (Miller and Kaspar, 1994; Zhao *et al.*, 1993) but 0.1% sodium benzoate did decrease O157:H7 levels and survival rates in the cider. Studies performed after the mayonnaise outbreaks showed similar results, where O157:H7 inoculated into mayonnaise were not able to grow, but could survive and persist depending on holding temperature (Hathcox *et al.*, 1995). Various levels of O157:H7 inoculated into mayonnaise samples and held at 25 °C became undetectable within days but, if held at 7 °C, they could survive for up to 35 days or longer (Weagant *et al.*, 1994; Zhao *et al.*, 1993).

Previous studies showed that *Salmonella* cells exposed to HCl at pH 5.8 also adapted to acidity and developed increased resistance to inactivation by organic acids found in cheese (Leyer and Johnson, 1992). Analysis of O157:H7 showed very similar results in that acid-adapted cells had increased resistance to the lactic acid produced during sausage fermentation and these cells also showed enhanced survival in both salami and cider as compared with non-adapted cells (Leyer *et al.*,

1995). Interestingly, the property of acid tolerance seems to be fairly prevalent, as one study showed that most O157:H7 strains isolated from bovine carcasses had been acid-adapted and had the capacity to survive in low pH environments (Berry *et al.*, 2004).

These various studies conclusively demonstrated that O157:H7 can be induced or adapted to tolerate low pH conditions and, although they do not appear to grow in low pH, their ability for survival in acidity is especially prolonged at refrigeration temperatures.

27.2.1 Regulatory, economic and market aspects

The US FDA 'Food Code' (FDA, 2009) provides scientifically based guidelines and protocols that are used by almost all food control jurisdictions to regulate the safety of foods served at the retail and food service industries, like restaurants and grocery stores, as well as institutional food services such as cafeterias and nursing homes. In the Food Code, Chapter 3, Part 5 – Limitation of Growth of Organisms of Public Health Concern, the pH value of 4.6 is significant because *Clostridium botulinum* cannot grow and produce toxin below this value. Therefore, it has been used as one of the criteria to define potentially hazardous foods and, from a safety point of view, foods that have pH <4.6 have often been deemed as low-risk. However, that perception, as well as some food processing and production practices, has been altered by the O157:H7 outbreaks that implicated acidic foods.

Juices

Most juices manufactured in the US are pasteurized; however, some specialty juice manufacturers and seasonal producers, such as those locally making cider and fruit juices, do not pasteurize their products. The 1991 cider outbreak revealed interesting observations: that apple cider is a viable vehicle for O157:H7 transmission, that not all preservatives are effective for O157:H7, and that low pH alone is insufficient to control O157:H7 in cider.

Consistent with these observations, unpasteurized apple juice caused another large O157:H7 outbreak in 1996 in Washington (USA) and British Columbia (Canada) that affected over 70 persons with 14 cases resulting in HUS and one fatality (Cody *et al.*, 1999). At about the same time, a large *Salmonella* outbreak that infected 62 persons in 21 states in the US was traced back to unpasteurized orange juice sold at an amusement park. As a result of these juice-related outbreaks, the FDA proposed to mandate pasteurization of all juices, but the proposal was rejected by dissenting opinions from the industry and consumers. Instead, the FDA enacted a food-labeling rule requiring health risk warning labels on containers of unpasteurized or untreated juices (FDA, 1998) and, later, implemented a juice hazard analysis and critical control points (HACCP) regulation that required the industry to use intervention steps that will reduce target pathogen levels by 5 logs in the processing of unpasteurized juices (FDA, 2001).

Mayonnaise

In accordance with Federal regulations, commercial mayonnaise made with unpasteurized eggs must have pH of ≤ 4.1 and 1.4% acetic acid in the aqueous phase. Most manufacturers have since switched to using pasteurized eggs and also, to lower the tartness of reduced-calorie dressings, the acetic acid level was dropped to 0.7%. Studies have shown that mayonnaises that meet these criteria and are made with good manufacturing practices are microbiologically safe. So, the implication of mayonnaise in the O157:H7 outbreak in 1993 was unexpected. Studies performed following the outbreak showed that mayonnaise does not support the growth of O157:H7 (Hathcox *et al.*, 1995) and, although the organism can persist in mayonnaise held at refrigeration temperatures (Hathcox *et al.*, 1995; Weagant *et al.*, 1994), it does not survive well at 25 °C, the normal processing and storage conditions for unopened mayonnaise (Hathcox *et al.*, 1995). These findings are consistent with the position that this product is generally safe microbiologically and that the 1996 outbreak was caused by improper handling that resulted in cross-contamination (Jackson *et al.*, 2000).

Cheese and other dairy foods

Fermentation of dairy products as a means of preservation has been commonly practiced by all the cultures in the world for centuries. For instance, yogurt is thought to have originated among the nomadic tribes of Eastern Europe and Western Asia, when they discovered that milk stored in animal skins would acidify and coagulate and could be kept for a long time. It is now well known that the acid produced from milk fermentation helped to preserve it from spoilage and from the growth of pathogens, so it was a practical means for maintaining product stability and safety.

In cheese making, the fermentation of lactose sugar in milk by lactic acid bacteria will lower the pH to ~4.6, but in the subsequent ripening processing the pH may rise so that the final product typically has a pH range of 5.1–5.7. The rise in pH is especially prominent in soft mold-ripened cheeses, where mold activity can raise the pH to neutral levels (pH 7.4). In the US, soft and fresh cheeses are required to be made using pasteurized milk. However, with some hard cheeses, such as cheddar and parmesan, and semi-soft cheeses like Gouda, raw milk is allowed to be used provided that the final product has been held at ≤ 1.7 °C for at least 60 days to allow curing, which also reduces pathogen levels (FDA, 2010). This requirement was established many years ago using another pathogen as a model and is still in effect and used by cheese manufacturers in various countries (Honish *et al.*, 2005).

Cattle are a well-known reservoir of O157:H7, so this pathogen has often been found on meats and products derived from cattle, including raw milk. Consumption of contaminated raw milk has been implicated in several outbreaks of O157:H7 worldwide, which raised the safety concerns of raw milk cheeses. Studies performed to investigate the survival of O157:H7 in the manufacture and curing of raw milk cheddar cheeses showed that the pathogen inoculated into milk can persist to the final product and be detected even after 158 days (Reitsma and

Henning, 1996; Schlesser *et al.* 2006). Similar studies done on aged Gouda and stirred-curd cheddar cheeses confirmed that, even at the low inoculum level of 20 cfu/g, O157:H7 can still be detected after 270 days (D'Amico *et al.*, 2010). These studies clearly showed that the 60-day aging requirement for raw milk cheeses is, by itself, inadequate to eliminate O157:H7 and that other intervention steps are needed to ensure the safety of raw milk cheeses.

Fermented meats

As with cheese, fermentation of meats is a common means of food preservation practiced by many cultures around the world, perhaps as far back as 1500 BC. Fermented meats are made by adding salt, sugar, spices and starter cultures to ground meats. The fermentation of sugar by bacteria produces lactic acid, which drops the pH to 4.8, and the subsequent drying/curing reduces moisture content of the product. It was a common belief that foods processed under low pH and low A_w and having high salinity content did not support pathogen growth; thus products like salami and sausages did not require cooking and are considered RTE. Hence, the 1994 salami outbreak puzzled many experts and prompted several hypotheses on the causes of the outbreak. One speculation was that O157:H7 was present on the raw meat but survived due to lapses in processing. Others suspected post-processing contamination, and yet another speculation was that O157:H7 can survive normal salami processing conditions (Tilden *et al.*, 1996). The first two hypotheses were effectively ruled out by investigations which showed that the implicated salami was properly made and handled and was not widely contaminated (Tilden *et al.*, 1996). To test the last hypothesis, many studies examined the survival of O157:H7 in fermented meats and found that, although O157:H7 inoculated into sausage batter will not grow, their tolerance for acidity enabled them to survive the fermentation and dry-curing processes used in sausage making (Glass *et al.*, 1992). Consistently, follow-up investigations of an O157:H7 outbreak with sausages in Sweden in 2002 (Sartz *et al.*, 2008) showed that the lack of a heat-treatment step, delayed addition of the starter culture that allowed O157:H7 to proliferate, and a short curing period were the factors that contributed to the outbreak. In response to the findings that sausage-processing conditions may be inadequate, the US Department of Agriculture (USDA) developed regulatory guidelines requiring manufacturers to include processes that will ensure a 5-log reduction of O157:H7 during the production of fermented meats.

Contaminated raw meat was the most likely source of O157:H7 in the salami and sausage outbreaks. The natural reservoir of O157:H7 is the intestinal tracts of cattle; and fecal contamination of meats often occurs during slaughter. In an effort to reduce such contamination, the USDA implemented intervention steps, including the use of acid washes on carcasses to reduce microbial load and to lower the pH to inhibit bacterial growth. While effective, it was interesting to note from a study that the majority of O157:H7 strains isolated from commercially slaughtered carcasses exhibited the ability to tolerate high acidity (Berry *et al.*, 2004).

27.3 The case history continues

The 1991 apple cider outbreak raised our awareness that acidic foods can serve as transmission vehicles for O157:H7, a fact that has since been more than substantiated by the many acidic vehicles that have caused O157:H7 outbreaks. In response to each of these incidents, new rules and intervention strategies were implemented and significant processing changes were also made by the industry. However, sporadic outbreaks of O157:H7 continue to implicate acidic foods (Hussein and Sakuma, 2005). For instance, Genoa salami and salami outbreaks in Canada in the late 1990s affected between 39 and 143 persons (Moore, 2004). Also in Canada, a Gouda cheese outbreak in 2002 affected 13, including two children who developed HUS (Honish *et al.*, 2005), and the raw milk-Gouda cheese outbreak in 2010 in the Western US made ~50 sick, to name a few.

27.4 Resolutions and outcomes

The implication of various acidic foods in O157:H7 outbreaks has had substantial impact on our food safety regulatory framework and on the manufacturing practices of the food industry. These outbreaks have also stimulated considerable scientific research, which showed that O157:H7 will not grow in most acidic foods, but can tolerate acidity and persist for a long time, especially at refrigeration temperatures. In some instances, the acidic foods themselves were safe, but were implicated in the outbreaks due to improper handling and cross-contamination. However, in other cases, these outbreaks pointed out inadequacies in rules and food processing conditions so that perhaps a review of existing regulations and guidelines, for example those relating to the manufacture of raw milk cheeses and fermented meats, may be warranted.

27.5 Commentary

This chapter only focused on the implication of acidic foods in outbreaks of O157:H7. There are, however, many other EHEC serotypes that can cause severe human illness and some of these serotypes have already caused outbreaks implicating acidic foods. For example, O26:H11 in fermented beef sausages caused an outbreak in Denmark in 2007 (Ethelberg *et al.*, 2009) and an O103:H2 outbreak of HUS in Norway in 2006 was linked to mutton sausages, suggesting that these non-O157 EHEC serotypes may also have the ability to tolerate acidic conditions. However, additional research will be required to determine if these non-O157 EHEC serotypes are as resilient as O157:H7 in acid environments and, if so, may be implicated in more acidic food outbreaks in the future.

27.6 Acknowledgment

The author thanks Dr S. Monday and Dr J. Sheehan for critical reading of this paper.

27.7 Critical questions for discussion

1. Should the 60-days requirement to ripen raw milk cheeses be extended?
2. What intervention steps should be implemented in the making of raw milk cheeses to reduce health risks?
3. Should preservatives or other intervention steps be used in fresh unpasteurized cider to ensure safety?
4. What intervention steps should be implemented in the making of fermented meat products to reduce health risk?
5. Is the acid spray wash commonly used in the US on carcasses to reduce microbial load inducing acid tolerance in O157:H7?
6. Do the non-O157 EHEC serotypes have the same ability as O157:H7 to tolerate acidity and will they be implicated more often in acidic food outbreaks?
7. What is the likelihood that low-pH foods, other than those mentioned above, can sustain survival of EHEC and cause infections?
8. What is the pH of RTE salsa, fruit salads and other similar products that are made with a variety of fresh produce and fruits?
9. Some frozen fruit preparations can have pH ~3.0 or less. What is the likelihood of these products causing O157:H7 infections?

27.8 References

BERRY E.D., BARKOCY-GALLAGHER G.A. and SIRAGUSA G.R. (2004). Stationary-phase acid resistance and injury of recent bovine *Escherichia coli* O157 and non-O157 biotype I *Escherichia coli* isolates. *J Food Prot*, **67**: 583–90.

BESSER R.E., LETT S.M., WEBER J.T., DOYLE M.P., BARRETT T.J. *et al.* (1993). An outbreak of diarrhea and hemolytic uremic syndrome from *Escherichia coli* O157:H7 in fresh-pressed apple cider. *J Am Med Assoc*, **269**: 2217–20.

CENTERS FOR DISEASE CONTROL AND PREVENTION (1995). *Escherichia coli* O157:H7 outbreak linked to commercially-distributed dry-cured salami – Washington and California, 1994. *Morb Mortal Weekly Rep*, **44**: 157–60.

CENTERS FOR DISEASE CONTROL AND PREVENTION (2000). Outbreak of *Escherichia coli* O157:H7 infection associated with eating fresh cheese curds – Wisconsin, June 1998. *Morb Mortal Weekly Rep*, **49**: 911–13.

CODY S.H., GLYNN M.K., FARRAR J.A., CAIRNS K.L., GRIFFIN P.M. *et al.* (1999). An outbreak of *Escherichia coli* O157:H7 infection from unpasteurized commercial apple juice. *Ann Intern Med*, **130**: 202–9.

D'AMICO D.J., DRUART M.J. and DONNELLY C.W. (2010). Behavior of *Escherichia coli* O157:H7 during the manufacture and aging of Gouda and stirred-curd cheddar cheeses manufactured from raw milk. *J Food Prot*, **73**: 2217–24.

ETHELBERG S., SMITH B., TORPDAHL M., LISBY M., BOEL J. *et al.* (2009). Outbreak of non-O157 Shiga toxin-producing *Escherichia coli* infection from consumption of beef sausage. *Clin Infect Dis*, **48**: e78–e81.

FENG P. (1995). *Escherichia coli* serotype O157:H7: Novel vehicles of infection and emergence phenotypic variants. *Emerg Infect Dis*, **1**: 16–21.

FOOD AND DRUG ADMINISTRATION (1998). Food labelling: Warning and notice statement: Labelling of juice products; Final rule. CFR 21 Part 101. *Federal Register*, **63**(130): 37029–56.

FOOD AND DRUG ADMINISTRATION (2001). Juice HACCP Final rule. 21 CFR Part 120. *Federal Register*, **66**(13): 6137–202.

FOOD AND DRUG ADMINISTRATION (2010). 21 *CFR Part 133 – Cheese and Related Cheese Product, Subpart B—Requirements for Specific Standardized Cheese and Related Products, Section 133.113. Cheddar cheese*. Available at http://www.accessdata.fda.gov/scripts/cdrh/cfdocs/cfcfr/CFRSearch.cfm?fr=133.113.

FOOD AND DRUG ADMINISTRATION (2009). *FDA Food Code*. Available at http://www.fda.gov/Food/FoodSafety/RetailFoodProtection/FoodCode/FoodCode2009/ucm186451.htm part3-5.

GLASS K.A., LOEFFELHOLZ J.M., FORD J.P. and DOYLE M.P. (1992). Fate of *Escherichia coli* O157:H7 as affected by pH or sodium chloride and in fermented, dry sausage. *Appl Environ Microbiol*, **58**: 2513–16.

HATHCOX A.K., BEUCHAT L.R. and DOYLE M.P. (1995). Death of enterohemorrhagic *Escherichia coli* O157:H7 in real mayonnaise and reduced-calorie mayonnaise dressing as influenced by initial population and storage temperature. *Appl Environ Microbiol*, **61**: 4172–7.

HONISH L., PREDY G., HISLOP N., CHUI L., KOWALEWSKA-GROCHOWSKA K. *et al.* (2005). An outbreak of *Escherichia coli* O157:H7 hemorrhagic colitis associated with unpasteurized gouda cheese. *Can J Public Health*, **96**: 182–4.

HUSSEIN H.S. and SAKUMA T. (2005). Prevalence of Shiga toxin-producing *Escherichia coli* in dairy cattle and their products. *J Dairy Sci*, **88**: 450–65.

JACKSON L.A., KEENE W.E., MCANULTY J.M., ALEXANDER E. R., DIERMAYER M. *et al.* (2000). Where's the beef? The role of cross-contamination in 4 chain restaurant–associated outbreaks of *Escherichia coli* O157:H7 in the Pacific Northwest. *Arch Intern Med*, **160**: 2380–5.

LEYER G.J. and JOHNSON E.A. (1992). Acid adaptation promotes survival of *Salmonella* spp. in cheese. *Appl Environ Microbiol*, **58**: 2075–80.

LEYER G.J., WANG L.L. and JOHNSON E.A. (1995). Acid adaptation of *Escherichia coli* O157:H7 increases survival in acidic foods. *Appl Environ Microbiol*, **61**: 3752–5.

MILLER L.G. and KASPAR C.W. (1994). *Escherichia coli* O157:H7 acid tolerance and survival in apple cider. *J Food Prot*, **57**: 460–4.

MOORE, J.E. (2004). Gastrointestinal outbreaks associated with fermented meats. *Meat Sci*, **67**: 565–8.

MORGAN D., NEWMAN C.P., HUTCHINSON D.N., WALKER A.M., ROWE B. *et al.* (1993). Verotoxin producing *Escherichia coli* serotype O157 infections associated with the consumption of yogurt. *Epidemiol Infect*, **111**: 181–7.

REITSMA C.J. and HENNING D.R. (1996). Survival of enterohemorrhagic *Escherichia coli* O157:H7 during the manufacture and curing of cheddar cheese. *J Food Prot*, **59**: 460–4.

SARTZ L., DE JONG B., HJERTQVIST M., PLYM-FORSHELL L., ALSTERLUND R. *et al.* (2008). An outbreak of *Escherichia coli* O157:H7 infection in southern Sweden associated with consumption of fermented sausage; aspects of sausage production that increase the risk of contamination. *Epidemiol Infect*, **136**: 370–80.

SCHLESSER J.E., GERDES R., RAVISHANKAR S., MADSEN K., MOWBRAY J. *et al.* (2006). Survival of a five-strain cocktail of *Escherichia coli* O157:H7 during the 60-day aging period of cheddar cheese made from unpasteurized milk. *J Food Prot*, **69**: 990–8.

STEELE B.T., MURPHY N., ARBUS G.S. and RANCE C.P. (1982). An outbreak of hemolytic uremic syndrome associated with ingestion of fresh apple juice. *J Pediatr*, **101**: 963–5.

TILDEN J. JR., YOUNG W., MCNAMARA A.M., CUSTER C., BOESEL B. *et al.* (1996). A new route of transmission for *Escherichia coli*: Infection from dry fermented salami. *Am J Public Health*, **86**: 1142–5.

WEAGANT S.D., BRYANT J.L. and BARK D.H. (1994). Survival of *Escherichia coli* O157:H7 in mayonnaise and mayonnaise-based sauces at room and refrigerated temperatures. *J Food Prot*, **57**: 629–31.

ZHAO T., DOYLE M.P. and BESSER R.E. (1993). Fate of enterohemorrhagic *Escherichia coli* O157:H7 in apple cider with and without preservatives. *Appl Environ Microbiol*, **59**: 2526–30.

28

Hygiene control in the dry food products industry: the roles of cleaning methods and hygienic indicators

J. L. Kornacki, Kornacki Microbiology Solutions, Inc., USA

Abstract: Manufacturers of dry foods in the United States find themselves in a difficult position. Wet cleaning and sanitization is generally necessary to establish a verifiable, effective microbicidal step that could be recognized by regulatory authorities, thereby minimizing the scope of any potential pathogen-related recall. Wet cleaning is also necessary to effect allergen removal. However, dry cleaning poses less contamination risk to the product. The following case studies illustrate this point. They also show the value of microbiological hygienic indicators and demonstrate how the industry has used them to solve problems and enhance food safety.

Key words: dry foods, dry cleaning, wet cleaning, hygienic indicators.

28.1 Introduction to the cases

Recognition of microbiological risks associated with dry (low a_w) foods is increasing, as evidenced by recent events including a large multi-state recall of hydrolyzed vegetable protein (HVP) (FDA 2010a), several peanut butter recalls in 2007, 2009 and 2011 (CDC 2007; FDA 2009a; FDA 2011), spice associated recalls (FDA 2009c, 2010b,c) and dry milk recalls (FDA 2009b). The type of products represented above, as with many other dry products, are often ingredients and consequently can drive further recalls of the products to which they have been added.

Water activity is a measurement of the amount of chemically (and microbiologically) available moisture. A food may have high moisture content but a relatively low water activity, e.g. jams and jellies, which have a fairly long shelf life. Foods with low water activity (e.g. a_w <0.60), are not subject to microbiological growth and or microbial spoilage. Consequently, these foods were wrongly (and sometimes are still) perceived to pose little microbiological public health risk. A series of articles was published in 2009 by the Grocery Manufacturers Association (GMA) addressing the complex issues related to production of dry foods (Scott *et al.* 2009a, 2009b, Chen *et al.* 2009). In large measure, these articles recognized that microbiological public health risks with dry foods are principally associated with *Salmonella* contamination, although other organisms may emerge.

28.1.1 Cleaning and sanitation in the dry products industry

Those who manufacture foods with higher water activity were and are still driven by economic constraints to design equipment in a hygienic manner. This is based upon the incidence and speed at which these products spoil, since spoiled products have reduced saleability in the marketplace. The result is that the economic and perceived public health risk associated with food with high water activities did not exist for foods with low water activity. Hence, much of the equipment for manufacturing foods with low water activity was not designed in a manner that could be readily cleaned in the traditional sense, e.g. with wet chemicals (Chen *et al.* 2009), and dry cleaning techniques, such as brushing, scraping, wiping, etc., have been used in the dry food industry for a long time.

Changes in regulations and recognition of risks associated with food allergens, however, have resulted in food processing equipment used to manufacture dry food, that was previously dry cleaned, now being subject to wet cleaning procedures to remove traces of allergens. This has had the result that the microbiological risk is increased in such equipment as it is not designed to accommodate water. Such equipment is typically manufactured in such a manner that during wet cleaning equipment entraps moisture, which is not easily removed. The only additional things necessary for such entrapped wet areas to become microbiological growth niches are the presence of an organism (present in non-sterile environments), nutrition (even in minute amounts) and incubation time (present in inaccessible areas) at a favorable temperature range. In most instances these factors will be present. Hence, the presence of moisture will result in microbial growth niche development in areas inaccessible for cleaning. Another factor to be considered is that microbial populations in dry food residues are likely to be injured. Hydration of these residues during wet cleaning and sanitization will result in repair (resuscitation) of such populations, as one of our case studies will demonstrate.

So now we have a situation with increased risk of contamination to the dry food product. Reduction of the business risk associated with contamination across lots requires the application of a microbicidal cleaning step, which most often means more applications of wet cleaning and sanititization, as will be discussed in Section 28.2.2, the efficacy of which it is essential to verify.

There is therefore an enhanced need to appropriately verify the sanitary quality of equipment and the product in the context of microbiological risk assessment in such a manner as not to create a liability to the facility. This can be done with hygienic indicators.

28.1.2 To test or not to test? The up- and downsides of hygienic indicators

Reviews of hygienic indicator tests (testing for hygiene indicator organisms) and their meanings and uses can be found in Kornacki (2010a, 2011). Numerous studies have revealed a decided lack of correlation between pathogen 'indicator' tests when compared on specific individual manufacturing production 'lots' (e.g. index organisms; Kornacki and Johnson 2001). However, hygienic indicators may have some value with regard to microbiological pathogen risk for several reasons. For example, on a pre-operational surface they can be used to assess the efficacy of cleaning and sanitization. If a given heat-sensitive process control indicator (e.g. coliforms destroyed by milk pasteurization) is found in thermally processed foods (milk in this example), then this provides evidence of post-pasteurization environmental contamination. If the environment is also contaminated with *Salmonella* it could be said that some risk exists. However, the presence of coliforms in pasteurized milk in this instance does not indicate that the milk is contaminated with *Salmonella*; only that such potential exists. However, a better hygienic indicator test would be one that detects a wide variety of microbes, including potential pathogens of concern and non-pathogens. In this instance positive results indicate that hygienic practices should be improved, whereas negative results suggest that pathogens are not present. This concept has been used with great effectiveness in a number of case studies. Some companies elect to use various ATP monitoring systems, but it should be recognized that these provide a better measurement of organic soil on a given surface and hence provide a measure of cleaning efficacy as opposed to microbial contamination, which provides a measure of sanitation efficacy (which can be a function, in part, of cleaning efficacy). ATP measurements often do not correlate with microbiological testing of surfaces (Kornacki, 2010a).

28.2 Significance of the cases

28.2.1 Microbial safety of dry foods

Data indicate that US recalls associated with spices appear to be rising dramatically. The FDA has noted an increased number of recalls due to bacterial contamination of dry spices in recent years (Vij *et al.* 2006). *Salmonella* was the predominant microbial cause of these recalls in the US, showing up in all but one of 21 recalls traced between October 1969 and December 2003. One recall was for the presence of *Listeria monocytogenes*, an organism seldom tested in dry foods. These recalls were associated with basil leaves, bay leaves (*L. monocytogenes*), ground black pepper, cerase spice, ground cumin, ground oregano, paprika, powdered red pepper, ground sage, sesame seeds, and ground thyme (Vij *et al.* 2006). The majority of these recalls were from paprika (four recalls) and ground oregano (three recalls). Spices are often added to foods that undergo no further heat treatment. In the last 4 years of the reported study, 16 of the 21 (76%) recalls occurred, with the remaining five occurring during the preceding 30 years. This increase was attributed to increased surveillance by the FDA and the Florida Department of Agriculture in response to discovery of contaminated spices at a Florida firm in 2001. The FDA's current focus on spices was addressed in a presentation recently by Doyle (2011).

A focus on *Salmonella* seems appropriate since there are over one million cases estimated annually in the US (Scallan *et al.* 2011; Lynch and Tauxe 2009), most of which are from unknown sources. Investigations of these situations are made more difficult because, under the right conditions, only a few cells may be required to initiate illness (Scott *et al.* 2009a). Hence, the magnitude of this risk from dry foods is unknown. If one traces the emergence of non-typhoid (essentially foodborne) Salmonellosis in the US over 86 years (1920–2006) it appears to increase dramatically beginning in the mid-1950s (see Fig. 28.1). How much of this reported illness can be traced to imported dry foods is presently unknown. An example follows. *Salmonella* serotype Agona emerged in 1969–70 in the US, the UK, the Netherlands and Israel. Each time the organism was isolated from imported fishmeal, a dry ingredient to animal feed, then domestic animals and finally humans (Clark *et al.* 1973). This organism was the eighth, and second most common, serotype isolated in the US and the UK, respectively, by 1972. It was more recently responsible for two foodborne illness-associated recalls in the United States associated with dry breakfast cereal in 1998 and 2008 (MMWR 1998; http://www.cdc.gov/mmwr/preview/mmwrhtml/00053368.htm, http://www.cdc.gov/*Salmonella*/agona/).

There is some evidence that microbiological risks other than *Salmonella* also exist or may be shown one day to exist in dry foods. These include *Cronobacter* spp. (formerly, *Enterobacter sakazakii*) outbreaks among neonates

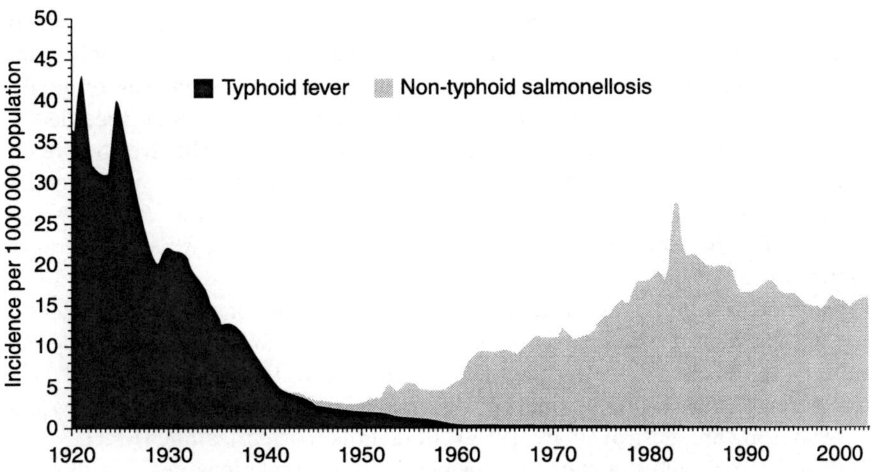

Fig. 28.1 Emergence of non-typhoid *Salmonella*: reported infections USA, 1920–2006, CDC national surveillance data. (Source: Lynch and Tauxe 2009.)

due to non-sterile powdered dry infant formula, which can range from septicemia to lethal meningitis with high mortality (Gurtler *et al.* 2005). One of the spice recalls above was due to *L. monocytogenes*, and risks with dry ready-to-eat foods from high numbers of spores, such as those produced by *Bacillus cereus* or *Clostridium perfringens*, also should not be ignored. Indeed, many dry foods were at one time high-moisture products that were dried. Hence, the microflora of those products prior to drying should be factored into our perceptions and investigations of microbiological risk. Thus an enterotoxin-producing strain(s) of *Staphylococcus* spp. may have grown in a food and produced a heat resistant enterotoxin(s). Populations of the organism may have been destroyed in the drying process, but the heat-resistant toxin(s) may remain in the dried food. Another putative pathogenic microbe on the horizon is *Mycobacterium avium* subsp. *paratuberculosis* (MAP), reported to be present at a low percentage in pasteurized fluid milk (Ellingson *et al.* 2005). This does not necessarily prove that these unusual organisms survive appropriate pasteurization temperatures, as there is a chance that post-pasteurization contamination or a failure to de-clump these very hydrophobic organisms before pasteurization may have occurred (Lund *et al.* 2002).

The composition of the product may affect the outcome of the pre-drying pasteurization process. As an example, exceedingly large numbers of *Salmonella* are destroyed by milk pasteurization in seconds (high-temperature short-time (HTST) pasteurization) and minutes ('vat' pasteurization), but very little reduction occurs after hours in milk chocolate at similar temperatures (Geopfert *et al.* 1970, Krapf and Gantenbein-Demarchi 2009). The heat resistance of different pathogens also varies. *Salmonella* are typically less heat resistant than *Listeria* spp., except perhaps *S.* Senftenberg 775W (in high moisture foods), which was shown to be similar in heat resistance to a ten-strain cocktail of *L. monocytogenes* in a review of the literature by Johnson *et al.* (1990b). *Mycobacterium* spp. are widely regarded for their heat resistance. MAP, in my view, are actually likely to be more resistant to pasteurization than *Listeria*, as milk pasteurization was originally established to destroy another *Mycobacterium* (tuberculosis), then regarded as the most heat-resistant pathogen that may occur in raw milk. This was before the discovery of *C. burnettii* in raw milk, which the pasteurization is now designed to kill (Johnson *et al.* 1990a). It is also well known that spores of *B. cereus* will survive milk pasteurization, some of which are able to grow at refrigeration temperatures.

The potential presence of human pathogens (putative or otherwise), whether due to their heat resistance or post-pasteurization contamination, in pasteurized milk is troublesome, given the lack of microbial lethality associated with commercial spray-drying operations. It should be noted that only 3.6–5.2 $\log_{10}$ of *Listeria* spp. are destroyed by HTST fluid milk pasteurization (Mackey and Bratchell 1989), and that *Listeria monocytogenes* populations are destroyed by spray drying to a much smaller extent (1–1.5 $\log_{10}$ reduction) (Doyle *et al.* 1985). This is because microbial heat resistance increases as the amount of chemically available moisture decreases (Goepfert *et al.* 1970) and the moisture decrease during spray drying occurs 'almost immediately' (Clark 1998). The problem of pathogen survival following drying processes is not limited to milk. A study that Kornacki Microbiology Solutions, Inc. performed (Kornacki 2006, 2011) with a surrogate organism and with *Salmonella* with a wheat-based product at two water activities is also reflective of the thermal protection afforded to a microbial population after a drying step (Fig. 28.2 and 28.3). In these examples, the lethality of a process to *Salmonella* and the surrogate represented in Fig. 28.2 and 28.3 as *Enterococcus* sp. NRRL B-2354 was greater at the lower temperature in the higher water activity cooked product (Fig. 28.2). In the lower water activity 'oven treated' product at higher temperatures, e.g. 93–116 °C, there was no appreciable lethality (Fig. 28.3).

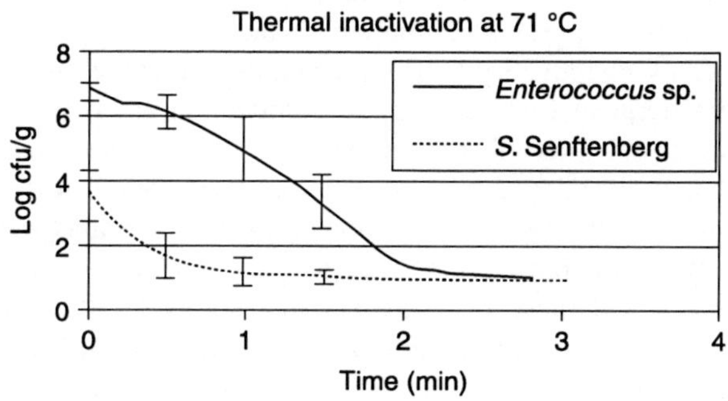

Fig. 28.2 Heat resistance of *Enterococcus* sp. (formerly *Pediococcus* sp.) NRRL B-2354 in cooked wheat (a high a_w product) (Kornacki 2006).

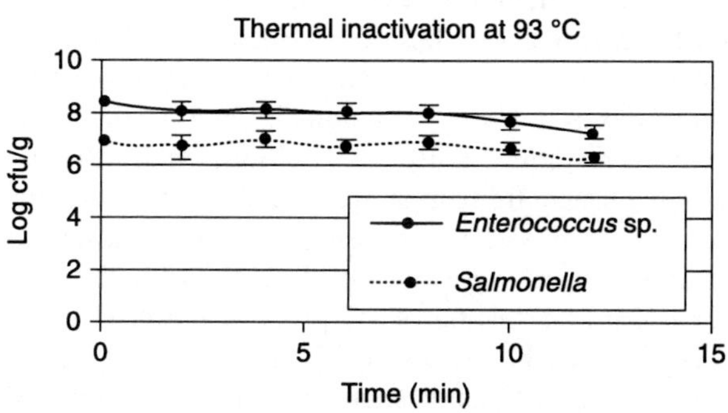

Fig. 28.3 Comparison of inactivation of surrogate and *Salmonella* cocktail in oven-treated low a_w product at 93 °C. Similar data were recorded at 104 and 116 °C. In the lower water activity product at higher temperatures, e.g. 93–116 °C, there was no appreciable lethality.

28.2.2 Regulatory and market aspects

The US-FDA regards all product manufactured on common production lines (or shared portions of lines) after a validated microbicidal step between validated wet clean-ups (including sanitization) to be adulterated if on one of the production days any 'lot' tests positive from such lines. FDA's perspective does not regard individual production days or lot assignments made by the production facility. Consequently, a contamination event on a dry food processing line that occurs on day 35 of a 35-day production cycle, between validated clean-ups, implicates all product produced on days 1–35. This makes it particularly important to have regular validated clean-ups (including sanitation); this is considered a 'sanitation break point' (Kornacki 2010c). However, sanitation break points are not achievable apart from by a biocidal treatment, and traditional approaches have utilized aqueous cleaners and sanitizers. Consequently, unless a wet cleaning and sanitization approach is taken, a dry-cleaning break point between lots is likely not to be recognized as a microbial sanitation break by regulatory agencies in the United States. Consequently, dry food manufacturers now feel increased pressure to perform more frequent wet cleaning and sanitization procedures to establish a sanitation break point. This allows companies to reduce their liability to those products on common lines produced between 'validated' wet cleaning and sanitation regimes. In the context of dry food production, frequent wet cleaning, by introducing moisture into production lines, potentially increases the contamination risk for dry products if these lines are not thoroughly dried in the early post-cleaning stages prior to food production on such lines. Thus, there is a great need for redesigning or retrofitting equipment to accommodate water (Chen *et al.* 2009).

28.3 Case studies

28.3.1 Case study 1: *Bacillus cereus* in dry foods

In a number of situations in recent years, we have been concerned about the potential for *B. cereus* contamination of RTE products with low water activity. Ingestion of 100 000/g of *B. cereus* has been associated with illness (Behling *et al.* 2010). Furthermore, spores of *B. cereus* are known to survive cooking processes; and germination of spores to high populations in foods or the intestinal tract may also release toxin(s) resulting in illness (Behling *et al.* 2010). The products involved were transformed from high water activity raw products and passed through atmospheric ovens and dryers, resulting in a low water activity finished product. The manufacturers were not comfortable with the liability involved with a direct approach (e.g. testing of the product and product contact surfaces for *B. cereus*). (This is often the case in the US, unless product is already implicated in a recall and the line(s) subsequently shut down for investigation. More US-based companies are becoming proactive about their risk of food contamination, though, so approaches need to be developed that are acceptable to the industry.) Consequently, the non-food contact processing environment was tested for the incidence of *B. cereus*, but in-line samples, finished product, selected product, and near-product contact surfaces were tested for their mesophilic aerobic spore (MAS) count as described below.

B. cereus is widespread in the environment and a natural contaminant of the raw materials we studied, so its presence in the environment was expected. However, vast numbers associated with germination, growth and sporulation would not be expected unless there was a breakdown in hygiene control. In processes and products in two facilities producing different dry products separated by hundreds of miles, a high incidence (>90%) of general production environment samples were positive for the presence of *B. cereus*. This incidence was much less in several other similar facilities. However, of some concern in these two cases, approximately 50% had levels in excess of 100 000 per sampled area (approximately 0.09 m^2), reflecting a need for greater hygiene controls in the general plant environment.

This, by itself, does not prove that product was contaminated with *B. cereus* (much less with dangerous levels of this contaminant). To assess this potential, we sampled and tested both in-line and finished product. In both facilities, product from a number of sites had MAS counts in excess of 100 000 per site, indicative of the need for better approaches to spore inactivation on such surfaces and of the potential that some of these sites may have *B. cereus*. A few product samples in both instances had MAS levels of 100 000 or more, which, if *B. cereus* were present in the product, would indicate a potential health risk if consumed directly without being diluted (unlikely in these cases) or if added to a food in which the organisms could grow. Consequently, we were able to make recommendations for testing finished product, spore control in the general processing environment and disinfection of equipment surfaces that did not involve an excessive amount of added moisture.

28.3.2 Case study 2: staphylococcal enterotoxin-producing organisms

Staphylococcal enterotoxin (SET)-producing strains of *Staphylococcus* are a concern in situations wherein they may grow and produce a heat-stable toxin in a pre-processed food, followed by population die-off due to various processing treatments. Levels of SET-producing strains of one million/g or more have been associated with production of measurable SET and illness. A similar approach to that described in the case above was employed. In this instance, in-line samples prior to a microbicidal step were tested in a similar fashion. However, in this instance in-line and product samples were plated onto Baird-Parker agar (BPA). This medium, apart from being a confirmation agar, is a hygiene indicator that will recover *Micrococcus* spp., *Staphylococcus* spp. (both enterotoxin and non-enterotoxin producers), *Bacillus* spp., *Listeria* spp. (Kornacki *et al.* 2003a, 2003b, 2011) and perhaps other microbes. Nevertheless, if the plates have no colonies, then it can be said that SET-producing *Staphylococcus* strains were not present in the sample. In this example, high numbers of BPA-recovered organisms were found in higher moisture in-line samples with growth-permissive water activity at a step in the process prior to a microbicidal drying step. Such results were used to make recommendations related to challenge studies and process changes to control the growth of SET-producing strains should they be present in the product before drying.

28.3.3 Case study 3: detection of Enterobacteriaceae strains with similar physiology to *Salmonella*

Similarly, Kornacki reported on an enrichment assay procedure that detects hydrogen sulfide-producing Enterobacteriaceae strains with similar physiology to *Salmonella* (Kornacki 2010b). Identification of hydrogen sulfide-producing thermophilic Enterobacteriaceae (HTEB) positive isolates by ABC Research (Gainesville, FL) from some samples collected by Kornacki revealed the presence of *Citrobacter freundii*, *Escherichia hermannii*, *Enterobacter ammigenus*-2, *Morganella morganii*, *Salmonella enterica* subsp. *arizona*, and *Proteus mirabilis*.

The use of this assay on equipment and environmental surfaces has provided useful information that has promoted food safety in the dry products industry. For example, performing both *Salmonella* and HTEB assays on the same sample is a valuable method for locating sites that have potential to become *Salmonella* niches when such sites test negative for *Salmonella*. Thus it provides the industry with the potential to eliminate such niches before they occur. In several instances swabbing near product contact surfaces for HTEB and Enterobacteriaceae resulted in the discovery that some post-operational sites were positive for HTEB and had high populations of Enterobacteriaceae, a finding that resulted in a powerful incentive for enhanced equipment breakdown for cleaning and development of an environmental *Salmonella* testing program and hold and test procedures on a dry product previously unrecognized as a potential risk for *Salmonella* contamination (Fig. 28.4).

28.3.4 Case study 4: an investigation related to wet cleaning and sanitization

The following case study is illustrative of the dangers associated with wet cleaning in a dry food processing environment. In this example, swabs were taken from dry product handling equipment after use in the dry state, after wet cleaning, and after gentle misting with sanitizer. In this study at least ten samples were taken at each of these time points, with the result that counts increased an average of 1.4 $\log_{10}$ cfu after wet cleaning and another 1.7 $\log_{10}$ cfu after gentle mist sanitization. Hence, wet cleaning and sanitization of this otherwise dry surface resulted in over 3 $\log_{10}$ increase in microbial load. This may have been because of hydration and revival of injured microbial populations. This result clearly revealed the need for great improvement in the approach to cleaning and sanitization and the inherent dangers in wet cleaning and sanitization approaches as applied to equipment designed to handle dry products. They also enabled us to make appropriate recommendations for approaches to cleaning and sanitization and appropriate monitoring of these surfaces.

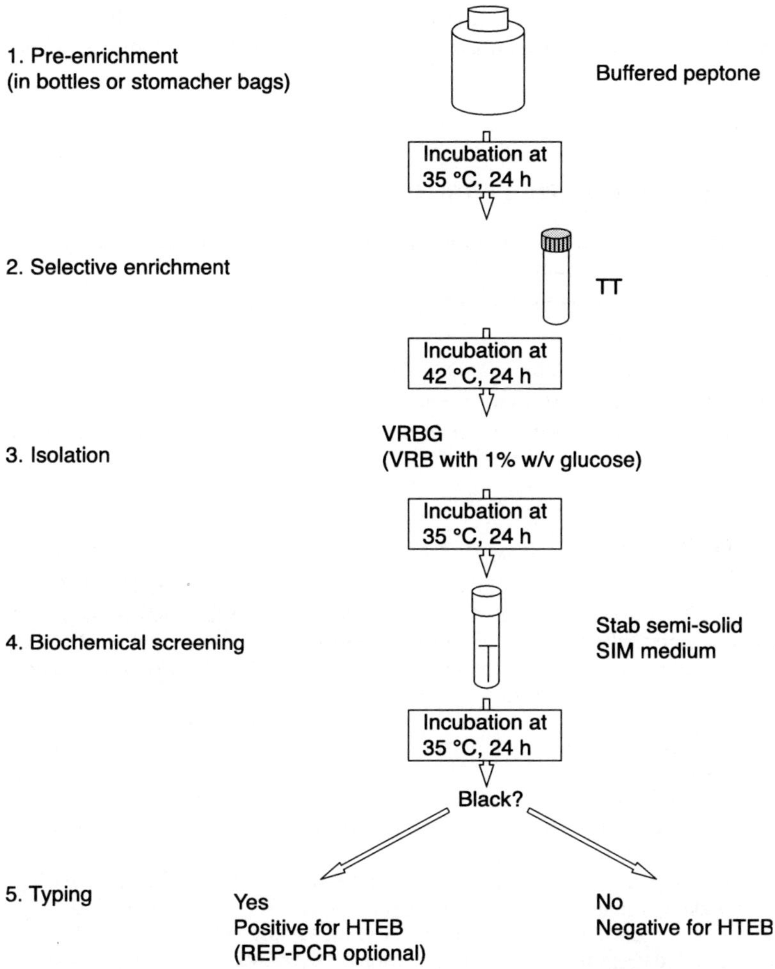

Fig. 28.4 Hydrogen sulfide-producing thermophilic Enterobacteriaceae (HTEB) assay for physiologically similar organisms to *Salmonella*. In several instances swabbing near product contact surfaces for HTEB and Enterobacteriaceae resulted in finding that some post-operational sites were positive for HTEB and had high populations of Enterobacteriaceae, a finding that resulted in a powerful incentive for enhanced equipment breakdown for cleaning and development of an environmental *Salmonella* testing program and hold and test procedures on a dry product previously unrecognized as a potential risk for *Salmonella* contamination.

28.4 Commentary: where do we go from here?

Awareness of microbiological risk associated with dry foods is growing, but the industry often lacks the appropriate tools to address and mitigate this risk. Improvements in facility and equipment design, research on dry methods of equipment and facility decontamination, and appropriate testing of the environment and product are needed to improve safety of dry foods and those products to which dry foods are added.

28.5 Critical questions for discussion

1. Why do wet cleaning and wet sanitization create more risk in dry food processing environments?
2. Why is there a microbiological risk from low moisture foods when bacteria cannot grow in them?
3. Does a hygiene indicator tell you if a pathogen is present?
4. What does a hygiene indicator test actually indicate?
5. Besides *Salmonella*, what other organisms may be significant in dry foods?

28.6 References

BEHLING R.G., EIFERT J., ERICKSON M.C., GURTLER J.B., KORNACKI J.L. *et al.* (2010). Selected pathogens of concern to industrial food processors: Infectious, toxigenic, toxico-infectious, selected emerging pathogenic bacteria. In J. L. Kornacki (ed.). *Principles of Microbiological Troubleshooting in the Industrial Food Processing Environment*, pp. 5–61. Springer, New York.

CDC (2007). *Salmonellosis – Outbreak investigation, February 2007*. Available at http://www.cdc.gov/ncidod/dbmd/diseaseinfo/salmonellosis_2007/outbreak_notice.htm. Accessed 29 August 2011.

CHEN Y., SCOTT V.N., FREIER T.A., KUEHM J., MOORMAN M. *et al.* (2009). Control of *Salmonella* in low-moisture foods II: Process validation and environmental monitoring. *Food Protection Trends*, July, 435–45.

CLARK G.M., KAUFMANNE A.F., GANGAROSA J. and THOMPSON M.A. (1973). Epidemiology of an international outbreak of *Salmonella* Agona. *The Lancet*, 1 September, 490–3.

CLARK W.S. JR. (1998). Concentrated and dry milks and wheys. In E. H. Marth and J. L. Steele (eds), *Applied Dairy Microbiology*, pp. 65–80. Marcel Dekker, New York.

DOYLE M.P., MESKE L.M., and MARTH E.H. (1985). Survival of *Listeria monocytogenes* during the manufacture and storage of dry milk. *J. Food Prot*, **48**: 740–2.

DOYLE M.P. (2011). Microbiological safety of spices: A paradigm shift. *Presentation in Symposium S37, Microbiological Safety of Dry Spices: A Hot Topic . . . literally. 2011 Annual Meeting of the International Association for Food Protection*. Milwaukee, WI.

ELLINGSON J.L.E, ANDERSON J.L., KOZICZKOWSKI J.J., RADCLIFF R., SLOAN S.J. *et al.* (2005). Detection of viable *Mycobacterium avium* subsp. *paratuberculosis* in retail pasteurized whole milk by two culture methods and PCR. *J Food Prot*, **68**(5): 966–72.

FDA (2009a). *Peanut Products Recall. June 19, 2009 (Update)*. Available at http://www.fda.gov/Safety/Recalls/MajorProductRecalls/Peanut/default.htm. Accessed 4 October 2011.

FDA (2009b). *Plainview Milk Cooperative Ingredient Recall Product List, Information current as of noon July 28, 2009 (update)*. Available at http://www.accessdata.fda.gov/scripts/Milk/. Accessed 29 August 2011.

FDA (2009c). *Adams Extract & Spice, LLC Announces Recall Because of Possible Health Risk. August 07 (update)*. Available at http://www.fda.gov/Safety/Recalls/ucm176343.htm. Accessed 29 August 2011.

FDA (2010a). *Hydrolyzed Vegetable Protein Product Recalls*. Available at http://www.fda.gov/Safety/Recalls/MajorProductRecalls/HVP/default.htm. Accessed 29 August 2011.

FDA (2010b),. *Wholesome Spice Recalls 25 Lb. Boxes Of Crushed Red Pepper Because Of Possible Health Risk. February 25 (Update)*. Available at http://www.fda.gov/Safety/Recalls/ucm202113.htm. Accessed 29 August 2011.

FDA (2010c). *Red & Black Pepper Spice Recalls Linked to the Salmonella Montevideo Outbreak Investigation, March 30, 2010 (Update)*. Available at http://www.fda.gov/Food/NewsEvents/WhatsNewinFood/ucm206052.htm. Accessed 29 August 2011.

FDA (2011). *Unilever Announces Recall of Skippy Reduced Fat Peanut Butter Spread due to Possible Health Risk Limited Recall of 6 best-if-used-by dates. March 4 (Update)*. Available at http://www.fda.gov/Safety/Recalls/ucm245897.htm. Accessed 29 August 2011.

GEOPFERT J.M., ISKANDER I.K., and AMUNDSON C.H. (1970). Relation of the heat resistance of *Salmonellae* to the water activity of the environment. *Appl Microbiol*, **19**(3): 429–33.

GURTLER J.B., KORNACKI J.L., and BEUCHAT L.R. (2005). *Enterobacter sakazakii*: A coliform of increased concern to infant health. *Int J Food Microbiol*, **104**: 1–34.

JOHNSON E.A., NELSON J.H., and JOHNSON M. (1990a). Microbiological safety of cheese made from heat-treated milk, Part I. Executive summary, introduction and history. Part I. *J Food Prot*, **53**: 441–518.

JOHNSON E.A., NELSON J.H., and JOHNSON M. (1990b). Microbiological safety of cheese made from heat-treated milk, Part II. Microbiology. *J Food Prot*, **53**: 519–40.

KORNACKI J.L. (2006). Examples of surrogate validation and use in the industry. In *IAFP Symposium S07 Surrogate Microorganisms; Selection, Use and Validation*. IAFP Annual Meeting, Calgary, 14 August. Alberta, Canada.

KORNACKI J.L. (2010a). How do I sample the environment and equipment? In J. L. Kornacki (ed.), *Principles of Microbiological Troubleshooting in the Industrial Food Processing Environment*, pp. 125–36. Springer, New York.

KORNACKI J.L. (2010b). An indicator approach to enteric contamination of at risk foods. *Presentation given as part of symposium S22 Food and Food Environment Test Considerations in View of Changing Regulations*. IAFP Annual Meeting, Anaheim, CA.

KORNACKI J. L. (2010c). Where do I start? (beginning the investigation). In J. L. Kornacki (ed.), *Principles of Microbiological Troubleshooting in the Industrial Food Processing Environment*, pp. 117–23. Springer, New York.

KORNACKI J.L. (2011). Indicator organism assays: Chaos, confusion and criteria. *Food Safety Magazine*, February/March. Available at http://foodsafetymagazine.com/article.asp?id=3945&sub=sub1. Accessed 20 December 2011.

KORNACKI J.L. and JOHNSON J.L. (2001). Enterobacteriaceae, coliforms and *Escherichia coli* as quality and safety indicators. In F. P. Downes and K. Ito (eds), *Compendium of Methods for the Microbiological Examination of Foods*, 4th ed. APHA, Washington DC.

KORNACKI J.L., GURTLER J.B., YAN Z., and COOPER C.M. (2003a). Evaluation of several modifications of an ecometric technique for assessment of media performance. Abstract T01. *International Association for Food Protection Annual Meeting*, 10–13 August, New Orleans.

KORNACKI J.L., GURTLER J.B., YAN Z., and COOPER C.M. (2003b). Evaluations of several modifications of an ecometric technique for assessment of media performance. *J. Food Prot*. **66**(9): 1727–32.

KRAPF T. and GAINTENBEIN-DEMARCHI C. (2009). Thermal inactivation of *Salmonella* spp. during conching. *LWT-Food Science and Technology*. Doi:10.1016/j.lwt.2009.10.009.

LYNCH M.F. and TAUXE R.V. (2009). Salmonellosis: Nontyphoidal. In P. S. Brachman and A. S. Evans (eds). *Bacterial Infections of Humans: Epidemiology and Control*, 4th ed., pp. 677–98. Springer, New York.

LUND B.M., GOULD G.W., and RAMPLING A. (2002). Pasteurization of milk and the heat resistance of *Mycobacterium avium* subsp. *paratuberculosis*: a critical review of data. *Int J Food Microbiol*, **77**: 135–45.

MACKEY B.M. and BRATCHELL N. (1989). A review: The heat resistance of *Listeria monocytogenes*. *Lett Appl Microbiol*, **9**: 89–94.

MMWR (1998). Multistate outbreak of *Salmonella* serotype Agona infections linked to toasted oats cereal – United States, April–May 1998. *Morb Mortal Wkly Rep*, **47**(22): 462–4.

SCALLAN E., HOEKSTRA R.M., ANGULO F.J., TAUXE R.V., WIDDOWSON M-A. *et al.* (2011). Foodborne illness acquired in the United States – Major pathogens. *Emerg Infect Dis*, **17**(1): 7–15.

SCOTT V.N., CHEN Y., FREIER T.A., KUEHM J., MOORMAN M. *et al.* (2009a). Control of *Salmonella* in low-moisture foods I: Minimizing entry of *Salmonella* into a processing facility. *Food Prot Trends*, June, 342–53.

SCOTT V.N., CHEN Y., FREIER T.A., KUEHM J., MOORMAN M. *et al.* (2009b). Control of *Salmonella* in low-moisture foods III: Process validation and environmental monitoring. *Food Prot Trends*, August, 493–508.

VIJ V., AILES E., WOLYNIAK C., ANGULO F.J., and KLONTZ K.C. (2006). Recalls of spices due to bacterial contamination monitored by the U.S. Food and Drug Administration: The predominance of *Salmonellae*. *J Food Prot*, **69**(1): 233–7.

29

Boosting exports of fresh meat by using faster laboratory methods in Denmark

C. Löfström, M. H. Josefsen and J. Hoorfar, Technical University of Denmark, Denmark and F. Hansen, Danish Technological Institute, Denmark

Abstract: As Denmark is one of the largest pork producers worldwide, exports of pork meat play an important role in the Danish economy. During the slaughter process fresh meat may become contaminated with enteric pathogens that can pose a risk to public health. Certain countries have strict legislative demands when importing meat, especially fresh minced meat due to its very short shelf-life. This case study describes the successful development and implementation of a rapid *Salmonella* detection method at the largest slaughterhouses in Denmark.

Key words: meat, pathogens, *Salmonella*, detection, export, rapid methods.

29.1 Introduction to the case

Denmark is one of the top ten pork producers in the world, with a total production of approximately 1.9 million tons in 2009 (Agriculture & Food, 2010). There has always been a pressing need to use the most efficient and cost-effective procedures available throughout the production chain to enable these large production volumes (Hinrichsen, 2010). *Salmonella* contamination of fresh meat presents a problem for public health. Occasionally, meat becomes contaminated with pathogens, such as *Salmonella*, for instance by cross-contamination between fecal matter and carcasses during the slaughter process. Moreover, *Salmonella* has the ability to survive under various unfavorable conditions (Podolak *et al.*, 2010) and can be difficult to eradicate from process equipment (Korber and Lawrence, 2004). Consequently, it is a challenge to produce *Salmonella*-free meat.

However, low or zero prevalence of *Salmonella* in meat products is required by an increasing number of countries, including some of the EU Member States. To assure the absence of *Salmonella* there is obviously a need for appropriate detection methods. Standard methods, including the methods from the International Organization for Standardization (ISO 6579:2002) and the Nordic Committee of Food Analysis (NMKL 71:1999) are based on the culture of *Salmonella* in different types of growth media to isolate the bacteria, followed by its identification by biochemical and serological tests (for a review, see Löfström *et al.* (2009a)). This is a cumbersome process taking at least 3 days to complete.

29.1.1 How did it all begin?

This case started in 2005 with a request from pork producers for a faster method for detection of *Salmonella* in different sample types related to the meat production chain. The meat producers had already implemented the fastest alternative detection methods on the market at that time, but these methods still took at least 24 h to perform. This resulted in unfortunate delays in the release of meat from the slaughterhouses to the customers. This in turn led to increased costs for storage rooms as well as decreased shelf-life of the products at retailers.

The requirements for the alternative method were that it should be:

- Rapid (<12 h in total),
- As specific and sensitive as the tests and standard methods currently used,
- Cost-effective with a low cost per sample analyzed (approximately <€4),
- User-friendly to enable use on-site at abattoirs,
- Approved according to the legislation in place for both domestic sale and export.

The prevalence of *Salmonella* in Danish pork, beef and poultry is low and hence most samples analyzed will be negative (Anonymous, 2010). The test should therefore be designed for fast screening of *Salmonella*-negative samples. As no test was available on the market at that time to meet all these demands, a research collaboration was established to develop the test.

29.1.2 What were the challenges involved?

To be able to develop a suitable test there were several challenges that had to be overcome. These were related to technical as well as operational aspects (Eijkelkamp *et al.*, 2009). It was decided that a polymerase chain reaction (PCR) method detecting DNA from *Salmonella* could be the best starting point to meet the above criteria (Malorny *et al.*, 2009).

Low numbers of Salmonella *present in samples*

First, the numbers of *Salmonella* in samples from the meat production chain were expected to be low (<10–100 CFU/sample) and the bacteria were considered to be partly injured due to the unfavorable environment encountered during meat production (Hoorfar and Baggesen, 1998). Such a low number is hard to detect directly from a sample as only a small fraction of the sample is actually analyzed (i.e. 100 µl for the spread plate technique or 2–10 µl for PCR). It is like finding a needle in a haystack. To increase the concentration of *Salmonella* cells in the samples the first challenge was therefore to design a procedure that could either multiply the *Salmonella* cells in the sample (make more needles in the same haystack) or concentrate them (keep the needle in a smaller haystack).

Design of sample preparation steps

Second, the steps taking place after the sampling had to be designed. This was not trivial because a variety of sample types had to be considered. These included fresh meat, gauze swabs used to sample the carcasses and poultry fecal samples.

With real-time PCR, it is possible, at least in theory, to detect one single copy of a gene, or in this case one single *Salmonella* cell. However, although PCR is a sensitive technique, inhibitory substances present in the sample might prevent the target from being amplified. Therefore one needs to obtain a relatively clean sample through design of suitable sample preparation steps (for a review see Hedman *et al.* (2012)).

Taking this into account, the multiplication strategy was chosen including a culture enrichment step, combined with a series of optimized sample preparation steps. These included easy-to-perform steps such as centrifugation and a semi-automated procedure to extract DNA from the samples based on magnetic beads that bind the DNA (see Fig. 29.1).

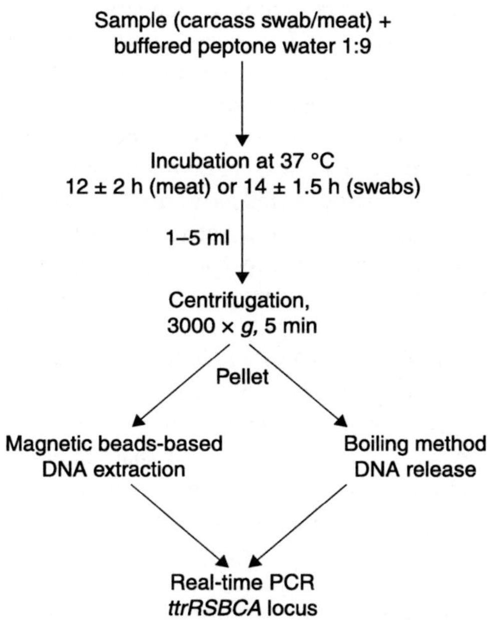

Fig. 29.1 Overview of the real-time PCR method used for detection of *Salmonella* in meat and carcass swab samples. There are two alternatives ways to release the DNA before PCR: magnetic beads-based method and boiling method.

Implementation on-site at abattoirs
Apart from the challenges involved with the design and optimization of a suitable method, there were also other factors that played a role in the final selection of the test. Among these can be mentioned the possibility of successfully implementing the developed method. It was decided that it should be possible for laboratories that are not so familiar with PCR testing to perform the analysis.

Approval by authorities and other factors
To enable the meat producers to use the method for official controls there was also a need to obtain approval for it from authorities and regulators. Last but not least, from a business point of view, the methods should be cost-effective.

29.2 Significance of the case

Salmonella is recognized as one of the most important foodborne pathogens worldwide, and human infections have wide health and socioeconomic consequences. Infection often presents as an acute gastrointestinal illness. The onset of diarrhea usually occurs within 6–48 h after ingestion of the contaminated food and, besides diarrhea, common symptoms include abdominal pain or cramps, fever, chills, nausea, vomiting, pain in the joints, headache, myalgia, general malaise and weight loss. In uncomplicated cases, fever usually resolves within 48–72 hours, and the average time to recovery is 1–2 weeks.

In patients with underlying illnesses or severe symptoms, prescription of antibiotics can be necessary; however, antimicrobial drug resistance against clinically important drugs may lead to reduced efficacy of early treatment and limitation of therapeutic choice. Complications include endocarditis or arterial infections, intestinal perforation, and abdominal, soft tissue and urogenital infections. Up to 15% of gastroenteritis cases are followed by reactive arthritis. The acute phase mortality has been estimated to be slightly above 1% in large parts of the industrialized world, but recent studies suggest that mortality after *Salmonella* infection may be underestimated (for a more comprehensive description, see Josefsen *et al.* (2011)).

The main sources of foodborne *Salmonella* vary between different parts of the world and depend on the prevalence of *Salmonella* in animals used for food production, food production chains, consumption habits and food preparation practices (Malorny *et al.*, 2011). However, meat, particularly pork and poultry, is considered an important source of foodborne salmonellosis in many countries (Denny *et al.*, 2007, http://www.cdc.gov/*Salmonella*).

29.2.1 Regulatory aspects

Some years ago, meat exporters, kit manufacturers and authorities requested the European Commission to accept the use of alternative methods, which could release meat for export within 24 h (termed 24 h methods hereafter). This resulted in an EU Directive on the authorization of certain alternative methods to be used in microbiological testing of meat intended for Finland and Sweden (Anonymous, 2005b).

Furthermore, the requirements for approval of the alternative methods for *Salmonella* are more specifically stated in the Commission Regulation (EC) No. 2073/2005 on microbiological criteria for foodstuffs (Anonymous, 2005a). In this directive it is stated that alternative methods to be authorized for special guarantees for *Salmonella* control must have been validated against the most recent editions of the ISO 6579 or NMKL 71 methods. Moreover, proprietary methods must be certified by a third party in accordance with the protocol set out in the standard EN ISO 16140 or other internationally accepted protocols. Examples of such protocols include AFNOR, AOAC – OMA, MicroVal and NordVal (Qvist, 2011).

29.2.2 Market and economic issues

Traditionally, pigs are slaughtered during the dayshift, and left for cooling overnight. Next morning the pig carcass is ready for cutting and deboning by the dayshift starting at 7 a.m. However, when using traditional *Salmonella* screening methods or even the 24 h methods that have been on the market for the last decade, the results from the *Salmonella* analyses are not ready by that time. Thus the slaughterhouse faces the risk of having to recall a large batch of pork cuts if the samples reveal contamination with *Salmonella.*

Special EU legislation is in force for certain EU countries (Sweden and Finland) regarding import of pork meat from other EU countries. This legislation allows Finland and Sweden to reject any batch of pork meat containing *Salmonella.* In order to avoid rejection of Danish meat, it must be certified as being free of *Salmonella* before being exported.

Formerly, meat for export to Sweden and Finland was certified using the 24 h methods, the rapid methods available at the time. The test took 24 h to complete, producing a result in the afternoon of the day after slaughter. Trucks were loaded during the morning, but had to wait until the batch of the pork meat tested negative. If the result was positive, that particular batch could not be exported to Sweden or Finland, and the truck had to be reloaded with another batch of pork that had tested free from *Salmonella.* This was, of course, a costly and inconvenient procedure, and delayed export of the meat for several hours.

Applying a more rapid test for *Salmonella* (developed in our project), pigs slaughtered during the day shift could be analyzed during the night shift and the result would be ready the next morning at 7 a.m., allowing the slaughterhouse to select a certified, *Salmonella*-free batch of meat for export. The cutting and deboning could then be initiated early in the morning without risk of recalling the meat, and the very costly and time-consuming process of reloading trucks meant for the Scandinavian markets when samples tested positive could be avoided.

However, if a more rapid test is to be cost-effective the running costs need to be markedly lower than those for the already existing 24 h methods that have formerly been implemented for the rapid release of the meat.

29.3 The case history continues

Despite the successful design of a real-time PCR method (Josefsen *et al.*, 2007, Löfström *et al.*, 2009b, 2010), it was not possible to meet the demand of reducing the total analysis time to 12 h for any of the matrices. The total analysis time for meat was 14 h, for carcass swabs 16 h and for poultry feces 20 h (see Fig 29.1). Nonetheless, it was decided to proceed with the next important step: obtaining approval for use at the production sites.

29.3.1 Validation studies

As this test was developed in Denmark, the decision was taken to use the international validation protocol of NordVal, the Nordic Organization for Validation of Alternative Methods (NordVal, 2009). This protocol is divided into two parts: a comparative study performed at the expert laboratory, followed by a collaborative test where identical samples are analyzed by a number of laboratories to assure reproducibility of the method.

A first validation study was performed in 2007 for meat samples followed by a second study in 2008 on carcass swabs and fecal samples (Löfström *et al.*, 2009b, 2010). In both studies the method was found to perform well and no differences were found when it was compared with the reference method. The tests were therefore granted NordVal approval in 2007 and 2008 and could be used for testing at the abattoirs (NordVal, 2008).

29.3.2 Implementation on-site

The test was now approved for use, but another challenge was yet to be encountered: the implementation at the routine laboratories on-site at the abattoirs. Several workshops were arranged by the expert laboratory to train the staff at the routine laboratories. This strategy enabled a successful introduction of the new method at the production laboratories.

29.3.3 Troubleshooting

Although the test was designed from the start with the end-user in mind, there were still problems with its implementation, challenges that were not considered from the beginning of the project. One of these was the ability to obtain quality-controlled reagents for the PCR step. The method was designed as a non-proprietary method, meaning that it is open for everyone to use without restrictions. Therefore no kits with ready-to-use reagent mixtures were available. To solve this, a contract was then taken out with a Danish biotechnology firm that was able to produce a ready-to-use reagent mix that could be delivered directly to the production laboratories.

Another obstacle was the relatively high cost of the magnetic beads-based DNA extraction method. The meat producers needed the method to be easier to perform, to reduce the total analysis time as well as the hands-on time and also to lower the investment and running costs of the analysis. Therefore an alternative strategy was evaluated based on a simple step that involved boiling the sample taken after culture enrichment to release the DNA and get rid of some of the interfering substances. The purified sample was then analyzed by real-time PCR to detect *Salmonella* (see Fig. 29.1).

The boiling strategy turned out to be a success: the method was found to perform as well as the more complicated magnetic beads-based method for meat and carcass swab samples (Löfström *et al.*, 2012). A NordVal validation study was conducted in 2010 and the method was granted NordVal approval at the end of that year (NordVal, 2010). The modified method is currently being used regularly by several of the major slaughterhouses in Denmark.

29.4 Commentary and personal reflections

The test strategy designed in this project is now being used by other expert laboratories for other major foodborne pathogens. With the continuing advance of probe chemistries, PCR technologies and nanochips, the possibility of faster laboratory tests is expanding. However, we should remember that there is a large 'technology cemetery' in which many fancy techniques are buried. These have all faced the problems of lack of robustness, very high costs and lack of feasibility for routine use.

29.5 Acknowledgments

This work was supported by the EU 6th Framework Programme under the BIOTRACER project, project number 036272 and the Danish Directorate for Food, Fisheries and Agri-Business (DFFE) grant 3414-04-01032. Pia Engelsmann, Kirsten Michaëlis, Julia Christensen and Liselotte Folling are acknowledged for excellent technical assistance.

29.6 Critical questions for discussion

1. How does *Salmonella* contamination of meat occur?
2. Why is *Salmonella* contamination of meat an important issue?
3. Give some examples of the expectations and demands that one can have when choosing an alternative detection method.
4. Are these expectations and demands different depending upon who you represent? Give some examples.
5. Are there any legislative differences between a result obtained by the method developed in this case study and a given standard culture method?

29.7 References

AGRICULTURE & FOOD (2010). *Statistics 2009 Pigmeat*. Copenhagen. Available at http://www.agricultureandfood.co.uk/Statistics/~/media/agricultureandfood-co-uk/Statistics/Statistics_2009.ashx.

ANONYMOUS (2005a). Commission Regulation (EC) No 2073/2005 of 15 November 2005 on microbiological criteria for foodstuffs. *Official Journal of the European Union*, **L 338**: 1–29.

ANONYMOUS (2005b). Commission Regulation (EC) No 1688/2005 of 14 October 2005 implementing Regulation (EC) No 853/2004 of the European Parliament and of the Council as regards special guarantees concerning *Salmonella* for consignments to Finland and Sweden of certain meat and eggs. *Official Journal of the European Union*, **L 271**: 17–19.

ANONYMOUS (2010). *Annual Report on Zoonoses in Denmark 2009*. National Food Institute, Technical University of Denmark, Søborg.

DENNY, J., BOELAERT, F., BORCK, B., HEUER, O., AMMON, A. *et al.* (2007). Zoonotic infections in Europe: trends and figures – A summary of the EFSA-ECDC annual report. *Euro Surveill*, **12**: 3336.

EIJKELKAMP, J., AARTS, H., and FELS-KLERX, H. (2009). Suitability of rapid detection methods for *Salmonella* in poultry slaughterhouses. *Food Anal Meth*, **2**: 1–13.

HEDMAN, J., LÖVENKLEV, M., WOLFFS, P., LÖFSTRÖM, C., KNUTSSON, R. *et al.* (2012). Pre-PCR processing strategies. In T. Nolan and S.A. Bustin (eds.) *PCR Technologies: Current Innovations*, 2nd ed. CRC Press, Boca Raton, FL. In press.

HINRICHSEN, L. (2010). Manufacturing technology in the Danish pig slaughter industry. *Meat Sci*, **84**: 271–5.

HOORFAR, J., and BAGGESEN, D.L. (1998). Importance of pre-enrichment media for isolation of *Salmonella* spp. from swine and poultry. *FEMS Microbiol Lett*, **169**: 125–30.

JOSEFSEN, M.H., KRAUSE, M., HANSEN, F., and HOORFAR, J. (2007). Optimization of a 12-Hour TaqMan PCR-based method for detection of *Salmonella* bacteria in meat. *Appl Environ Microbiol* **73**: 3040–8.

JOSEFSEN, M., LÖFSTRÖM, C., OLSEN, K., MØLBAK, K., and HOORFAR, J. (2011). *Salmonella*. In D. Liu (ed.) *Molecular Detection of Foodborne Pathogens*, pp. 1023–37. CRC Press, Boca Raton, FL.

KORBER, D.R., and LAWRENCE, J.R. (2004). Biofilm formation. In K. J. Werner (ed.) *Encyclopedia of Meat Sciences*, pp. 59–68. Elsevier, Oxford.

LÖFSTRÖM, C., HOORFAR, J., SCHELIN, J., and MALORNY, B. (2009a). *Salmonella*. In D. Liu (ed.) *Molecular Detection of Foodborne Pathogens*, pp. 447–58. CRC Press, Boca Raton, FL.

LÖFSTRÖM, C., KRAUSE, M., JOSEFSEN, M.H., HANSEN, F., and HOORFAR, J. (2009b). Validation of a same-day real-time PCR method for screening of meat and carcass swabs for *Salmonella*. *BMC Microbiol*, **9**: 85.

LÖFSTRÖM, C., HANSEN, F., and HOORFAR, J. (2010). Validation of a 20-h real-time PCR method for screening of *Salmonella* in poultry faecal samples. *Vet Microbiol*, **144**: 511–14.

LÖFSTRÖM, C., HANSEN, F., MANSDAL, S., and HOORFAR, J. (2012). *Salmonella* detection in meat: Comparative and collaborative validation of a non-complex and cost-effective pre-PCR protocol. *J AOAC Int*, **95**: 100–7.

MALORNY, B., HUEHN, S., DIECKMANN, R., KRAMER, N., and HELMUTH, R. (2009). Polymerase chain reaction for the rapid detection and serovar identification of *Salmonella* in food and feeding stuff. *Food Analytical Methods*, **2**: 81–95.

MALORNY, B., BHUNIA, A., AARTS, H.J.M., LÖFSTRÖM, C., and HOORFAR, J. (2011). *Salmonella* in pork, beef, poultry, and egg. In J. Hoorfar (ed.) *Rapid Detection, Characterization, and Enumeration of Foodborne Pathogens*, pp. 179–94. ASM Press, Washington, DC.

NORDVAL (2008). NordVal certificate no 031. Søborg, Denmark. Available at http://www.nmkl.org/NordVal/Sertifikater/NO31.pdf.

NORDVAL (2009). *Protocol for the Validation of Alternative Microbiological Methods: NordVal*. NordVal/NMKL, Oslo, Norway. Available at http://www.nmkl.org/NordVal/NordValprotocolmarch2009.pdf.

NORDVAL (2010). *NordVal certificate no 041*. Oslo, Norway. Available at http://www.nmkl.org/NordVal/Sertifikater/NO31.pdf.

PODOLAK, R., ENACHE, E., STONE, W., BLACK, D.G., and ELLIOTT, P.H. (2010). Sources and risk factors for contamination, survival, persistence, and heat resistance of *Salmonella* in low-moisture foods. *J Food Prot*, **73**: 1919–36.

QVIST, S. (2011). International validation, ring trial, and standardization of rapid methods. In J. Hoorfar (ed.) *Rapid Detection, Characterization and Enumeration of Foodborne Pathogens*, pp. 157–161. ASM Press, Washington, DC.

30

Using predictive microbiology to benefit the Australian meat industry

J. Sumner, University of Tasmania, Australia, I. Jenson, Meat and Livestock Australia, Australia and T. Ross, University of Tasmania, Australia

Abstract: Since 2002, predictive food microbiology models have become an everyday part of meat processing in Australia. In early February 2011, tropical cyclone Yasi, the most intense cyclone to land on the Australian continent in 100 years, struck the north east coast of Australia. Many services were interrupted, including electricity supply. The integrity of perishable foods under refrigerated storage was compromised, including a large volume of fresh meat at a processing plant. Fortunately, the concept of predictive microbiology is well established and accepted in the Australian meat industry. This case describes how the operators of one plant used predictive modelling to avert considerable economic loss and food wastage following the loss of electricity supply for refrigeration.

Key words: predictive microbiology, commercial application, meat safety, Refrigeration Index, *Escherichia . coli*.

30.1 Introduction to the case

In the mid-1990s the Australian meat industry began an investment in predictive microbiology, the outcomes of which have changed radically how the regulation of processing, retailing and distribution of meat and meat products has evolved. As a work in progress, Sumner and Krist (2002) summarized how predictive microbiology was leading to the development of models for predicting *E. coli* growth in uncooked comminuted fermented meat (UCFM) and in chilling, storage and distribution of red meats. The authors completed their summary with ‘It is anticipated that the model will be integrated into a software package that will interface seamlessly with electronic data loggers to provide an estimate of the potential growth of *E. coli* during the period of interest.’

Since 2002, predictive models have become an everyday part of meat processing in Australia, particularly in the fermented meat industry and in the chilling of red meats after slaughter and dressing. It is in the latter area where predictive microbiology provided essential support for a company (here called Plant A) that was soon to be overtaken by Australia's worst cyclone and rain event in over a century.

30.1.1 Development of a model to predict the growth of *E. coli* in meat and the 'Refrigeration Index' (RI)

Traditionally, meat carcasses have been slaughtered, dressed and chilled before being deboned into specific cuts, plus trimmings. Deboning carcasses, either freshly slaughtered (hot boning) or after short chilling (warm boning), provides logistical advantages in reduced handling and chilling costs. After deboning, meat (27.2 kg) is packed into cartons under controlled chilling. The downside is that the centre of a carton chills slowly, potentially allowing growth of *Salmonella* and Shiga toxigenic *E. coli* (STEC). In Australia, the original regulatory approach prescribed the time allowed for meat in the centre of a carton (initially 25–35 °C) to reach 7 °C, an approach that proved impossible for establishments that hot-boned carcasses and chilled the cartons using large blast freezers, as opposed to plate freezers.

Accordingly, Meat and Livestock Australia (MLA; an industry-owned company that is jointly funded by government and industry and that has responsibility for marketing, and research and development, on behalf of the Australian red meat industry) commissioned the University of Tasmania to develop a model to predict the growth of *E. coli* in meat. The model predicts growth rate and lag time of *E. coli*, based on product pH, lactate, water activity and temperature over time. The development and validation of the model are described in Ross (1999), Mellefont *et al.* (2003) and Ross *et al.* (2003), and reviewed in McMeekin (2007). The model, and its integration into software called the Refrigeration Index (RI), was incorporated into Australia's Export Control (Meat and Meat products) Orders 2005 for all meat, whether boned from hot or chilled carcasses.

The model calculates potential growth of *E. coli* during chilling and processing at the site of greatest microbiological concern. At temperatures less than 7 °C (the minimum temperature for growth of *E. coli*) no growth is predicted. The site of microbiological concern is the slowest cooling point that may be contaminated with *E. coli*. For a carcass, this is the carcass surface, while for packaged product it is the centre of the package. The potential for *E. coli* growth, based on the time and temperature history, and for appropriate values of pH, lactic acid concentration and water activity, is determined using the model and expressed as the RI. The temperature changes over time are recorded with electronic data loggers. The pH, water activity and lactic acid values are assumed constant at agreed values considered representative for different kinds of meat.

The specific values of the RI are the criteria against which product acceptability is assessed. For meat processes to be considered acceptable, processing establishments needed to demonstrate in routine production that:

- The RI of the process is no more than 1.5,
- 80% of RI values are not more than 2.0,
- No RI is more than 2.5.

A simple computer software tool that receives data from electronic temperature data loggers was developed to allow time–temperature data to be analysed and the RI calculated (see Section 30.7, Further Reading).

30.2 Significance of the case

Since its introduction the RI has been used by abattoirs and boning rooms as a routine tool for controlling their processes. However, if loss of refrigeration control occurs, the RI can be used to demonstrate to regulators whether meat safety has been compromised or not, without recourse to microbiological testing, which is often impractical due to the time required to obtain the results. It was in February 2011 that the RI was used to its fullest extent, when Australia's most damaging cyclone and rain event in over 100 years hit Queensland's coastal areas.

30.3 The case history continues – extreme rainfall and tropical cyclone Yasi

30.3.1 The genesis of tropical cyclone Yasi

During December 2010 and January 2011, over 50 Queensland cities and towns were inundated by flooding (BOM, 2011a), including many in which meat processing plants operate. Sections of the state capital of Brisbane were evacuated due to rainfall of over 630 mm in 48 h (a once-in-100-years event) and a flood level that was the highest in over 25 years (BOM, 2011b). In late January 2011, Tropical Cyclone (TC) Yasi began as a tropical low-pressure weather system northwest of Fiji and began tracking towards Australia (BOM, 2011c). To the north of Vanuatu the system quickly intensified to cyclone category and was named Yasi (Fijian for sandalwood) on 30 January by the Fiji Meteorological Service. TC Yasi maintained a westward track and rapidly intensified to Category 2 on 31 January and then to Category 3 later the same day. The cyclone began to accelerate towards the tropical Queensland coast and, when it crossed the coast early on Thursday 3 February, was upgraded to a category 5 tropical cyclone, the highest storm intensity rating. Category 5 cyclones are characterized by winds of more than 280 $km.h^{-1}$ and widespread destruction (e.g. trees uprooted, cars turned over, boats washed ashore, houses destroyed).

30.3.2 The plant prepares for the coming power loss

At Plant A, cattle were slaughtered on Day 1 and the carcasses underwent the usual overnight active chilling cycle. The Bureau of Meteorology confirmed its prediction of an impending extraordinary weather event. Foreseeing potential problems, including the loss of electricity supply for refrigeration, the operators of Plant A took proactive steps to be able to assess the integrity of the considerable volume of meat in the plant after the event. On Day 2, conditions were unsafe for staff to travel to work and a skeleton crew at the plant did its best to take steps to manage the impending crisis. Significant actions and events were as follows:

- Day 2: refrigeration was switched off at 11:30 p.m. and data loggers were placed with carcasses in every chiller,
- Day 3: Power to Plant A was lost at 01:35,
- Day 4: Power was returned to Plant A at 18:06,
- Day 4: Refrigeration was reactivated at 18:20,
- Day 7: Data loggers were downloaded and temperature records interpreted using the RI,
- Day 7: The carcasses killed on Day 1 were processed in the boning room.

30.3.3 The aftermath

In all, refrigeration at Plant A was lost for 55 h and, when power was returned, QA staff recovered the data loggers from each chiller, spot-checked carcass surface temperatures and set about assessing the damage.

The refrigeration systems for Chiller #6 had an ammonia leak and could not be reactivated for some time; carcasses in this chiller were condemned.

Data loggers from the other chillers were downloaded and the RIs calculated. An example is presented in Fig. 30.1, which indicates an RI of 0.96. It is typical of what happened in all six chillers – a normal chilling rate over the first 20 h before power was lost, followed by a gradual rise in temperature over the following 55 h. Carcasses were warmer than 7 °C (the minimum growth temperature for Gram-negative pathogens such as *Salmonella* and Shiga-toxic *E. coli*), for much of the time and ranged up to 10–13 °C across all chillers.

When the RIs for product in all chillers were calculated they did conform with the criteria specified in the Export Control (Meat and Meat Product) Orders 2005 (see Section 30.1.1 above). This gave the company the incentive to continue processing the carcasses rather than sending them for processing into meat and bone meal, which would result in considerable economic loss.

Consequently, the product was boned, cartoned and frozen while the company pondered its disposition.

30.4 Resolution and outcomes

While the RIs indicated regulatory compliance, the company would also need to convince customers that the product was acceptable for human consumption, so the Quality Assurance staff also begin organoleptic and microbiological assessment for quality indicators.

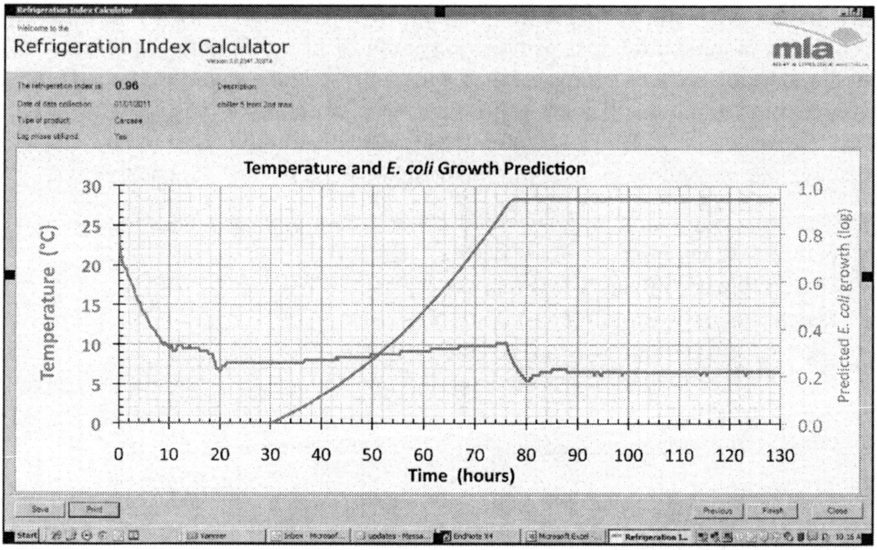

Fig. 30.1 A copy of an output from the Refrigeration Index for one of the chillers in Plant A during the time of cyclone Yasi. The graph shows the temperatures experienced by the meat including the period when electricity supply was interrupted for ~55 h starting after about 20 h. The chart also shows the predicted *E. coli* growth, which is the basis of the RI calculation. Growth was predicted to be possible, after resolution of lag phase, after ~32 h into the process, and until refrigeration was restored at ~75 h into the process.

A team of independent, experienced meat technologists thawed a large number of cartons, sampling the meat by taking 38 mm core samples for microbiological testing. After testing they reported, 'No off odours were detected from any of the samples and each had a natural and normal range of meat and fat colours. There was no evidence of odour or discolouration that could be attributed to excessive bacterial growth. The samples bloomed as expected for frozen meat', and the microbiological profile indicated:

- The mean aerobic plate count was 6.16 $\log_{10}$ cfu.g^{-1},
- *E. coli* was present on 38.3% of samples (limit of detection 10 g^{-1}) with positive samples having a mean 1.62 (SD 0.81) $\log_{10}$ cfu.g^{-1},
- *Staphylococcus aureus* was detected on six of 30 samples (20%) with positive samples having a mean of 2.12 (SD 0.18) $\log_{10}$ cfu.g^{-1},
- Lactic acid bacteria were detected on only one of 30 samples (5.36 $\log_{10}$cfu.g^{-1}),
- *Brochothrix thermosphacta* was not isolated (limit of detection 100 cfu.g^{-1}).

All microbiological quality indicators (Aerobic Plate Count (APC), *E. coli*, Enterobacteriaceae) were higher than the company's normal levels, an outcome not unexpected given the length of time product was warming from 7 to 13 °C.

The company was advised that these results were similar to those obtained for retail minced (ground) beef in Australia (Phillips *et al.*, 2008). Given the additional time that would elapse between further processing this meat and its presentation for retail sale, the company sought to sell the meat for human consumption, following heat treatment, or for non-human consumption.

On the basis of the RI, the regulator accepted that the meat was microbiologically safe for consumption but was concerned that the overall microbiological quality would result in a reduced shelf-life. The regulator decided that, although the counts were higher than normal, the meat could be used for human consumption, following suitable heat treatment and requiring that any purchaser was made aware of the history and status of the product.

The outcome was of great economic value to the company since the product would be sold at a reasonable price rather than being downgraded to pet food use or for meat and bone meal for animal feed supplementation. The company estimated that being able to sell the meat for human consumption in processed foods represented a saving of approximately US$500 000 for the company.

30.5 Commentary

This case study is an excellent example of the value of the application of science in decision-making, and of the time and money that can be saved in food microbiology by adopting a well-validated predictive microbiology approach. While this case is an extreme example, the RI is used routinely in the Australian meat industry to assure process integrity. It is also used to evaluate the microbiological significance of loss of temperature control, often with the result that product that has undergone loss of temperature control can be shown to remain microbiologically sound and need not be discarded, with savings both to the industry and also to consumers through increased industry productivity.

This case study demonstrates the value of:

- Predictive microbiology tools to assist in decision-making,
- Deployment of tools across an entire industry sector,
- Agreement between industry and regulators on the use of scientific data for decision-making,
- A technology that prepares food processors to manage unexpected events.

Predictive microbiology for the safety assessment of the meat was an aid to the company's decision to continue processing the carcasses but, ultimately, microbiological testing was required to provide data for other microbiological parameters, i.e. the available predictive microbiology tools provided assurances about the safety of the product, but not the quality. The development and application of a wider range of predictive models might have been able to avoid testing completely, and may be a subject for additional research.

30.6 Critical questions for discussion

1. What behaviours of microorganisms in foods might be predicted by mathematical models?

2. How many foodborne pathogens have predictive growth models? How can they be accessed and how are they used?
3. Are there other examples of predictive models being used by food regulators?
4. What information is required to be able to use predictive microbiology with confidence to assess the quality and safety of foods? Could predictive models replace the need for microbiological testing of product?
5. In this case, only temperature was measured. What values might have been assigned to the other model parameters (pH, water activity, lactate)?

30.7 Further reading

The Refrigeration Index and supporting reports and documentation can be accessed at www.foodsafetycentre.com.au/refrigerationindex.php or http://www.redmeatinnovation.com.au/innovation-areas/food-safety/predictive-modelling-tools.

General information about predictive microbiology can be found at the Predictive Microbiology Information Portal Home Page at http://portal.arserrc.gov/ and ComBase at http://www.combase.cc/.

The reader is also directed to the following texts and reviews:

MCDONALD, K. and SUN, D.W. (1999). Predictive food microbiology for the meat industry: A review. *Int J Food Microbiol*, **52**:1–27.

MCKELLAR, R.C. and LU, X. (eds) (2004). *Modelling Microbial Responses in Foods*. CRC Press, Boca Raton, FL.

MCMEEKIN, T.A., OLLEY, J., ROSS, T. and RATKOWSKY, D.A. (1993). *Predictive Microbiology. Theory and Application*. Research Studies Press, Taunton.

ROSS, T. and MCMEEKIN, T.A. (2003). Modeling microbial growth within food safety risk assessments. *Risk Anal*, **23**: 179–97.

30.8 References

BOM (Australian Government Bureau of Meteorology). (2011a). *Map of flood affected cities and towns in Queensland December 2010 and January 2011*. Available at http://www.bom.gov.au/hydro/flood/qld/fld_reports/qld_flooded_towns_2011.pdf. Accessed 17 October 2011.

BOM (Australian Government Bureau of Meteorology). (2011b). *Flood summary for the Brisbane River at Brisbane City*. Available at http://www.bom.gov.au/hydro/flood/qld/fld_reports/brisbane_fact_sheet_2011.pdf. Accessed 17 October 2011.

BOM (Australian Government Bureau of Meteorology). (2011c). *Severe Tropical Cyclone Yasi. 30 January–3 February 2011*. Available at http://www.bom.gov.au/cyclone/history/yasi.shtml. Accessed 14 October 2011.

MCMEEKIN T.A. (2007). Predictive microbiology: Quantitative science delivering quantifiable benefits to the meat industry and other food industries. *Meat Sci* **77**: 17–27.

MELLEFONT, L.A., MCMEEKIN, T.A. and ROSS, T. (2003). Performance evaluation of a model describing the effects of temperature, water activity, pH and lactic acid concentration on the growth of *Escherichia coli*. *Int J Food Microbiol*, **82**: 45–58.

PHILLIPS, D., JORDAN, D., MORRIS, S., JENSON, I. and SUMNER, J. (2008). A national survey of the microbiological quality of retail raw meats in Australia. *J Food Protect*, **71**: 1232–6.

ROSS, T. (1999). *Predictive Food Microbiology Models in the Meat Industry*. Meat & Livestock Australia, Sydney, Australia.

ROSS, T., RATKOWSKY, D.A., MELLEFONT, L.A. and MCMEEKIN, T.A. (2003). Modelling the effects of temperature, water activity, pH and lactic acid concentration on the growth rate of *Escherichia coli*. *Int J Food Microbiol*, **82**: 33–43.

SUMNER, J. and KRIST, K. (2002). The use of predictive microbiology by the Australian meat industry. *Int J Food Microbiol*, **73**: 363–6.

Part VI

Food adulteration and authenticity

31

Forensic investigation of a sabotage incident in a factory manufacturing nut-free ready meals in the UK

M. J. Walker, LGC Ltd, UK

Abstract: Deliberate sabotage (tampering) of food occurs from time to time and is frequently dealt with by management with or without the involvement of police, but generally with minimum media coverage. The following case study describes the forensic science of a sabotage incident in a medium-sized food factory.

Key words: allergen, cross-contamination, analysis, forensic, incident, recall, sabotage.

31.1 Introduction

Amy (not her real name) was a student in her first year at university. The end-of-term dinner was a glittering occasion with printed menu, formal dress and speeches. Amy was allergic to almonds and was well aware of the dangers of eating the wrong food. She was always careful and felt in control. But, unknown to her, the shortbread she ate towards the end of the meal contained almonds, a fact not revealed on the menu. Amy started to feel unwell and suspected she might have eaten something that contained almonds but she didn't want to make a fuss.

So she sat on, embarrassed to say anything on the grand occasion and hoped the reaction would pass. It didn't. Amy died later that night.

Food allergy is an adverse reaction mediated by the immune system, and its prevalence appears to be increasing. The number of children with peanut allergy in the UK appears to have increased in recent years with estimates as high as one in 55 children affected. There are about ten deaths (and potentially many more near misses) every year in the UK in food allergy-related incidents (Wight and Hattersley, 2007). Many of these tragically affect young people, teenagers at the start of their adult lives. It is vitally important that food prepared for those who have a food allergy is free from the allergen of interest, and many food manufacturers have geared up to provide nut-free food, recognising that nut allergy is one of the most severe and prevalent in the Western world.

Deliberate sabotage (tampering) of food occurs from time to time and is frequently dealt with by management with or without the involvement of police, but generally with minimum media coverage. Objects used in sabotage incidents range from the innocuous (paper) to the highly offensive (condoms) and have included glass, needles, blades, staples and mercury.

The following case was highly unusual in that it involved sabotage, by peanuts, of the production area of a plant manufacturing nut-free ready meals for major retailers. As previously highlighted, this could have had extremely serious consequences for nut-allergic individuals. The case illustrates the difficulties of allergen management in a food manufacturing plant, the challenges facing the investigation of a sabotage incident, and the role of the expert witness in a criminal trial.

31.1.1 How did it all begin?

The product on the line was a nut-free ready meal manufactured for several of the retail multiple stores. Management and staff had worked hard to achieve the quality standards needed to supply such a potentially sensitive product to the multiples. There was training, signage, stringent hand-washing, audit of suppliers, and testing of incoming raw materials and of finished product. All the prerequisite programmes were in place along with hazard analysis and critical control points (HACCP). The environment was carefully controlled and monitored for the allergens of interest. The audits had been tough but the company had come through with credit and was now a trusted supplier of nut-free meals.

The morning shift had taken over the production line. A worried production supervisor asked to see the production manager. 'I've a cleaner here who says she has found some peanuts in the production area.'

31.1.2 Immediate actions

Over the next few hours more peanuts were found at different locations, and at 4:00 p.m. managers decided to stop production and search the area. More peanuts were found, seized and bagged. An urgent meeting of the senior team took place. What could be said with certainty about the day's production, and previous batches? Was it possible that product contaminated with peanuts had already been released? What would happen if a nut-allergic consumer ate one . . . and died? It was decided that it was prudent to inform the central competent authority, the Food Standards Agency, and the hard decision was taken to inform the company's customers. A full product recall swung into action. The future of the company was on a knife edge.

31.2 Significance of the case

Children with a peanut allergy are faced with food and social restrictions due to the potentially life-threatening nature of food allergies, for which there is no cure or treatment. This inevitably impacts upon their quality of life. One study reported that, in comparison, children with a peanut allergy had a poorer quality of life than children with insulin-dependent diabetes mellitus (Avery *et al.* 2003). There was more fear of an adverse event and more anxiety about eating, especially when eating away from home. Children with peanut allergy felt more threatened by potential hazards within their environment, felt more restricted by their allergy regarding physical activities, and worried more about being away from home. However, they felt more safe when carrying epinephrine kits and were positive about eating at familiar restaurants. Their anxiety may be considered useful in some situations, promoting better adherence to allergen avoidance advice and rescue plans (Avery *et al.*, 2003).

Food allergy hospital admissions are rising steeply (Gupta *et al.*, 2003) with concomitant costs to the health services.

This case is the only example, so far as the author is aware, not only of a significant food recall incident based on sabotage using a major food allergen, but also of a criminal trial resulting from the incident.

31.2.1 Regulatory aspects

Existing and more recent legislation has an impact on the use of food allergens (European Community, 2002; legislation.gov.uk, 1990; legislation.gov.uk, 2007). General food law (both EU and UK) makes it an offence to sell food that is unsafe for, or not of the nature, substance or quality demanded by, consumers. Foods that have incorrect allergen labelling or are not correctly described (such as unjustified claims of 'free from') are therefore illegal. The UK Food Standards Agency has published comprehensive best practice guidance on allergen cross-contamination and 'may contain' labelling (Food Standards Agency, 2006). For prepacked foods, if any of 14 specified allergens are incorporated in their formulation, the law requires, with certain limited exemptions, a labelling declaration. On 30 January 2008 the European Commission issued a proposal for a new food information regulation (European Commission, 2008). The proposal maintains the derogation for Member States to adopt national rules for the provision of information for food sold non-prepacked and pre-packed for direct sale. However, the Commission proposal has extended mandatory allergen labelling to apply to both food sold non-prepacked and pre-packed for direct sale, including catering establishments. The UK Food Standards Agency has published best practice guidance on the provision of allergen information for non-prepacked foods (Food Standards Agency, 2008). There has also been a series of cases in the UK courts (Gowland and Walker, 2011) involving civil litigation with substantial damages awarded, health and safety and criminal food law prosecutions involving food allergens.

31.2.2 Market and economic issues

The provision of allergen-free food can be regarded as a 'public good' and there is a market for allergen-free foods; but food manufacturers find it difficult and expensive to deal with risk management issues around offering such food. After 2004 the number of allergen food incidents per year in the UK rose steeply to 86 in 2009 and 79 in 2010. Of these, 49 and 34 respectively were alerts indicating that product was withdrawn (FSA Incidents 2009, 2010). While no formal figures are available, a single recall may cost the food company involved a substantial sum. It is known that the incident reported here is reported to have cost the company around £1 million. Thus, while the annual cost to the UK of allergen recall incidents is unknown, it can be estimated to have cost up to perhaps £34 million in 2010 (personal communication). Food recall incidents such as that described in the current case are very serious, not only for the manufacturer involved, but for all sectors of the food industry supplying allergen-free or allergen-labelled product.

From the point of view of the manufacturer, disposal of recalled product has the immediate effect of financial loss for that product (raw materials, staff costs, storage costs, etc.), but also a long-term effect of the inability to supply markets and reputational damage, thus losing market share.

31.3 The case history continues

The seriousness of the incident, already high, escalated.

31.3.1 The investigation

It quickly became obvious that this was not a lapse in procedures. Someone had to have done this deliberately. No other explanation fitted the fact that whole, salted peanuts were found scattered in multiple locations in the food-handling areas. Senior management considered the situation: who had a grudge or motive? Who had access? A possible suspect was a fitter who had been asked to remove some inappropriate material from his workshop wall. He had access to all parts of the plant, and the cleaner who had first spotted peanuts said he had been in the vicinity not long before.

His locker was searched. His overalls and factory casual clothing were seized and searched. No peanuts were found in the pockets and the clothing was locked in the Production Manager's office along with the peanuts that had been found in the food-handling area.

31.3.2 The police

It was now obvious that this was not a simple disciplinary matter and the management decided to call in the police, who arrived, took statements and bagged up the evidence.

31.3.3 Forensic science

The fitter was arrested and the exhibits (his clothing and the seized peanuts) were sent off for forensic examination to our laboratory at LGC Ltd, which is the forensic supplier to the police force dealing with the inquiry. It was hoped that DNA would be recovered from the peanuts, pinpointing who had handled them. However, despite extensive swabbing and PRC, no recoverable DNA was found and this avenue of investigation proved fruitless.

Founded in 1842 as the Laboratory of the Government Chemist, LGC provides the full range of forensic services from crime scene examination through to DNA, toxicology, questioned documents, e-crime, ballistics and specialist services such as pathology, entomology, archaeology and palynology. Moreover, LGC is not an ordinary forensic supplier, since in addition LGC deals with food analysis and hosts the Government Chemist function. The main remit of the Government Chemist team is to be the independent technical appellate arm of formal food control. When prosecution and defence laboratories disagree on their findings in a food prosecution the Government Chemist can be called in by either party, or the court, to carry out a definitive analytical investigation. Thus the Food Chemistry team is forensically aware and accustomed to giving expert testimony in the courts. Moreover, food law in the UK is criminal law and any investigation must be carried out to the criminal burden of evidential probity – beyond reasonable doubt.

Rather than the trail going cold, LGC Forensics asked their food science colleagues to look for traces of peanut protein in the pockets of the overalls and clothing belonging to the defendant. The analysis and evidence recovery were undertaken in a dedicated restricted-access 'special projects' containment suite. This is designed to guard against any possibility of laboratory cross-contamination.

When working in the containment suite analysts must gown-up in disposable (colour-coded) forensic coveralls. There is positive-pressure air-handling and environmental swabbing to ensure the area remains free of allergens. There is maximum use of disposable equipment, extending to the recycling of blenders to non-sensitive areas of other work if they have been used with a positive sample or control. Separate safety cabinets are used for different aspects of the work. The analysts must take care over their own diet to ensure that a major food allergen has not been recently consumed when working in the containment suite.

The analytical approach used was enzyme-linked immunosorbent assay (ELISA). Swab samples were taken using a proprietary allergen swab kit. The swabs were extracted with buffer and the resulting extracts were analysed in duplicate using a commercial peanut assay kit, which has been validated by Association of Analytical communities (AOAC) (Walker *et al.*, 2008). For quality control purposes a swab blank, swab QC material and kit standards were included in each batch.

Following protocols designed to preserve the chain of custody of evidence, the exhibits in the case were transferred to the containment suite, photographed and documented (protocols involve recording who transfers the exhibits to whom and when and where they are stored with a written, signed record).

No peanuts had been found in the pockets of two exhibits of clothing allegedly belonging to the defendant. Therefore, experiments were carried out to determine if traces of peanut protein were present on the inside surfaces of the pockets. This was done by carefully swabbing the inside surfaces of the pockets and extracting the swabs for subsequent ELISA. In keeping with Government Chemist forensic procedures, all key steps in the process were witnessed by a second scientist.

31.3.4 Forensic evidence

The exhibits consisted of five sets of overalls, one allegedly from the defendant and four from other workers at the plant. A pair of tracksuit bottoms, a high-visibility jacket and a pair of industrial boots allegedly belonging to the defendant were also submitted along with plastic bags that had contained some of the exhibits. Swabs from the garments allegedly associated with the defendant returned positive results for peanut protein. In all, 24 swabs were taken from the defendant's garments and 21 were positive for peanut protein. The overalls belonging to four other workers were swabbed 25 times, all with negative results.

31.3.5 The trial

The defendant was charged with possessing materials for contaminating goods with intent to cause public alarm, injury and economic loss and with threats to kill. The Crown Court jury trial lasted 3 weeks. The forensic results detailed above formed a major part of the trial and were given in evidence by the author assisted by two colleagues.

Part way through the trial, the defence requested experiments to assess the possibility of contact transfer of traces of peanut protein after handling peanuts with subsequent handling of garments and the effects of hand washing. The defence hypothesized that anyone coming into contact with peanuts and then subsequently handling the exhibits could transfer traces of peanut protein onto the exhibit, thus contaminating them.

Instructed by the Crown, the author set up experiments to see if handling a peanut transferred sufficient peanut protein to the fingers for it to be picked up from fabric.

The results and outcome of the tests indicated that, even after brief contact with a peanut, peanut protein was readily transferred to clothing and easily detected by analysis even after ten successive finger/fabric contacts. It was also found that only rigorous hand washing stopped the transfer. The defence argued that, because it was the same management team who found the peanuts and who seized the clothing, it was reasonable to suggest that the management team, in searching the pockets of the clothing, had themselves contaminated the clothing with peanut protein.

Much other evidence was heard in the trial and is not reproduced here.

31.4 Resolution and outcomes

After over 8 hours of deliberations, the jury indicated they could not reach a verdict on which all of them were agreed and were discharged by the Judge. Consequently the defendant was found not guilty.

31.5 Commentary and personal reflections

The case raises some key points.

- Food factory management were (and generally remain) unaware of the precautions required in dealing with an allergen sabotage incident.
- There are well-known procedures in crime scene examination that must be followed in order to preserve DNA evidence. These require the personnel collecting the evidence to ensure that they do not themselves cross-contaminate the evidence, by wearing forensic coveralls, gloves and masks. The evidence must be individually securely bagged and tagged so as not to cross-contaminate other evidence and to preserve the evidential chain of custody. Essentially, allergen sabotage evidence must be collected to the same standards. This is not expected to be widely known to police, crime scene investigators or food plant managers.
- Even minimal handling of peanuts risks widespread subsequent transfer to other surfaces.
- Hand washing needs to be to a high standard to remove peanut protein deposited on fingers.

These points have implications for food manufacturers and possibly for dermal exposure and sensitization of infants.

Those with allergies have a right to the same quality of life as their non-allergic counterparts and, as prevalence increases, so too does the need for efficient and effective allergen management in the food industry generally and especially in the catering sector. If you have completed this chapter you have gained some knowledge of the problems in this area and perhaps stimulated your interest to learn more. As you take your studies further, as a qualified food science professional, you will have the opportunity to bring real benefits to people with food allergies, and perhaps save a life.

Finally, it is the duty of every expert witness to assist the court no matter which side has commissioned the work; however, it is unusual for a single expert to give evidence for both the prosecution and the defence in a criminal trial. The fact that all the scientific work carried out in the case was conducted according to Government Chemist referee analysis protocols went a long way to ensuring that the court was satisfied with the impartiality of the evidence.

31.6 Acknowledgements

Indu Patel, Kirstin Gray, Malvinder Singh and Pete Colwell are thanked for their input to the case. The permission of the Government Chemist, Dr Derek Craston, to publish this material is gratefully acknowledged.

31.7 Critical questions for discussion

1. Describe the principles of ELISA and the advantages and disadvantages of this technique, focusing on its application to food allergens. Compare and contrast

ELISA with DNA and mass spectrometry approaches to allergen detection and quantification.
2. What are the major food allergens and what laws govern their labelling on prepacked and non-prepacked foods?
3. Discuss the management of allergens in the context of the food supply chain (harvest, transport, storage, processing, labelling and retail).
4. (a) What is a 'prerequisite programme' – give examples; (b) how would you validate cleaning in a food plant to remove allergens?
5. Discuss the terms (a) chain of custody of evidence', (b) 'beyond reasonable doubt', (c) 'expert witness', (d) 'examination in chief' and 'cross examination' in the context of the UK or USA legal systems.
6. What is 'anaphylaxis'? Discuss in the context of food allergens.

31.8 Further reading

For further information on this and other food forensic cases please see G. Taylor, *Forensic Enforcement, The Role of the Public Analyst* (2010), RSC Publishing, London.

See http://www.safefoodallergy.ning.com for further information and an opportunity to network with those with an interest in food allergy and food intolerance.

31.9 References

AVERY, N.J., KING, R.M., KNIGHT, S. and HOURIHANE, J.O. (2003). Assessment of quality of life in children with peanut allergy. *Pediatric Allergy and Immunology*, **14**: 378–82.

EUROPEAN COMMISSION (2008). *Commission proposal to overhaul EU food labelling rules*. Available at http://ec.europa.eu/food/food/labellingnutrition/foodlabelling/proposed_legislation_en.htm. (Accessed May 2011.)

EUROPEAN COMMUNITY (2002). Regulation (EC) No 178/2002 Of The European Parliament and of The Council of 28 January 2002 laying down the general principles and requirements of food law, establishing the European Food Safety Authority and laying down procedures in matters of food safety. *Official Journal of the European Communities*, **L 31**, 1–24, Articles 14 and 16.

FOOD STANDARDS AGENCY (2006). *Guidance on Allergen Management and Consumer Information*. Available at http://www.food.gov.uk/multimedia/pdfs/maycontainguide.pdf. (Accessed May 2011.)

FOOD STANDARDS AGENCY (2008). *Providing Allergen Information for Non Pre-packed Foods*. Available at http://www.food.gov.uk/foodindustry/guidancenotes/labelregsguidance/nonprepacked. (Accessed May 2011.)

FSA INCIDENTS (2009). *Annual Report of Incidents 2009*. Available at http://www.food.gov.uk/multimedia/pdfs/incidents09.pdf.

FSA INCIDENTS (2010). *Annual Report of Incidents 2010*. Available at http://www.food.gov.uk/multimedia/pdfs/publication/annualreportofincidents2010.pdf.

GOWLAND, H.M. and WALKER, M.J. (2011). Food allergy, a summary of recent cases in the criminal and civil courts of the UK. *Clinical and Translational Allergy*, **1**(Suppl 1): O1. Available at http://www.ctajournal.com/supplements/1/S1. (Accessed January 2011.)

GUPTA, R., SHEIKH, A., STRACHAN, D. and ANDERSON, H.R. (2003). Increasing hospital admissions for systemic allergic disorders in England: Analysis of national admissions data. *BMJ*, **327**: 1142–3.

LEGISLATION.GOV.UK (1990). *Food Safety Act 1990, (C.16) Sections 8, 14 and 15*. Available at http://www.legislation.gov.uk/ukpga/1990/16/contents.

LEGISLATION.GOV.UK (2007). *The Food Labelling (Declaration of Allergens) (England) Regulations 2007 with Devolved Equivalents*. Available at http://www.legislation.gov.uk/2007?title=The%20Food%20Labelling%20%28Declaration%20of%20Allergens%29.

WALKER, M.J., COLWELL, P., ELAHI, S., GRAY, K. and LUMLEY, I. (2008). Food Allergen Detection: A Literature Review 2004–2007. *J Assoc Public Analysts* (Online), **36**: 1–18.

WIGHT, M. and HATTERSLEY, S. (2007). *Food Standards Agency Board Paper INFO 07/09/01* and references therein. Available at http://www.food.gov.uk/multimedia/pdfs/board/info070901.pdf.

32

Lessons from the large-scale incident of animal feed contamination with dioxins in Germany in 2011

J. Zentek and F. Knorr, Freie Universität Berlin, Germany, A. Mader, Freie Universität Berlin, Germany and Federal Institute for Risk Assessment, Germany and H. Schafft, Federal Institute for Risk Assessment, Germany

Abstract: During a recent incident in Germany, it was reported that animal feeds were contaminated with dioxins. Although the established traceability system enabled the source of the contamination to be identified, the incident only came to light after most of the contaminated material had already been consumed by livestock. This case describes the event and its consequences, which were significant, and discusses the lessons that can be learned.

Key words: animal feed, contamination, tracing, dioxins, feed, oil, fat.

32.1 Introduction to the case

Approximately 2300 tonnes of fat contaminated with dioxins was distributed to 25 compound feed producers in the year 2010 in Germany. Fatty acids meant for industrial uses (i.e. non-food uses, such as biodiesel) were processed into fats for animal feeds by a single company. This alternative use was unauthorized. Although the producer was aware of the contamination of the material with dioxins, countermeasures were not carried out, nor were the authorities informed. The dioxin load in the compound feeds was finally revealed by routine tests conducted by feed manufacturers who had used the contaminated fats as feed ingredients. The feed manufacturers immediately informed the responsible authorities.

On 27 December 2010, the first alert was published by one of the German federal states, Schleswig-Holstein, via the Rapid Alert System for Feed and Food (RASFF). The estimated quantity of compound feed contaminated with dioxins in the year 2010 was up to 150 000 tonnes. In total, approximately 25.4 mg of dioxins were introduced to the feed chain, according to a mathematical model based on facts published by the Federal Ministry of Food, Agriculture and Consumer Protection, as well as the Directorate General for Health and Consumers of the European Commission. Compound feed manufacturers that had processed the contaminated fat were quickly identified. In the first days of January 2011, feed manufacturers in 12 German counties were affected, resulting in the delivery of contaminated batches to approximately 4760 farms. Some meat and egg samples were found to have levels of dioxins that were higher than those allowed under EU law. Although no acute health implications were identified for consumers, the products had to be disposed of. The economic impact was considerable due to the reduced consumption of animal-derived food and trade restrictions.

32.2 Significance of the case

Polychlorinated dibenzodioxins and polychlorinated dibenzofurans are often summarized under the term 'dioxins'. Many different isomers (congeners) exist, depending on the physical and chemical conditions during the combustion processes. These substances are not produced intentionally but are formed as by-products of chemical reactions such as heating of carbon compounds in the presence of inorganic or organic chlorine. Dioxins escape into the environment as combustion gases or production wastes. They accumulate in soils and sediments, but can also occur in the food chain, especially in the fatty tissue of meat, as well as in milk and eggs. They are characterized by a very long half-life. Some dioxins have very high toxicity, but the isomers differ significantly in their biological effects. Another group of widespread chemicals that are regularly mentioned in context with dioxins are polychlorinated biphenyls (PCBs). Due to the fact that some of the PCB compounds have toxicological properties that are similar to dioxins, they are often termed 'dioxin-like' PCBs.

Normally, dioxin levels are low in feedstuffs (Schwind *et al.* 2007). Dioxins can be introduced into the feed chain from contaminated ingredients and additives such as certain clays, roughage, oils and fats. They can occur when combustion processes take place in the presence of chlorine, either in natural events such as fires or volcanic eruptions or in industrial manufacturing processes (Gude *et al.* 2008). Therefore, most incidences of contamination can be considered unavoidable, but they may also occur due to the deliberate use of contaminated materials. As the result of various incidents in recent years, regulations have been implemented to limit the contamination of feed and food. Animals and humans are primarily at risk from contaminated food and feed.

32.2.1 Regulatory aspects

Due to the occurrence of various incidents involving dioxin-contaminated feed in the food chain in recent years, relevant statutory rules have been tightened in Europe and elsewhere. Dioxins are among the 'undesirable substances' in animal feed mentioned in Directive 2002/32/EC, which includes maximum limits for dioxins in feed materials, and strict guidelines exist for feed producers. The European Commission recommendation 2006/88/EC provides advice on the reduction of the presence of dioxins, furans and polychlorinated biphenyls in feed and foodstuffs.

32.2.2 Economic and market aspects

The dioxin scandal had severe consequences for the German agriculture and food industry, leading to a significant reduction in the consumption of meat and eggs. Conventionally produced products were particularly affected, as many consumers increasingly consumed organic products. From a total of 217 400 livestock farms in Germany in 2010, 12 000 farms were organic, according to Federal Statistical Office of Germany (DESTATIS). About 1% of 27.8 million pigs and 3% of 116.7 million poultry were raised organically.

In addition to these major changes in the internal market, export restrictions contributed to the significant repercussions on the German meat and egg market. Five and a half million tons of pork and 10 billion eggs for consumption were produced in 2010 and 2009, respectively. To give some idea of the trade volumes involved, 1.5 million tons of pork (2010) and 1.9 billion eggs (2009) were exported, according to DESTATIS. In 2010 some countries, including China and South Korea, imposed import bans on meat products from Germany. This led to a significant decline in prices that threatened the existence of many farms. According to the statistical evaluation of the Johann Heinrich von Thuenen-Institute, the price of pork decreased from about €1.42 within 3 weeks of the crisis to about a quarter of the original price, with a rapid improvement thereafter. The price of eggs decreased to about one-third of the original price with a price recovery occurring only after more than 2 months. Beyond the direct costs of the dioxin crisis, permanently higher costs due to increased controls and increased costs for surveillance can be expected. This will presumably lead to a financial burden on the producers of meat and eggs. More importantly, there will presumably also have been a decline in consumer confidence in the national production system.

32.3 The case history continues

After the outbreak of the dioxin crisis, testing laboratories were suddenly under considerable pressure. Dioxin analyses take 3–5 days to perform as rapid screening methods are not yet available, and cost €400–500 on average per sample. Due to the high demand for feed and food analysis during the crisis, many laboratories quickly reached the limits of their capacities. In order to be better prepared for future emergencies, state and private laboratories must have larger capacities. The need for analyses will also increase due to the higher demand for routine investigations. It cannot be expected that the increased sample number will lead to further cost reductions as the high-end laboratory equipment and laboratory professionals still need to be maintained. Overall, the indirect costs of the dioxin crisis are considerable.

32.4 Resolution and outcomes

Due to the recent incidents in Germany, an action plan was presented by the Ministry of Agriculture to further restrict the use of high-risk feed components. As a consequence of the dioxin scandal, the Ministry proposed higher standards for the safety measures in the animal feed industry and tightened the controls. The measures include increased reporting and monitoring obligations, liability insurance for businesses and products, and particularly the mandatory reporting of dioxin results by private laboratories. In the future, a strict separation of fats for technical purposes and fat for feed production is mandatory.

Over several months, the dioxin crisis had considerable negative economic consequences. Due to the various factors that must be considered when estimating the financial impact of the crisis, no final figures are currently available. The public has also become much more aware of the relationship between the safety of feed and food. By improving analytical capabilities and controls, the likelihood that such cases can be detected more efficiently has increased. However, it is also clear that recklessness or criminal intent of individuals could always lead to similar cases. With the new rules on disclosure requirements, such problems are now less likely to occur. The overall picture shows that the feed chain is linked very closely with the food chain and that neglecting feed safety has serious consequences for the agricultural sector and the food industry (Kamphues and Schulz 2006).

32.5 Commentary

The safety of animal feed is an essential prerequisite for the production of high-quality food of animal origin. The farm-to-fork approach is established in Europe but it must be subjected to continual refinement. This is only possible if there is appropriate expertise at all levels of the feed and food production chain. The safety of food of animal origin can best be achieved by integrating expertise from different areas, using reliable and time-efficient analytical methods, understanding the temporal and spatial distribution of feed and food, and assessment of the carry-over of undesirable substances from feed to food of animal origin (Fernandes *et al.* 2011). The present case shows very clearly that solutions are possible only when there is cooperation between the private sector and governmental surveillance authorities. Knowledge of the complex network within the food production chain and the awareness of all participants involved in the chain are also prerequisites for the efficient production of safe food. In addition to fats and oils, other feed ingredients and additives, and also roughage, may be significant sources of dioxins in animal feed, which must not be ignored (Taube and Kamphues 2009; EFSA 2010; Schwind *et al.* 2010).

32.6 Critical questions for discussion

1. To which category of substances do dioxins belong?
2. What factors determine whether or not undesirable substances are carried over from feed to food?
3. What are the analytical methods used to detect dioxins, and why are they chosen?
4. List cases in the past in which feedborne contamination led to a carry-over of contaminants to animal-derived food for humans.
5. Where do dioxins come from and under what conditions are they generated?
6. What possibilities exist to reduce the presence of dioxins in feed and foodstuff, and what is the role of the quality control manager?
7. What are the current legislative rules on dioxin in feed and dioxin limits in animal nutrition in the EU and the WTO?
8. What other undesirable substances are of concern as potential feed contaminants?

32.7 References

EFSA (2010) Results of the monitoring of dioxin levels in food and feed. *EFSA Journal*, **8**: 1385.

FERNANDES A.R., FOXALL C., LOVETT A., ROSE M., and DOWDING A. (2011). The assimilation of dioxins and PCBs in conventionally reared farm animals: occurrence and biotransfer factors. *Chemosphere*, **83**: 815–22.

GUDE K., TAUBE V., BRUNS-WELLER E., SEVERIN K., SCHULZ A.J. *et al.* (2008). Contamination of feed by dioxins and dl-PCB and their relevance regarding food safety. *Übersichten zur Tierernahrung*, **36**: 93–144.

KAMPHUES J., and SCHULZ A.J. (2006). Dioxins: Risk management by agriculture and feed industry – options and limits. (Special issue. Dioxins) (in German). *Deutsche Tierärztliche Wochenschrift*, **113**: 298–303.

SCHWIND K. H., DÄNICKE S., and JIRA W. (2010). Survey of dioxins, dioxin-like PCBs and marker PCBs in German feeds of plant origin. *Journal für Verbraucherschutz und Lebensmittelsicherheit*, **5**: 413–20.

SCHWIND K. H., JIRA W., and KLEINHENZ S. (2007). The levels of dioxins and dioxinlike PCBs in feedstuffs and food. *Fleischwirtschaft*, **87**: 94–7.

TAUBE V., and KAMPHUES J. (2009). Dioxins: contamination of feed and subsequently of food derived from cattle? (In German). *Übersichten zur Tierernährung*, **37**: 167–78.

33

Investigating the adulteration of UHT milk in Brazil

C. F. Cardoso and A. G. Cruz, Cidade Universitária 'Zeferino Vaz', Brazil, U. M. Pinto, Universidade Federal de Ouro Preto, Brazil and J. A. F. Faria, Cidade Universitária 'Zeferino Vaz', Brazil

Abstract: Brazil is among the top ten producers of milk in the world. Milk may be adulterated with H_2O_2NaOH to reduce bacterial flora and the pH subsequently adjusted with NaOH. Whey may be added as another adulterant. Despite a police operation, reports of illegal adulteration of ultra-heat treated (UHT) milk from various regions in Brazil are still received. A new organization, 'Centro Integrado de Monitoramento da Qualidade dos Alimentos – CQuali Leite', aims to integrate actions from all the participants mentioned above in order to control food safety and quality and strengthen prevention measurements to fight quality issues in foods, including fraud.

Key words: Brazil, pasteurized milk, long-life milk, ultra-heat treated (UHT) milk, whey, quality standards, fraud, adulteration.

33.1 Introduction to the case

When one considers per capita consumption of milk, Brazil occupies 65th position worldwide, with consumption of about 161 l/inhabitant per year according to the Food and Agriculture Organization (FAO, United Nations); however, Brazil is among the top ten producers of milk in the world (Soares, 2010; Anonymous, 2011). In the 1990s, Brazilian annual national milk production almost doubled, from 11.6 billion to 21.3 billion l. The type of milk that was most often consumed in Brazil during that period was 'type C milk', a category that includes pasteurized milk, ultra-heat treated (UHT) milk and whey drinks (De Martino, 2001; ABLV, 2011a). Most of the milk that was collected from farms in that period was not refrigerated and thus milk acidification was a problem. UHT milk is currently the type of milk most consumed by the Brazilian population, with a sharp increase in consumption from 20.2% of all milk consumption in 1994 to 74.6% in 2009 (ABLV, 2011a). In the same time period we have also seen a shift from the canned milk system to the bulk tank system with refrigeration at the farm and during transportation to the dairy plant, a practice already adopted a long time ago in the developed countries.

There is variation in the quality of both the pasteurized and the UHT milk and it is not uncommon to find milk that does not meet official quality standards (Souza, 2006; Carvalho *et al.*, 2007). Indeed, controlling Brazilian milk quality is a constant issue for the relevant authorities and health and safety experts (Roma Júnior *et al.*, 2006; Carvalho *et al.*, 2007). Two Brazilian governmental agencies, the Ministério da Agricultura e Abastecimento (MAPA) and the Agência de Vigilância Sanitária (ANVISA), are responsible for the handling and inspection of milk sold in the country. Their main objective is to control milk quality, including milk authenticity (MAPA, 2011). Factors such as vast geographical distances together with a huge number of milk processing plants and a small number of inspection staff have favored an unfortunate increase in the incidence of milk adulteration. The main driving forces for this practice are increases in profit through gains in volume and extended shelf life; however, adulteration of milk can be potentially harmful to human health and the economy (Souza, 2006; Roma Júnior *et al.*, 2006; Carvalho *et al.*, 2007; Mendes *et al.*, 2010). Fraudulent practices include the addition of water, whey, stabilizing agents, preservatives such as hydrogen peroxide (H_2O_2) and acidity regulators such as sodium hydroxide (NaOH), nitric acid and urea to milk (Carvalho *et al.*, 2007; Mendes *et al.*, 2010).

In 2007, the Brazilian Federal Police launched an operation called *Ouro Branco* (White gold), which had the objective of limiting fraud in the pasteurized and long-life (UHT) milk industries. Alleged incidences of the addition of H_2O_2, NaOH and whey to milk were investigated (Folha de São Paulo, 2011a, 2011b, 2011c, 2011d, 2011e). This operation was performed in the whole of the Brazilian territory, but focused mainly on the states that produced the most milk, such as Minas Gerais, Paraná, Goiás and São Paulo (Folha de São Paulo, 2011a, 2011b). According to official denouncements, individuals with some technical knowledge adulterated milk using techniques that made it unlikely that the inspection agencies would discover the fraud during their periodical analyses (Folha, 2011a, 2011b). Adding to this problem are incidences of adulteration involving whey, which is a low-cost dairy product (Recio *et al.*, 2000; de la Fuente and Juárez, 2005). The methods currently employed for detecting adulteration are tested in laboratory conditions on simple mixtures of milk and whey. In practice, though, adulterated products may contain many other ingredients and these other ingredients may have little or no reactivity on the tests specifically developed for the detection of whey (Furtado and Vilela, 1996).

33.2 Significance of the case

H_2O_2 is added to milk with the intention of partially eliminating its bacterial flora, including those that can spoil milk, and ultimately delaying the deterioration process. However, the addition of this compound leads to a drop in pH, which is often subsequently adjusted for by the addition of NaOH, a practice that makes it unlikely that the addition of H_2O_2 will be detected (Souza, 2006; Mendes *et al.*, 2010; Nascimento *et al.*, 2010).

Because whey is a by-product of the cheese making process, it is frequently available in the dairy industry, at virtually no cost, making fraud involving whey attractive (Recio *et al.*, 2000). The addition of whey to milk also solves a problem for the industry, which normally has to discard this product, thus increasing the level of organic matter in waste treatment plants. Adulteration with whey brings about quality problems related to the alteration of the basic milk components, the increased risks associated with microbial contamination, and economic losses for the consumers and other milk processors that follow the legislation (Carvalho *et al.*, 2007).

33.2.1 Regulatory aspects

Brazilian legislation is clear when it comes to milk adulteration, as stated by Act 543 from the Regulamento de Inspeção Industrial e Sanitária de Produtos de Origem Animal (RIISPOA), which declares that milk must not contain any added extraneous substances (Brasil, 1983, 2002). Indeed, it is compulsory for dairy processors to carry out a fraud control procedure daily that consists of analysis of the raw milk received from producers for adulterants (Brasil, 2002). Brazilian legislation (Brasil, 2007) classifies milk as fraudulent when:

- Water has been added,
- One of its constituents (except fat) has been extracted,
- Preservatives or any other extraneous material have been added,
- The milk is not of the kind declared on the package's label,
- Raw milk is sold as pasteurized.

Therefore, the addition of H_2O_2 and NaOH to milk is prohibited. If whey is added to milk, the final product should be called 'lactic drink' and should not be sold as pasteurized or UHT milk (Dracz, 1996).

Both Brazilian and international legislation prohibits the addition of whey to milk destined for human consumption (Brasil, 1983, 1991). Ordinance 1725/79 from the European Union establishes that powdered milk should not contain added whey of any type. In Brazil, normative instruction no. 69 of 13 December 2006 states the criteria that should be used to evaluate fluid, concentrated, powdered and reconstituted milk, based on an analytical method called the CMP index (casein-macropeptides). This index is used to determine if milk has been adulterated with whey, as described in the normative instruction no. 68 of 12 December 2006. This normative instruction declares that the CMP index should be lower than 30 mg/l if milk is destined for direct consumption (Brasil, 2006).

33.2.2 Economic aspects

Due to the nationwide media coverage of the case and activities of the Public Sector (Federal and State), a branch of a multi-national company had its activities interrupted for a period of time and a milk cooperative was permanently closed. Both industries processed long-life milk (Folha de São Paulo 2011c, 2011d). To the consumers who frequently bought products from this multi-national business this came as a surprise. It is important to mention that, prior to this episode, the multinational company produced and sold a very well-known brand, which was associated with high levels of quality and safety (Folha de São Paulo 2011c, 2011d).

Even after extensive disclosure of the frauds in the media, sales of long-life milk experienced no setbacks between 2007 and 2008; indeed, an increase of 5.42% was observed (ABLV, 2011a, 2011b). This might be because the Brazilian population was by then accustomed to consuming long-life milk as opposed to pasteurized milk because the latter has a very short shelf-life of about 7 days, due to the poor quality of raw milk. The market share of long-life milk increased moderately in the same time period from 74.6% to 74.7% (ABLV, 2011a, 2011b).

No cases of disease possibly caused by the ingestion of the fraudulent milk were disclosed to the press. However, even though the amounts of H_2O_2 and NaOH found in milk were considered to be low, it is known that the ingestion of these compounds may cause irritation to the mouth and stomach mucosa and in severe cases may even be lethal (ABLV, 2011a, 2011b).

33.3 The case history continues

As a result of operation *Ouro Branco*, 29 persons were arrested by the Brazilian Federal Police because of their involvement with the adulteration of milk with H_2O_2, NaOH and whey. Some of those arrested were chemical engineers (Folha de São Paulo, 2011a, 2011b, 2011c, 2011d, 2011e). Even though the case was publicized nationally, with ongoing reports of the case on the TV, the internet and in the press, and the arrest of many individuals, reports of illegal adulteration of UHT milk from various regions in Brazil are still received (Rosa-Campos *et al.*, 2011). The specific types of adulteration detected actually tend to vary according to the region. This means that actions by the regulatory agencies could be planned to target specific types of frauds in certain geographic locations in Brazil, resulting in the better use of human and financial resources (Souza *et al.*, 2011). The lack of human resources for inspection is a great obstacle to a more effective response and hinders long-term planning.

33.4 Resolution and outcomes

Not only legislation, but also the official laboratories find it difficult to keep up with the constant changes in, and increased sophistication of, the procedures used to adulterate milk. To strengthen the Brazilian response to problems of this kind, as an immediate action, the National Agency of Sanitary Inspections (ANVISA), Health Department (Ministério da Saúde), Department of Defense and Protection of the Consumer (Departamento de Proteção e Defesa do Consumidor – DPDC),

Justice Department (Ministério da Justiça) and Department of Agriculture (Ministério da Agricultura, Pecuária e Abastecimento – MAPA) created a new inter-departmental center: 'Centro Integrado de Monitoramento da Qualidade dos Alimentos – CQuali Leite'. This center is an inter-departmental branch run by the National Agency of Sanitary Inspection (Agencia Nacional de Vigilância Sanitaria – ANVISA), which aims to integrate actions from all the departments mentioned above in order to control food safety and quality and strengthen prevention measurements to fight quality issues in foods, including fraud. CQuali Leite inspects farms and production units and monitors the conformity to legislation of milk destined for consumption as pasteurized, UHT and powdered products. It also aims to reduce informal sales of raw milk, which are still a big problem in Brazil.

33.5 Commentary and personal reflections

Fraud involving H_2O_2 and NaOH as additions to milk is carried out to decrease the high bacterial load in milk and to increase product shelf-life. This is one indication that milk in Brazil is still produced under sub-optimal hygienic conditions, and therefore has very high bacterial counts and is of low quality. This obviously does not justify any adulteration that may be practiced either by the industry itself or by farmers. The milk-producing and processing industries, universities and governmental agencies should work together to teach farmers about the advantages of producing milk of high quality and especially the benefits to consumer safety of more hygienic milk production. As more educational programs become available, we expect improvements in the quality of milk sold in Brazil and the elimination of fraudulent actions.

33.6 References

ANONYMOUS (2011). Consumo de leite cresce. *Tribuna do Norte*, 22 February 2011. Available at www.tribunadonorte.com.br/noticia/consumo-de-leite-cresce/174032. Accessed 17 March 2011.

ASSOCIAÇÃO BRASILEIRA DE LEITE LONGA VIDA (ABLV) (2009). Sobre o leite longa vida: Estatísticas. Available at http://www.ablv.org.br/Estatisticas.aspx. Accessed 11 April 2011.

ASSOCIAÇÃO BRASILEIRA DE LEITE LONGA VIDA (ABLV) (2010). Sobre o leite longa vida: Estatísticas. Available at http://www.ablv.org.br/Estatisticas.aspx. Accessed 11 April 2011.

BRASIL (2002). Ministério da Agricultura, Pecuária e Abastecimento. Instrução normativa n° 51, de 18 de setembro de 2002. Aprovar os Regulamentos Técnicos de Produção, Identidade e Qualidade do Leite tipo A, do Leite tipo B, do Leite tipo C, do Leite Pasteurizado e do Leite Cru Refrigerado e o Regulamento Técnico da Coleta de Leite Cru Refrigerado e seu Transporte a Granel, em conformidade com os Anexos a esta Instrução Normativa.

BRASIL (1983). Ministério da Agricultura, Pecuária e Abastecimento. Portaria n° 5, de 07 de março de 1983, aprova os Critérios de Inspeção do Leite e Produtos Lácteos, em estabelecimentos de laticínios registrados no Serviço de Inspeção Federal/SIPA – MA. Diário Oficial da União, 14 March 1983.

BRASIL (1991). Ministério da Agricultura, Pecuária e Abastecimento. Departamento de Inspeção de Produtos de Origem Animal. Portaria n° 124, de 23 de setembro de 1991. Aprova os métodos analíticos qualitativos e quantitativos de detecção de soro em leite. Diário Oficial da República Federativa do Brasil, p. 20561, 24 set. 1991. Seção I.

BRASIL (1997). Ministério da Agricultura, Pecuária e Abastecimento. Decreto n° 30.691, de 29 de março de 1952, alterado pelos Decretos n°s.1255, de 25 de junho de 1962, n° 1236, de 2 de setembro de 1994, n° 1812, de 8 de fevereiro de 1996, e n° 2.244, de 4 de junho de 1997. Regulamento da Inspeção Industrial e Sanitária dos Produtos de Origem Animal-RIISPOA. Brasília, DF.

BRASIL (2006). Ministério da Agricultura, Pecuária e Abastecimento. Departamento de Inspeção de Produtos de Origem Animal. Instrução Normativa n° 69, de 13 de dezembro de 2006. Institui critério de Avaliação da qualidade do leite in natura, concentrado em pó, reconstituídos, com base no método analítico oficial físico-químico denominado "Índice CMP", de que trata a Instrução Normativa n° 68 de 12 de dezembro de 2006. Diário Oficial da União, 15 de dezembro de 2006, Seção 1, Página 67. Ministério da Agricultura e Abastecimento.

CARVALHO, B.M.A., CARVALHO, L.M., ALCÂNTRA, L.A.P., and BONOMO, R.C.F. (2007). Detection methods of adulteration in milk by addition of cheese whey. *Rev. Eletrônica de Veterinária*, **8**(6): 1–7.

DE LA FUENTE, M., and JUÁREZ, M. 2005. Authenticity assessment of dairy products. *Critical Reviews in Food Science and Nutrition*, **45**: 563–85.

DE MARTINO, D.B. (2001). A escalada do leite longa vida. *Engarrafador Moderno*, **10**: 10–15.

DRACZ, S. (1996). Desenvolvimento de um método imunoenzimático para análise de queijo em leite. 57p. Dissertação (Mestrado em Medicina Veterinária) – Universidade Federal de Viçosa, Viçosa, MG.

FOLHA DE SÃO PAULO (2011a). Ministério Público Federal em Passos (MG) denuncia 29 por fraude no leite. Folha Online. 2008. Available at http://www1.folha.uol.com.br/folha/cotidiano/ult95u420387.shtml. Accessed 12 May 2011.

FOLHA DE SÃO PAULO (2011b). 28 pessoas vão responder por crime contra a saúde pública por adulteração de leite em MG. Folha Online. 2008. Available at http://www1.folha.uol.com.br/folha/cotidiano/ult95u480799.shtml. Accessed 12 May 2011.

FOLHA DE SÃO PAULO (2011c). Governo proíbe comercialização de leite longa vida de quatro empresas. Folha Online. 2007. Available at http://www1.folha.uol.com.br/folha/cotidiano/ult95u342230.shtml. Accessed 12 May 2011.

FOLHA DE SÃO PAULO (2011d). Parmalat diz que ministério liberou produção de fábrica no RS. Folha Online. 2007. Available at http://www1.folha.uol.com.br/folha/cotidiano/ult95u342395.shtml. Accessed 12 May 2011.

FOLHA DE SÃO PAULO (2011e). Promotoria de Goiás vai denunciar laticínios suspeitos de fraude. Folha Online. 2007. Available at http://www1.folha.uol.com.br/folha/cotidiano/ult95u347880.shtml. Accessed 12 May 2011.

FURTADO, M.A.M., and VILELA, M.A.P. (1996). Fraudes em leite de consumo: limites de detecção. *Leite & Derivados*, **29**: 14–18.

MENDES, C.G., SAKAMOTO, S.M., SILVA, J.B.A., JÁCOME, C.G.M., and LEITE, A.I. (2010). Physical-chemical analysis and research of fraud of informal milk commercialized in Mossoró, RN. *Ci Anim Bras*, **11**: 349–56.

MINISTÉRIO DA AGRICULTURA, PECUÁRIA E ABASTECIMENTO (MAPA) (2011). Available at http://extranet.agricultura.gov.br/sislegis/action/detalhaAto.do?method=abreLegislacaoFederal&chave=50674&tipoLegis=A. Accessed 18 May 2011.

NASCIMENTO, W.W.G., ALHADAS, M.M., BELL, M.J.V., and ANJOS, V.C. (2010). Detection of hydrogen peroxide and sodium hydroxide added in milk, through electrical conductivity measurements. In *27° Congresso Nacional de Laticínios*, 27 ed., pp. 1–7. Epanig: Juiz de Fora, MG.

RECIO, I., GARCÍA-RISCO, M.R., LÓPEZ-FANDIÑO, R., OLANO, A., RAMOS, M. (2000). Detection of rennet whey solids in UHT milk by capillary electrophoresis. *Internation Dairy Journal*, **10**: 333–8.

ROMA JÚNIOR, L.C., FISCHER, P.C., AMARAL, T.G.R., CASSOLI, L.D., and MACHADO, P.F. (2006). Influência da fraude com uréia sobre os componentes do leite durante armazenamento. In *II Congresso Brasileiro de Qualidade do Leite*, 2nd ed., pp. 1–6. Talento: Goiania, GO.

ROSA-CAMPOS, A.A., ROCHA, J.E.S., BORGO, L.A., and MENDONÇA, M.A. (2011). Physical-chemical evaluations and research fraud in integral pasteurized type C″ produced in Brasília, Federal District. *Revista do Instituto de Laticínios Cândido Tostes*, **379**: 30–34.

SOARES, T. (2010). Embalagem – mais competitividade ao setor de leite. *Engarrafador Moderno*, **198**: 26–31.

SOUZA, V. (2006). Características físico-químicas, microbiológicas, celulares e detecção de resíduos de antibióticos em amostras de leite de tanque comunitário. 69p. Dissertação (Mestrado em Medicina Veterinária) – Faculdade de Ciências Agrárias e Veterinárias, Universidade Estadual de São Paulo, Jaboticabal, 2006.

SOUZA, S.S., CRUZ, A.G., WALTER, E.H.M., FARIA, J.A.F., CELEGHINI, R.M.S. *et al.* (2011). Monitoring the authenticity of Brazilian UHT milk: A chemometric approach. *Food Chemistry*, **124**: 692–5.

34

Food adulteration with melamine on an international scale: field work and troubleshooting in Africa

D. Schoder, Veterinarians Without Borders Austria, Austria and University of Veterinary Medicine Vienna, Austria

Abstract: The globalized food market on occasion comes with a price to human health. A recent example was the epidemic of melamine poisoning in China, which affected more than 294 000 children, who suffered from urinary tract stones. At the end of September 2008 China's Ministry of Health made an official announcement that all affected milk powder batches had been discarded. At the same time the UN issued a worldwide alert, after which 31 countries in Asia, South America and Africa banned imports of Chinese milk products. Africa is dependent on imported milk powder goods, and the major part has no functioning food monitoring system. Was melamine-containing milk powder manufactured in China imported to Africa despite the ban?

Key words: melamine, milk, infant formula, adulteration.

34.1 Introduction to the case

In September 2008 intentional contamination of melamine in food became a real topic of international attention. On 5 September 2008 the dairy company Fonterra notified the New Zealand Government about the possible contamination of infant formulae with melamine in China. On 9 September news reports began circulating in mainland China, and on 10 September the news was broken internationally by Reuters. On 16 September 2008 infant formulae from 22 Chinese companies tested positive for melamine (Anonymous, 2008a). Subsequently, the United Nations issued a worldwide alert, the European Union banned Chinese-made baby food, and France went further by banning all foods containing Chinese milk as a precautionary measure. Finally, on 26 September, 16 days after the first international news item, 31 countries in Asia, South America and Africa (including Tanzania) banned the import of Chinese milk products (Parry, 2008).

Despite an official announcement by China's Ministry of Health at the end of September that all affected milk powder batches had been confiscated and burnt, locally sold contaminated milk powder and infant formulae, originating from China, were identified in several other countries (Kuehn, 2009). Interestingly, no single report referred to Africa, despite the fact that Africa is the world's leading importer of milk powder. In Africa, milk powder not only serves as a food aid for millions of malnourished Africans, but it is also recommended as a milk substitute for healthy infants of women suffering from HIV/AIDS.

Africa was left potentially vulnerable to the consequences of melamine-contaminated dairy products. Consequently, the aim of this study, as part of an effort by Veterinarians Without Borders Austria, was to determine whether melamine-containing milk powder manufactured in China was, despite import bans, imported to the African market (Schoder, 2010).

34.2 Significance of the case

34.2.1 What is melamine and why was it added to milk?

Melamine (1,3,5-triazine–2,4,6-triamine or $C_3H_6N_6$) is an industrial chemical that was developed in 1834 and is primarily used as a plastics stabilizer and fire retardant. It is a component in coatings, glues, dishes and kitchenware. Melamine is also rich in nitrogen, containing 67% nitrogen per mass unit. This has relevance to milk processing as the total nitrogen content has long been used to calculate the protein content using the non-specific Kjeldahl method (Kennaway, 1921). Consequently, melamine added to food commodities, such as milk, wheat gluten and rice protein, can artificially indicate increased protein content and increase the profit.

Effects of melamine on health

Melamine can cause renal failure by nephrolithiasis, the formation of insoluble crystals in renal tubules and/or the formation of calculi in the kidneys, urethra and bladder. These calculi are a mixture of melamine, protein, uric acid and phosphate, and as such are distinct from other kidney stones. They are radiolucent and therefore give a negative image on urinary tract radiographs. As a result, diagnosis of melamine-related urinary stones requires more sophisticated ultrasonographic equipment (Guan *et al.*, 2009).

The Ministry of Health of the People's Republic of China has confirmed that some 294 000 children suffered from urinary tract stones, 51 900 infants were hospitalized and – according to unofficial reports – up to 11 babies have died from melamine poisoning (Chen, 2009). According to all clinical studies published so far, even moderately melamine-contaminated infant formula has a significant clinical impact (Guan *et al.*, 2009; Wang *et al.*, 2009). Kidney stones were seen in nearly 7% of the children studied in Beijing who received moderately contaminated formulae (Guan *et al.*, 2009), and more than 20% of children residing in Taiwan and fed formulae with a melamine content exceeding 2.5 mg/kg had kidney stones (Wang *et al.*, 2009).

34.2.2 Regulatory aspects

International authorities, such as the World Health Organization (WHO) and regulatory bodies such as the US Food and Drug Administration (FDA) and the European Food Safety Authority (EFSA), have attempted to assess tolerable daily intakes (TDI) for melamine. A TDI is calculated by dividing a no-observed-adverse-effect level by a safety factor (SF). Based on appropriate choices of values, the FDA determined two TDI values, expressed as milligrams per kg body weight per day, at first 0.63 and then 0.063, while the WHO determined this to be 0.5 and then 0.2, as a result of further increasing the SF values in the calculation (Hsieh *et al.*, 2009).

According to a Chinese clinical trial, the patient group with the lowest melamine intake (>0–0.2 mg/kg of body weight), which is lower than the TDI for melamine recommended by the WHO, was still 1.7 times more likely to have nephrolithiasis compared with controls without exposure (Li *et al.*, 2009). The authors of this trial concluded that a TDI for melamine of 0.2 mg/kg body weight, as recommended by the WHO, seems unsuitable for young children, in whom a monotonous food intake of milk formula every day over months can be expected. In their opinion the TDI should be at least lower than 0.1 mg/kg. Finally, according to a Taiwanese study, even the ten-fold safety factor (TDI/10) of 0.063 mg melamine per kg body weight, introduced by the FDA, is not appropriate and the introduction of a new TDI of 0.0081 mg/kg of bodyweight has been strongly advised (Hsieh *et al.*, 2009).

34.2.3 Economic and market aspects

Global milk powder production

Between 2000 and 2007 China increased milk powder production from 615 000 to 1 300 000 tonnes and in so doing overtook New Zealand (660 000 tonnes) and the European Union (764 000 tonnes) in terms of production volume. Consequently, China has now become the world's largest milk powder producer (Anonymous, 2008b). Africa, on the other hand, is the world's leading importer, and up to 55% of the world's milk powder production is sold to the African continent each year (Food and Agriculture Organization of the United States, 2007).

Sub-Saharan milk powder supply and marketing systems

In Africa there are only a few milk powder production plants, and as a result of milk scarcity sub-Saharan Africa is dependent on imported milk powder, where the marketing systems can be categorized by the type of milk powder distributed and by the type of functions performed: branded products of international and multinational companies, and illegally imported unlabelled products, which are sold illegally on the market or local 'black' markets.

The categories by function are retail and modern distribution outlets, such as supermarkets, which sell exclusively imported branded products (products of the formal market), and street vendors and micro-retailers selling illegally imported unlabelled products (products of the informal market). Supermarket customers are mainly high-income earners (both expatriates and nationals), whereas street vendors and micro-retailers sell unlabelled milk powder products to low-income consumers for half the price, or even less.

34.3 The case history continues

34.3.1 Troubleshooting

In view of the limited financial resources of this case study, selection of a suitable place of investigation, the design of a cost-efficient sampling plan, and the application of a cheap and simple detection method were crucial factors for successful completion of this pilot study. It was thus important to find the most suitable place of investigation.

Dar es Salaam, the largest city and former capital of Tanzania, East Africa, was selected for three reasons. Firstly, Dar es Salaam is the largest seaport in East Africa and therefore the first destination in Africa for cargo ships from Asia. Secondly, Dar es Salaam is one of the fastest-growing African megacities and a centre of international trade. It supplies major parts of Africa via key trading routes and serves as a shipment hub for thousands of tonnes of milk powder to sub-Saharan, Central and East Africa. Thirdly, Tanzania was one of the nine African countries that banned Chinese milk products from being imported on 26 September 2008.

Designing a cost-efficient sampling plan

The aim of the study was to determine the incidence of melamine in milk powder and infant formulae imported to the (East) African market and the distribution of the contaminated products through both formal/legal and informal/illegal channels. Consequently it was necessary to include both main categories of milk powder in this study: branded products of international companies from the formal market, and illegally imported, unlabelled products from the informal market. Thus, branded samples were purchased from the retail and modern distribution outlets, including all ($n = 8$) supermarkets based in Dar es Salaam.

Sampling of the informal market was more difficult. Firstly, we had to collect reliable data about the number and the location of small kiosks selling milk powder on the local black market. Secondly, due to the national import prohibition of unlabelled milk powder in Tanzania and rigid inspections concerning the observance of the ban (especially after 26 September 2008), micro-retailers of the black market had become wary of showing their illegal products. For successful execution of the study we required the assistance of a local scout, who pretended that his newborn child urgently needed milk powder.

Finally, it was important to find a locally suitable detection method. The most commonly used techniques to detect melamine are liquid chromatography (LC) and gas chromatography (GC) combined with tandem mass spectrometry (MS/MS). However, chromatographic methods are relatively expensive, require special instrumentation and are laborious to perform. Due to melamine contamination of milk and foods, commercial enzyme-linked immunosorbent assay (ELISA) test kits have been developed for quantitative analyses as a potential solution to meet the needs of high-throughput sample screening. In our study, the ELISA testkit AgraQuant Melamine Sensitive Assay was selected as the method of detection for the analysis of melamine for three main reasons. Firstly, the assay was claimed to detect melamine concentrations ≥0.15 mg/kg and is therefore about equisensitive to mass spectrometric analyses, both GC/MS/MS and LC/MS/MS. Secondly, the test can be performed in microplates where a large number of samples (up to 80 samples) can be analysed simultaneously. Thirdly, the assay does not require expensive laboratory equipment or specially trained staff, and does not have high running costs.

34.3.2 The case history continues

The total area of Dar es Salaam is 1393 km^2 and it is divided into three administrative districts (Kinondoni, Ilala and Temeke), each broken into three–four divisions and 22–27 wards.

Two categories of samples were collected between October and December 2008, immediately after the melamine contamination of Chinese products became public: market brands of all international companies supplying the East African market, and illegally sold products from informal channels.

Branded products were selected on the basis of their immediate availability, product category (milk powder/infant formula), type of manufacturer, and production and expiry date. Branded products with the most recent production dates and expiry dates on or later than January 2009 were chosen for further examination.

With regard to the informal market, we estimated that there were about 300 kiosks and micro-retailers selling milk powder to low-income consumers in Dar es Salaam. These small kiosks were mostly located in open spaces and lacked basic infrastructure (concrete floors, roofs, drainage and facilities for sorting, storing and handling products). Temeke, the southern most district of Dar es Salaam, belongs to the city's largest slum area and showed the highest rate of unplanned kiosks and micro-retailers. Within the period of examination (October–December 2008) it was possible to inspect two-thirds (200 of 300) of all micro-retailers located in the Dar es Salaam region. With regard to the inspected kiosks, 26% (52 of 200) were located in Kinondoni, 21.5% (43 of 200) in Ilala, and 52.5% (105 of 200) in Temeke. The results of the ELISA assay were calculated using the Romer Log/Logit regression model (Table 34.1).

Table 34.1 Results of the milk powder and infant formula products tested by ELISA. The samples were taken in Dar es Salaam, Tanzania

No.	Product name	Manufacturer	Lot no.	Melamine (ppm ± SD[g])
1[b]	Cowbell® (Instant filled milk powder)	Promasidor Ltd, Zug, Switzerland	6191A	Neg
2[a]	Aspen™ – Nutritionals S-26/1 (Infant formula with iron)	Pharmacare Ltd, Sandton, South Africa	502472	Neg
3[a]	Nestlé® Lactogen®/1 (Starter infant formula with iron)	Nestlé (South Africa) (Pty) Ltd, Randburg, South Africa	73400179L1	Neg
4[c]	Wyeth® – S-26 Gold Stage 1 (Infant formula from birth to 6 months)	Wyeth Nutritionals, Askeaton, Co., Limerick, Ireland	8C295611	Neg
5[a]	Nestlé® NAN®/1 (Starter infant formula with iron)	Nestlé (South Africa) (Pty) Ltd, Randburg, South Africa	7325017981	Neg
6[a]	Nestlé® NIDO® (Full cream milk powder)	Nestlé (South Africa) (Pty) Ltd, Randburg, South Africa	80560181	Neg
7[c]	Daily Fresh™ (Full cream milk powder)	Daily Fresh Foods Ltd, UK	AM/CH/090702	0.485 (± 0.043)
8[a]	Nestlé® NAN®/1 (Starter infant formula with iron)	Nestlé (South Africa) (Pty) Ltd, Randburg, South Africa	8065017981	Neg
9[c]	Kerry Gold® (Full cream milk powder)	Irish Dairy Board, Dublin, Republic of Ireland	02K/400	Neg
10[e]	Imalaseko Supermarket Milk Powder[f]	Imalaseko Super Market, Dar-es-Salaam, Tanzania	NA[e]	Neg
11[b]	Cowbell® (Instant filled milk powder)	Promasidor Ltd, Zug, Switzerland	5141A	Neg
12[a]	Nestlé® Lactogen®/2 (Follow-up formula with iron and prebiotics)	Nestlé (South Africa) (Pty) Ltd, Randburg, South Africa	82040181 SWWBFM	Neg
13[a]	Elite® (Fat free instant milk powder)	Clover S.A. (Pty) Ltd, Roodepoort, South Africa	HO91-8-(3)-2	Neg
14[a]	Aspen™ Nutritionals – S-26 Progress/3 (Growing up drink for children)	Pharmacare Ltd, Sandton, South Africa	502535	Neg
15[c]	Marvel (Original dried skimmed milk)	Chivers Ireland Ltd, Dublin, Republic of Ireland	7268	Neg

(Continued)

Table 34.1 Continued

No.	Product name	Manufacturer	Lot no.	Melamine (ppm ± SD[g])
16[c]	Daily Fresh™ (Full cream milk powder)	Daily Fresh Foods Ltd, UK	AM/IMP/14B	Neg
17[c]	Cow & Gate®/2 (Nutrionally complete infant milk)	Cow & Gate Ireland, Dublin, Republic of Ireland	001251	Neg
18[a]	Elite® (Fat free instant milk powder)	Clover S.A. (Pty) Ltd, Roodepoort, South Africa	HO63-8-(3)-3	Neg
19[b]	Cowbell® (Instant filled milk powder)	Promasidor Ltd, Zug, Switzerland	4381B	Neg
20[a]	Nestlé® NIDO® (Full cream milk powder)	Nestlé (South Africa) (Pty) Ltd., Randburg, South Africa	81260181 SCEOIM	5.474 (± 0.127)
21[b]	Nestlé® NIDO® (Full cream milk powder)	Nestlé Foods Kenya Ltd, Nairobi, Kenya	72700169 A1	Neg
22[b]	Cowbell® (Instant filled milk powder)	Promasidor Ltd, Zug, Switzerland	3301B	Neg
23[b]	Nestlé® NIDO® (Full cream milk powder)	Nestlé Foods Kenya Ltd, Nairobi, Kenya	80720169 D1	Neg
24[d]	Luna (Instant full cream milk powder)	National Food Industries Co. Ltd, Jeddah, Kingdom of Saudi Arabia	104 NZMP DR 27	Neg
25[b]	Nestlé® NIDO® (Full cream milk powder)	Nestlé Foods Kenya Ltd, Nairobi, Kenya	80780169 D1	Neg
26[a]	Aspen™ – Nutritionals Infasoy/1	Pharmacare Ltd, Sandton, South Africa	502249	3.342 (± 0.157)
27[b]	Cowbell® (Instant filled milk powder)	Promasidor Ltd, Zug, Switzerland	3351 A	Neg

[a] Produced in Africa; [b] Only packaged in Africa; [c] Produced or packaged in Europe; [d] Produced or packaged in New Zealand; [e] Data not available (NA); [f] Store-brand of a Kenyan supermarket chain; [g] The final concentration of melamine contamination (ppm) of sample Nos 1 to 49 was calculated based on the optical density OD-450 values, using the Romer® Log/Logit spreadsheet and considering a dilution factor of 25. Since the amount of melamine in each standard (0 ppb; 20 ppb; 100 ppb; 1000 ppb) was known, assay results of the samples were measured by interpolation from the standard curve.

Source: Schoder, 2010.

34.4 Resolution and outcomes

Despite the national import prohibition of Chinese milk products and unlabelled milk powder in Tanzania, and rigid inspections concerning the observance of the ban, 11% (22 of 200) of the inspected micro-retailers sold milk powder on the local black market. A total of 49 milk powder batches were collected in Dar es

Salaam. Manufacturers could be identified for only 55% (27) of the 49 investigated batches.

Six percent (three of 49) of all samples and 11% (three of 27) of all international branded products tested revealed melamine at concentrations of up to 5.5 mg/kg of milk powder (Table 34.1). In total, branded products ($n = 27$) were manufactured and/or marketed by 11 international and multinational companies. A total of 16 different brands could be included in this study, 19% (five of 27) of which were not duly labelled and did not show any production date at all.

Taking the TDI of the recently published Taiwanese study of 0.0081 mg/kg into account, and considering a 5 kg infant, its tolerable amount of melamine would be 0.04 mg melamine per day. This amount would be reached after consuming 750 ml of reconstituted formula contaminated at a melamine level around 0.05 mg/l (ppm). For comparison, milk powder sample number 20, which was analysed in our study, was contaminated at a level of 5.5 mg/kg powder, corresponding to approximately 0.8 mg/l in the reconstituted product (assuming a seven-fold reconstitution factor) and results in 0.12 mg melamine intake per day. This is about twice the TDI/10 of 0.063 mg (US FDA) or 15 times the TDI of 0.0081 mg/kg suggested by Hsieh *et al.* (2009).

34.5 Commentary and personal reflections

Our study clearly demonstrated that, despite official regulations, melamine contamination of milk powder was a significant problem in Africa. When 6%, and up to 11% of random samples taken in the Dar es Salaam region proved positive for melamine in this pilot study, the potential total contamination could be alarming.

It is important to note that the Chinese authorities did take reasonable steps to destroy contaminated stocks of milk powder and to prevent its export. It is also not clear that all the contaminated milk powder necessarily originated from Chinese sources. However, it is reasonable to assume that, as the largest exporter of milk powder and as the source of a serious case of melamine contamination, at least a small part of the contaminated powder may have originated from illegal Chinese sources. This suggests the need for all the responsible regulatory agencies, including those in China as well as other countries, to do all they can to control illegal trading of potentially contaminated or adulterated foodstuffs.

It is also important to note that reports regarding clinical cases of melamine poisoning in Africa are still rare, largely because melamine-related urinary stones – a mixture of melamine, protein, uric acid and phosphate – are different from other kidney stones and difficult to detect in traditional clinical radiology. As mentioned above, the diagnosis of melamine poisoning requires technologically complex ultrasonographic equipment and considerable expertise (Guan *et al.*, 2009). However, due to poor technical facilities in the majority of sub-Saharan hospitals, clinical cases of infants and small children exposed to contaminated formulae are seldom diagnosed. On the basis of our study, and considering that moderate contamination with melamine carries a substantial health risk, it can be assumed that the number of affected children in Africa is substantially high.

In summary, continuous food control as well as the speedy introduction of reliable surveillance tools for monitoring melamine contamination in the least-developed countries is urgent.

34.6 Acknowledgements

This work was supported by Veterinarians Without Borders Austria. The author thanks Mr Awesu Mgango for his guidance to Dar es Salaam and great support with the sampling. Many thanks also to John Laffa, Benedict Lema and Abdoulla Zangana for excellent technical assistance.

This research work was awarded the Heinrich Stockmeyer Science Prize 2011.

34.7 Critical questions for discussion

1. How much melamine must food contain before it poses a risk to human health?
2. How can such cases be identified more quickly in future?
3. What extra precautions can exporting and importing countries take to prevent any illegal trade in contaminated foodstuffs?

34.8 References

ANONYMOUS (2008a). *China seizes 22 companies with contaminated baby milk powder.* Available at: http://news.xinhuanet.com/english/2008-09/17/content_10046949.htm. Accessed 30 May 2011.

ANONYMOUS (2008b). The world dairy situation 2008. *Bulletin of the International Dairy Federation*, **432**: Annex 1, p. 84.

CHEN, J. (2009). A worldwide food safety concern in 2008 – Melamine-contaminated infant formula in China caused urinary tract stone in 290 000 children in China. *Chin Med J*, **122**: 243–4.

FOOD AND AGRICULTURE ORGANIZATION OF THE UNITED STATES (2007). *Commodities. No 3. Import surges in developing countries: The case of dairy products. FAO briefs on import surges.* Available at: http://www.fao.org/es/esc/en/378/406/highlight_425.html. Accessed 30 May 2011.

GUAN, N., FAN, Q., DING, J., ZHAO, Y., LU, J. *et al.* (2009). Melamine-contaminated powdered formula and urolithiasis in young children. *N Engl J Med*, **360**: 1067–74.

HSIEH, D.P., CHIANG, C.F., CHIANG, P.H., and WEN, C.P. (2009). Toxicological analysis points to a lower tolerable daily intake of melamine in food. *Regul Toxicol Pharmacol*, **55**: 13–16.

KENNAWAY, E.L. (1921). The estimation of non-protein nitrogen in blood by a micro-Kjeldahl Method. *Biochem J*, **15**: 510–12.

KUEHN, B.M. (2009). Melamine scandals highlight hazards of increasingly globalized food chain. *JAMA*, **301**: 473–5.

LI, G., JIAO, S., YIN, X., DENG, Y., PANG, X. *et al.* (2009). The risk of melamine-induced

nephrolithiasis in young children starts at a lower intake level than recommended by the WHO. *Pediatr Nephrol* (Serial online). Available at http://www.springerlink.com/content/h8353152m3140536/. Accessed 30 May 2011.
PARRY, J. (2008). China's tainted milk scandal spreads around world. *BMJ*, **337**: a1890.
SCHODER, D. (2010). Melamine milk powder and infant formula sold in East Africa. *J Food Prot*, **73**: 1709–14.
WANG, I.J., CHEN, P.C., and HWANG, K.C. (2009). Melamine and nephrolithiasis in children in Taiwan. *N Engl J Med*, **360**: 1157–8.

35

Investigating cases of taste disturbance caused by pine nuts in Denmark

N. Z. Ballin, Danish Veterinary and Food Administration, Denmark

Abstract: Exotic edible nuts are increasingly consumed in the Western world, especially as part of a raw food diet or as a healthy snack. Recently, a quality problem has arisen due to taste disturbance caused by some types of pine nuts. The taste disturbance results in a bitter and metallic taste, and is now recognised as pine nut syndrome. The syndrome occurs 1–2 days after the ingestion of pine nuts from inedible species, and symptoms last up to 2 weeks. This paper is based on a Danish case that resulted in identification of the source of the problematic pine nuts and their withdrawal from the Danish market.

Key words: authentication, pine nuts, pine nut syndrome, *Pinus armandii*, species identification.

35.1 Introduction to the case

European consumption of pine nuts has increased during the last 10 years, with a significant increase in pine nut-related taste disturbance. This was first described in 2001 and involved an unpleasant bitter and metallic taste (Mostin, 2001). Mostin reported several cases of taste disturbance but several years passed before additional cases were described in the literature.

Today, many cases of pine nut intake and taste disturbance have been reported through the Internet in addition to a few scientific papers (Munk, 2010; Destaillats *et al.*, 2010; Flesch *et al.*, 2011). The taste disturbance is now referred to as pine mouth syndrome (PMS), pine nut syndrome (PNS), or the medical term dysgeusia. The taste disturbance has been associated with ingestion of pine nuts from the species *Pinus armandii* (Destaillats *et al.*, 2011). Because of the obvious consumer implications involved in the case, further investigations were initiated at international, European and national levels.

35.1.1 Pine nut species and origin

More than 100 different *Pinus* species exist worldwide, but, according to the Food and Agriculture Organization (FAO), only 29 species produce edible nuts. Thus, the number of edible nuts that are used commercially is only around 20. In other words, the majority of pine species produce inedible nuts. These inedible nuts are in some cases used for animal feed production or non-food industrial purposes.

China is one of the major suppliers of edible pine nuts on the international market. The most common edible Chinese pine nuts belong to the species *Pinus koraiensis* and *Pinus sibirica* and, to a lesser extent, *Pinus yunnanensis*, *Pinus pumila*, *Pinus tabuliformis*, and *Pinus wallichiana* (formerly known as *Pinus griffithii*). Other common edible pine nuts belong to the species of *Pinus pinea* and *Pinus gerardiana* that originate from the Mediterranean region and the area of the north-western Himalaya, respectively. Common inedible pine nuts belong to the species *Pinus armandii* and *Pinus massoniana* that originate from China (Table 35.1).

35.1.2 Market and economic issues

The majority of pine nuts consumed in Europe originate from non-Chinese countries, but a minority are imported from China. This leaves China with a great opportunity to increase its market share, especially in a period with increasing European pine nut consumption. Currently, China exports 300–400 tonnes of pine

Table 35.1 Pine nut species, their use, shape, and origin

Species*	Use*	Shape and approximate size (length and width in mm)†	Origin (market share)*
Pinus koraiensis	Edible	Triangular–obovoid (12–16 × 5–6‡)	China (60%)
Pinus sibirica	Edible	Obovoid (10 × 5–6)	China (12%)
Pinus yunnanensis	Edible	Slightly appressed, nearly ovoid or obovoid (4–5 × 2‡)	China (9%)
Pinus armandii	Inedible§	Obovoid (10–15 × 6–10)	China (8%)
Pinus pumila	Edible	Triangular–obovoid (7–10 × 5–7)	China (5%)
Pinus tabuliformis	Edible	Narrowly ovoid (6–8 × 4–5)	China (3%)
Pinus massoniana	Inedible	Narrowly ovoid (4–6 × 2‡)	China (2%)
Pinus wallichiana¶	Edible	Ellipsoid–obovoid (3–9 × 4–5)	China (1%)
Pinus pinea	Edible	Cylinder (13 × 5)‡	Mediterranean
Pinus gerardiana	Edible	Cylinder (17 × 5)‡	North-western Himalaya

* Based on information provided by Chinese officials to the European Commission (DG SANCO) 2011. † Adapted from Wu and Raven (1999). ‡ Own measurement. § *Pinus armandii* is reported as a food ingredient only after roasting and pasting. However, Chinese officials have promised to exclude export of the *Pinus armandii* species to the European community. ¶ An old name for *Pinus wallichiana* is *Pinus griffithii*, which is used by Chinese officials.

nuts each month, and one way to further increase that amount is to provide high-quality pine nuts. However, continuous problems with PNS caused by Chinese pine nuts might have a negative impact on European consumers. Consumers can avoid Chinese pine nuts that are small and round, in contrast to the non-Chinese pine nuts that are long and cylindrical (Table 35.1).

35.1.3 How did it all begin?

Initially, the link between taste disturbance and pine nut ingestion (Mostin, 2001) attracted little attention in Europe. However, the case evolved rapidly after a notification to the European Commission's Rapid Alert System for Food and Feed (RASFF) in August 2009 (news notification 09-573). This notification made for the first time information about pine nuts and taste disturbances available to the authorities in all member states.

A few months later followed the first Danish consumer complaint. The Danish Veterinary and Food Administration (DVFA) decided soon after to publish information about the phenomenon on their website, advising consumers to report cases of pine PNS.

35.1.4 What were the challenges involved?

The taste disturbance appears 1/2 days after ingestion of certain pine nuts, which makes it difficult to link the symptoms with pine nut intake. This incubation period probably postpones or limits the number of consumer complaints. However, since the RASFF news notification from 2009, work had been undertaken by the European authorities to act appropriately. In this respect, the challenges for the Danish food control authorities included five main areas:

1. An understanding of the seriousness of the case: The seriousness was difficult to establish as not all people who ingested the problematic pine nuts were affected.
2. Identification of the causative agent: The cause of PNS was unknown and could involve a pine nut intrinsic constituent, mycotoxins, or perhaps an external chemical from pine nut production.
3. Identification of the species involved in PNS: Visual identification of the species was impossible as the botanical features are very similar in several species originating from China. An effort to find adequate analytical authentication methods was made but only a very limited number of published methods were available.
4. To define the legal basis involved in batch withdrawals: Legal actions were difficult to establish before the European Union had gathered enough information to provide the member states with solid information on how to react.
5. To stop the import of problematic pine nuts: One way to address the problem is to trace the pine nuts back to their origin and discontinue further collaboration with the involved companies. However, in a global market, food commodities

are often redistributed many times and tracing can therefore be difficult and time consuming.

35.2 Significance of the case

In the Western world, an increasing number of consumers are turning toward vegetarian or raw food diets, where nuts constitute the main source of protein. Nuts from Africa and Asia are especially gaining popularity, because of health claims and their exotic image. To keep up and encourage this healthy development requires access to good quality ingredients and a redoubled effort by food control regulators.

35.2.1 Regulatory aspects

Decisions about the legal actions in relation to batch withdrawal were taken at a meeting in the European Standing Committee on the Food Chain and Animal Health, Section of Toxicological Safety of the Food Chain, January 2011. At this meeting, it was made clear that pine nut batches containing nuts from *Pinus armandii* are considered unfit for human consumption and are deemed to be unsafe. This argument complies with the European Regulation 178/2002 that lays down the general principles and requirements of food law. Article 14 states that food shall be deemed to be unsafe if it is considered to be unfit for human consumption. This provided the European authorities with arguments that enabled withdrawal of pine nuts from *Pinus armandii* based on a solid legal background.

35.3 The case history continues

Member states must report to the RASFF system if they have any information about a serious health risk derived from food or feed. The number of reported RASFF notifications does, therefore, indicate the extent of the food or feed problem on a European scale. Since the first RASFF news notification concerning pine nuts and taste disturbances from August 2009, 13 addenda were attached. Furthermore, 11 individual RASFF notifications were reported from France, the Czech Republic, Slovakia, Spain, Italy, Ireland, Finland, Malta, Belgium, and Denmark. Except a few addenda and notifications, all implicated pine nuts originated from China. The increased European focus on pine nuts prompted the DVFA to encourage consumers to report cases of PNS. The DVFA reported 20 consumer complaints in 2010 and 23 in 2011 (until 7 March). These consumer complaints initiated batch withdrawals from the Danish market. The withdrawals were based on the European Commission statement that pine nuts from *Pinus armandii* are considered unfit for human consumption (see Section 35.2.1, Regulatory aspects). During the first 6 months of 2011, seven batches were identified through consumer complaints and removed from Danish stores. In some of these batches, species identification confirmed the presence of *Pinus armandii*. Meanwhile, the DVFA informed importers of pine nuts about the pine nut problems and pointed out that all pine nut import must be accompanied by documentation of the exact species.

35.3.1 Troubleshooting

The work performed at the DVFA followed the described challenges and included:

1. An understanding of the seriousness of the case. The scientific papers published so far have unfortunately not disclosed much information about the seriousness of the symptoms involved in PNS. The author, therefore, initiated a small sensory trial to investigate the PNS phenomenon further. The trial, which involved the author and his colleagues, showed that six–eight pine nuts from a presumed problematic batch (later identified as *Pinus armandii*) were sufficient to cause PNS in four out of six people. The incubation period was 1/2 days and symptoms lasted for a maximum of 5 days. The predominant symptom was a constant bitter or metallic taste; one reported its intensification upon intake of specific foods such as red wine and coffee. These symptoms and the fact that not all were seriously affected are in agreement with other trials. An article has recently reported that as few as two pine nuts were sufficient to cause PNS and that symptoms might last up to 2 weeks (Flesch *et al.*, 2011).
2. Identification of the causative agent. The causative agent involved in PNS is still unknown. Analyses carried out at official laboratories in Europe have not revealed abnormalities in respect to pesticides, heavy metals, mycotoxins, or bacterial content. The role of specific fatty acids or their oxidative forms has been hypothetically suggested but more research is needed to resolve the matter.
3. Identification of the species involved in PNS. Most of the identification of pine nut species performed in Europe at the national control laboratories involved gas chromatographic analysis of fatty acids. The fatty acid profiles are species-specific and a comparison of the fatty acid profile of unknown pine nut species with the fatty acid profile of known pine nut species (Wolff *et al.*, 2000) can, therefore, establish the species identity in some cases (Destaillats *et al.*, 2010). However, DVFA soon discovered that interpretation of fatty acid data was difficult or impossible in cases of pine nut mixtures and, of course, in complex food products. Mixtures of pine nuts do, therefore, require a careful separation of nuts into subsamples that belong to the same species. This is often difficult as some species look very similar. As these problems were recognized other methods were explored. As very few methods were described in the literature, focus was turned toward DNA technologies for more species-specific identification. In addition to the fatty acid profiling performed at DVFA, a problematic batch was sent to a commercial laboratory, and DNA sequencing and DNA fragment length analysis helped establish the species identity.

4. To define the legal basis involved in batch withdrawals. The withdrawal of problematic batches of pine nuts needed to be justified legally. The legal considerations that took place in the European Commission were dependent on how the case evolved, and Denmark and other European member states provided the European Commission with relevant information about the case.
5. To stop the import of problematic pine nuts. Even with the legal procedure in place, import of problematic batches can only be stopped after the batch has been identified and traced back to the company involved in the mislabelling of pine nuts. This identification involved two strategies: batch identification through consumer complaints, and identification of *Pinus armandii* through analytical testing. Consumer complaints depend on consumers' awareness of PMS. Appropriate initiatives to increase consumer awareness are, therefore, currently discussed in DVFA and could involve more information about PMS being made public. The second strategy involves analytical testing of randomly collected samples at importers and stores. One such authentication project is currently being conducted and an additional project is in preparation. Yet another way to address the problematic import of *Pinus armandii* is to make the Chinese authorities aware of the problem. Official meetings were, therefore, held between the European Community and Chinese officials.

35.4 Resolution and outcomes

In 2010, the European Food and Safety Authority (EFSA) and the International Nut and Dried Fruit Council (INC) put pressure on the Chinese authorities to stop the export of pine nuts from *Pinus armandii*. The Chinese authorities did, thereafter, inform the European Commission that export of pine nuts from *Pinus armandii* would stop in September 2010. However, so far the problem continues.

Another issue concerns the quality of pine nuts. The Chinese authorities have given assurances that only pine nuts of Grade A will be exported as a product for direct human consumption (Table 35.2). Grades B and C are considered to be industrial products. These quality grades were further discussed at a meeting at the European Commission, and it was informed that quality grades B and C from species other than *Pinus armandii* might also cause PMS. It is, therefore, necessary that control focuses not only on the species of *Pinus armandii* but also on the specific quality grades. It is, of course, possible that batches of non-*Pinus armandii* nuts labelled quality grades B or C contain *Pinus armandii* in an amount sufficient to cause PMS, but this has not been confirmed. In addition, it has been considered that immature pine nuts from species regarded as edible might cause PNS.

35.5 Commentary and personal reflections

During the last few years, the scientific focus on identification of pine nut species has increased with the increase in PNS cases. Nevertheless, there is a need for

Table 35.2 Grade A specifications for Chinese pine nut kernels

Parameter	Specification
Quality	100% peeled
Size	As per each specific variety
Moisture	3% maximum
Colour	As each specific variety
Broken kernels	2% maximum (w/w)
Damaged kernels	1% maximum*
Loose shells	5 pieces/12.5 kg
Foreign material	0.05% maximum†
Free fatty acids	0.5% maximum (ISO 660)
Peroxide value	5 meq/kg maximum (AOAC 8b-90; ISO 3960)
Aflatoxin	4 ppb maximum, with B1 less than 2 ppb
Total plate count	100 000 cfu/g maximum (ISO 4833)
Coliform	1000 cfu/g maximum (ISO 4831; ISO 4832)
E. coli	10 cfu/g maximum (ISO 7251)
Salmonella	Negative (ISO 6579)
Yeast and mould	3000 cfu/g maximum (ISO 7954)

* Yellow peak, insect damage, polluted, and immature kernels (based on CIQ pine nut kernels concept: SN/T 0788-1999) and insect damage, polluted, diseased, and immature kernels are not exceeding 0.5% of the 1%. † Excluding metals, glass, hair, stones, and foreign objects.

The above mentioned specifications are category A quality, which is the only export quality to the overseas markets. Other specifications such as rancidity, broken, shrivelled, defect, brown center kernels, and immature kernels, are of grade B and grade C qualities. They are not suitable for human consumption (reproduced from information provided by the INC).

more research in the identification of pine nut species, especially in pine nut mixtures and in complex food products. When adequate analytical authentication methods become available, more information on pine nut species and PNS will be gathered. This information might answer the question, 'are species other than *Pinus armandii* involved in PNS?'

Pine nut syndrome seems to disappear after approximately 1 week with no apparent side effects, but the chemical agent responsible for PNS and the mechanism of action have not yet been established. Thus, more serious side effects cannot be ruled out completely. There is, therefore, an urge to identify the chemical nature of the compound responsible for PNS. The agent responsible for PNS might be an intrinsic constituent of *Pinus armandii*, but process conditions, pesticides and other external chemicals used for isolation of the raw nuts could be responsible (Destaillats *et al.*, 2011).

35.6 Critical questions for discussion

1. Considering the incubation period of PNS, suggest mechanisms of action.
2. Discuss other common pairs of raw food that appear identical but where one is inedible.

3. Are there other natural or unnatural compounds that cause bitter and metallic taste?
4. Suggest why PNS has not been described prior to 2001.
5. Could pine nut syndrome be caused by an external chemical, or is it more likely an intrinsic *Pinus armandii* constituent?
6. What further actions can be taken to prevent import of inedible pine nuts?

35.7 References

BALLIN N.Z. (2012). A trial investigating the symptoms related to pine nut syndrome. *J Med Toxicol*, DOI: 10.1007/s13181-012-0216-4.

DESTAILLATS F., CRUZ-HERNANDEZ C., GIUFFRIDA F., and DIONISI F. (2010). Identification of the botanical origin of pine nuts found in food products by gas-liquid chromatography analysis of fatty acid profile. *J Agric Food Chem*, **58**: 2082–7.

DESTAILLATS F., CRUZ-HERNANDEZ C., GIUFFRIDA F., DIONISI F., MOSTIN M. *et al.* (2011). Identification of the botanical origin of commercial pine nuts responsible for dysgeusia by gas-liquid chromatography analysis of fatty acid profile. *J Toxicol*. Available at http://www.hindawi.com/journals/jt/2011/316789/.

FLESCH F., RIGAUX-BARRY F., SAVIUC P., GARNIER R., DAOUDI J. *et al.* (2011). Dysgeusia following consumption of pine nuts: More than 3000 cases in France. *Clin Toxicol*, **49**: 668–70.

MOSTIN M. (2001). Taste disturbances after pine nut ingestion. *Eur J Emerg Med*, **8**: 76.

MUNK M.D. (2010). "Pine mouth" syndrome: cacogeusia following ingestion of pine nuts (genus: *Pinus*). An emerging problem? *J Med Toxicol*, **6**: 158–9.

WOLFF R.L., PEDRONO F., PASQUIER E., and MARPEAU A.M. (2000). General characteristics of *Pinus* spp. seed fatty acid compositions, and importance of Δ5-olefinic acids in the taxonomy and phylogeny of the genus. *Lipids*, **35**: 1–22.

WU Z.Y. and RAVEN P.H. (1999). *Pinacea*. In Z. Y. Wu (ed.), *Flora of China (Cycadaceae through Fagaceae)* Vol. 4, pp. 11–40. Science Press, Beijing, and Missouri Botanical Garden Press, St Louis.

36

Development and application of molecular tools to investigate the mislabeling of cod sold in Sweden

E. E. Nielsen, J. Hemmer-Hansen and D. Bekkevold, Technical University of Denmark, Denmark

Abstract: Labeling of fish and fish products with geographical origin is mandatory within the EU. Whereas control of illegal fishing has received substantial attention, relatively little work has been done on substitution and mislabeling in the different levels of retailing from boat to consumer. Substitution of high-quality fish with less valued products can provide high illegal revenues for retailers and pose serious health risks to consumers, but can also lead to the eradication of endangered fish stocks. This case study demonstrates that, by using molecular tools, individual fish fillets can be traced back to the sea area from where they were fished.

Key words: Atlantic cod, illegal, unreported and uncontrolled (IUU) fishing, DNA tests, tracing origin.

36.1 Introduction to the case

Since 2002 it has been mandatory by law to label and establish traceability procedures that allow food products to be followed in Europe at all stages of production (EC No 178/2002). For fish, accompanying certificates and labels should document the product (species), production method and catch area. The reason for the focus on area relates to the fact that fisheries worldwide are regulated through spatial management, allowing quotas to be set for each species within a specified geographical region. Still, illegal, unreported and uncontrolled (IUU) fishing poses a large and continuing threat to sustainable fisheries.

Recently, there has also been an increased focus from consumers on 'eco-labeling' of fish by organizations such as the Marine Stewardship Council (MSC), which certify fisheries that meet the criteria for sustainable harvest. In addition, fish of a particular species caught from different areas may differ significantly in flesh quality, caused by, e.g., different access to feed, feed content, or the physical conditions experienced by the individual and genetic factors (Børresen 1992).

Finally, fish from different areas may contain different levels of organic contaminants or heavy metals (mercury, cadmium, etc.), which may pose a health risk for the consumer. Accordingly, methods that unambiguously establish the geographic origin of fish and fish products independently of the accompanying paper trail are in increasingly high demand.

In the 2000s, it was estimated that up to 40% of the landings of Atlantic cod in the Eastern Baltic Sea were from IUU fishing (International Council for Exploration of the Sea (ICES, www.ICES.dk)), resulting in severe overfishing. Despite the apparently huge numbers of illegal cod being landed, it was not established how they entered the market and were sold to the European consumer.

In 2003, a group of journalists from the Swedish television channel TV4 decided to investigate whether black market fish were 'white-washed' by fish wholesalers or retailing companies and sold to Swedish consumers under false labels of geographical origin. However, it was not possible to establish the stock of origin using traditional traceability methods (Cadrin *et al.* 2005). TV4 journalists therefore contacted population geneticists at the Danish Institute for Fisheries Research (now National Institute of Aquatic Resources, DTU-Aqua), in order to inquire whether genetic methods could determine the origin of cod sold by Swedish fishmongers.

36.2 Significance of the case

In general, Eastern Baltic cod is considered to have inferior flesh quality compared with cod from North Sea and North Atlantic stocks. In addition, fish from the Eastern Baltic have higher levels of organic pollutants, such as dioxin mainly originating from tree pulp bleaching associated with paper production. This has led to recommendations in several countries around the Baltic Sea for limited consumption of Baltic Sea fish with high fat content such as salmon and herring, in particular by pregnant women. For fish with low flesh fat content, such as cod, the problem is mainly limited to eating cod liver and roe. For a thorough assessment of risks and management of dioxin-like compounds in Baltic Sea fish, see Assmuth and Jalonen (2005).

Collectively, these factors have led to the general perception that North Sea cod is of higher quality, and Baltic cod has therefore generally fetched prices about 20% lower. Accordingly, there is a large profit to be made from substituting and mislabeling Baltic cod with cod of North Sea or Northeast Atlantic origin. The Swedish television journalists estimated that the large fish retail companies in theory could make more than €500 000 extra annually through deliberate mislabeling, underpinning the potential for large economic revenue from illegal activities.

36.3 The case history continues

The response from the population geneticists at DTU was positive. A genetic method for tracing cod back to population of origin had recently been developed (Nielsen *et al.* 2001) using Northeast Arctic (Barents Sea), North Sea and Eastern Baltic cod as test cases. The method was based on the analysis of ten so-called DNA microsatellites (Fig. 36.1), which are short, repetitive stretches of DNA, typically with a repeat unit of 2–6 base pairs and a relatively high mutation rate. Due to these attributes microsatellites are highly variable DNA markers, well suited for the procedure of individual assignment (IA).

In brief, IA assigns an individual (fish) to the population sample (and thus geographic origin) where its genotype has the highest likelihood of occurring. The procedure consists of two steps. Firstly, baseline samples are collected from the populations/areas of interest (e.g. North Sea and Eastern Baltic Sea; see Fig. 36.2). Individuals from these baselines are then consecutively assigned back to the most likely population of origin (self-assignment) in order to test the statistical power of IA using individuals of known origin.

Secondly, individuals of unknown origin (case samples) are assigned to the most likely baseline population and the statistical certainty is evaluated in relation to the results of the self-assignment in step one. Under a conservative statistical model, 91% of our baseline of Eastern Baltic cod were correctly assigned back to

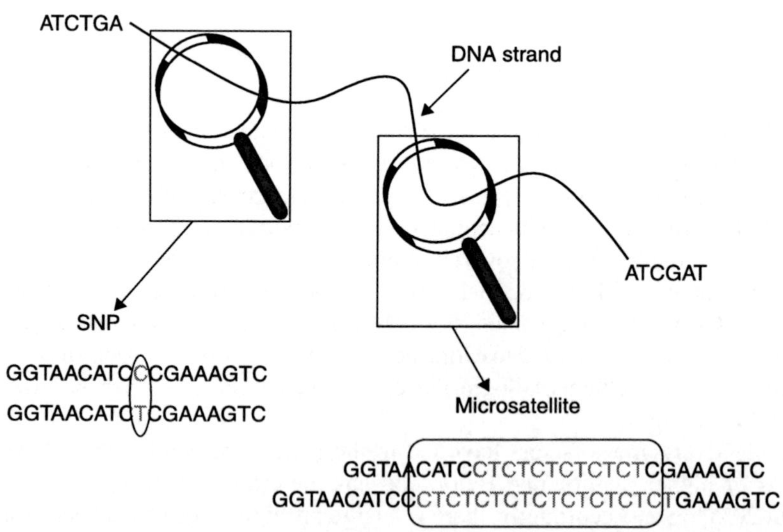

Fig. 36.1 A microsatellite is a segment of a DNA strand that consists of a variable number of tandem repeats. Here the repeat motif is CT. The two depicted variants (alleles) have six and ten repeats respectively and are flanked by segments of identical DNA. In contrast, a single nucleotide polymorphism (SNP) is a DNA sequence variation at a single position in the DNA. Here the two variants contain a C and T, respectively.

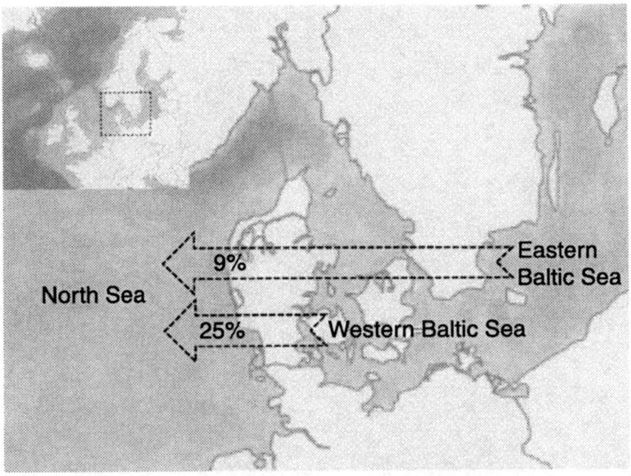

Fig. 36.2 Map showing the location of the geographical areas where the baseline samples for individual assignment were collected (North Sea, Baltic Sea and Western Baltic Sea). Illustrated are also the percentages of fish with known origin that were mis-assigned, showing the level of statistical uncertainty of the applied DNA-based tool.

that population, while 9% were mis-assigned to the North Sea population (Fig. 36.2).

The television journalists now had the tools they had been looking for. They collected at least two cod fillets (Fig. 36.3) from ten supermarkets/fish shops across Sweden. When possible, they visited all the shops twice in order to assess whether potential fraud was a single event or if it occurred repeatedly. In all cases they specifically bought North Sea cod and asked the shop assistant to confirm the origin. This was documented using a candid camera.

From each of 42 sampled fillets of presumed North Sea origin the journalists cut out 0.5 cc^2 of tissue, which was then preserved in ethanol and sent from Sweden to DTU-Aqua's molecular analysis laboratory. Approximately 2 weeks later the samples were processed and the results were clear. Twenty samples were assigned correctly to the claimed North Sea cod origin. Seventeen samples were assigned to Eastern Baltic cod. The last five samples had genotypes that were compatible with neither North Sea nor Eastern Baltic origin. Despite the fact that it was not possible to assign individual cod with 100% certainty to population of origin, the hypothesis that all cod originated from the claimed North Sea origin could be rejected (with probability ≈ 0).

In fact, it appeared that more than half of the fish were mislabeled and that the vast majority of these fish originated from the Eastern Baltic. Overall, seven shops delivered two or more tentatively mislabeled samples, while the remaining three shops all delivered four samples assigned to the correct area of origin. In several cases all four samples, collected on two separate occasions, were mis-assigned, suggesting more systematic mislabeling.

Fig. 36.3 Where do these cod come from? Visual inspection does not give much away. In contrast to barcoding, which is a categorical method where fish can only be one species or the other, Individual Assignment (AI) used here relies on a probabilistic model, the attributes of which can be difficult to grasp for the non-specialist. In relation to this, genetic differences among marine fish populations are generally not very large, resulting in low power for IA (source: courtesy of Finn Sivebæk).

36.4 Resolution and outcomes

Confronted with the evidence, all supermarkets and fish shops that allegedly had sold mislabeled fish admitted that mislabeling of cod and cod products in their shops was indeed possible. In addition, they all apologized and recognized the need for the correct label of origin for consumer protection. Finally, they all promised to initiate an investigation into where in the production chain the mislabeling had taken place.

A general tendency was that the majority of the shops/supermarkets claimed that the substitution had taken place at the fish buyer/wholesaler level, so that the fillets sold were already mislabeled before they entered the shops. The journalists pursued this issue and contacted the major fish wholesalers in Sweden, including one company that supplies more than 300 shops in Sweden. None of these companies agreed to give an interview.

Based on the investigation and interviews, a TV program was produced and broadcast on Swedish television in 2003. The program caused a big uproar from consumers in Sweden and neighboring countries, as the majority of fish wholesale companies are international. As an epilogue to the case, the Danish population geneticists were contacted by a Danish fish wholesale company, which was amongst the largest in Europe. This company was known to buy a large proportion of its cod in Poland, which borders the Eastern Baltic Sea. The company was among those that delivered cod to some of the investigated shops that sold mislabeled cod. The contact person claimed that their cod were caught in the transition area (Belt Sea) between the North Sea and the Baltic Sea, where there is a gradual change in genetic composition from North Sea to Baltic Sea cod (Nielsen *et al.* 2003).

According to the company it was not a question of mislabeling but of mis-assignment from our side. Fortunately, baseline samples from the Belt Sea were available (see Fig. 36.2). The assignment tests revealed that only two of the ten cod samples originating from this company were assigned to the claimed Belt Sea (labeled North Sea) origin, while eight individuals were assigned to the Eastern Baltic Sea. The probability that this was due to mis-assignment alone was 0.0002. In conclusion, it was still highly likely that the fish were accidentally or deliberately mislabeled.

36.5 Commentary and personal reflections

This case shows that there is a very high demand for methods to trace the geographical origin of fish, independently from the paper trail and certificates accompanying fish products en route from boat to consumers. If such tools are not developed and implemented, there is an ever ample opportunity for intentional or unintentional mislabeling, which potentially compromises product quality and consumer health.

DNA methods hold a clear advantage over other conventional identification tools. As DNA is found in all tissues in all organisms, DNA-based tools are relatively insensitive to type and amount of tissue as well as the degree of processing, and they are therefore good candidates for universal traceability tools. Whereas DNA barcoding for species identification has gained some general application (e.g. see the Fish Barcode of Life Initiative (FISH-BOL) at www.fishbol.org), the number of cases where geographic origin assignment of fish has been used in a forensic or certification process is still low, despite the fact that some of the DNA tools have been around for more than a decade.

There are a number of potential reasons for this. First of all, these methods are perhaps not so well known outside the scientific community. In contrast to barcoding, which is a categorical method where fish can only be one species or the other, individual assignment relies on a probabilistic model, the attributes of which can be difficult to grasp for the non-specialist. In relation to this, genetic differences among marine fish populations are generally not very large, resulting in low power for IA. Unambiguous assignment is essential for most forensic purposes, where deliberate illegal fishing or mislabeling has to be proved in a court of law beyond reasonable doubt. Finally, standardization of these methods for food organisms besides model species such as domestic cattle and pigs has been relatively difficult and time-consuming.

However, the development of new tools, such as thousands of single nucleotide polymorphisms (SNPs) under the support of EU 7th framework projects (e.g. FishPopTrace; http://fishpoptrace.jrc.ec.europa.eu), means that standardization of method has been much improved. Still, there is a need for authorities and labeling institutions to implement these methods.

A recent example from MSC eco-labeled Patagonian toothfish (*Dissostichus eleginoides*) highlights this point (Marko *et al.* 2011). An independent DNA-based test revealed that a number of retailed samples displayed genotypes incompatible with an origin from the certified fishery, thus undermining both consumer confidence in the product and the validity of the certification procedures.

36.6 Acknowledgments

We thank the European Community's Seventh Framework Programme (FP7/2007–2013) under grant agreement KBBE 212399 (FishPopTrace) for financial support.

36.7 Critical questions for discussion

1. Why is it important to have paper trail-independent methods for tracing food?
2. What constitutes a good laboratory traceability method?
3. What are the advantages of using DNA-based traceability tools?
4. What are the specific laboratory challenges in relation to tracing fish and fish products?
5. Do you know how IUU fishing and mislabeling are controlled in your country?
6. Do you think that eco-labelling of fish will be able to assure sustainable fishing and protect against mislabeling and associated health risks for the consumer? How will this be done?
7. What could the authorities do in order to assure better traceability and authenticity documentation?
8. If you were to develop new fish traceability tools, how would you like to see the current methods improved?

36.8 References

ASSMUTH T. and JALONEN P. (2005). *Risks and Management of Dioxin-like Compounds in Baltic Sea Fish: An Integrated Assessment*. Nordic Council of Ministers, Copenhagen.

BØRRESEN T. (1992). Quality aspects of wild and reared fish. In H. H. Huss, M. Jakobsen, and J. Liston (eds), *Quality Insurance in the Fish Industry*, pp. 1–17. Elsevier, Amsterdam.

CADRIN S.X., FRIEDLAND K.D. and WALDMAN J.R. (eds) (2005). *Stock Identification Methods: Application in Fisheries Science*. Academic Press, London.

MARKO P.B., NANCE H.A. and GUYNN K.D. (2011). Genetic detection of mislabeled fish from a certified sustainable fishery. *Curr Biol*, **21**: R622.

NIELSEN E.E, HANSEN M.M., RUZZANTE D.E., MELDRUP D. and GRONKJAER P. (2003). Evidence of a hybrid-zone in Atlantic cod (*Gadus morhua*) in the Baltic and the Danish Belt Sea revealed by individual admixture analysis. *Mol Ecol*, **12**: 1497–508.

NIELSEN E.E., HANSEN M.M., SCHMIDT C., MELDRUP D. and GRONKJAER P. (2001). Fisheries – Population of origin of Atlantic cod. *Nature*, **413**: 272.

37

Testing for genetically modified organisms (GMOs) revealed the real source of rice imported to Norway

A. Holst-Jensen, A. R. Ali, L. Emanuelsen, T. M. Grønbeck, S. B. Harbo, C. Skjæret and B. Spilsberg, Norwegian Veterinary Institute, Norway and H.-U. Waiblinger, State Institute of Chemical and Veterinary Analysis of Food Freiburg, Germany

Abstract: Genetically modified organisms (GMOs) are controversial. The presence of GMOs or derived products in food is regulated and foods, feeds and seeds are routinely tested for the presence of GMOs to reveal non-compliance with authorization and labeling regulations. The number and diversity of GMOs developed and commercialized is rapidly increasing. Foods are also often composed of ingredients from several species. These are just two of the factors responsible for the complexity of possible scenarios involving GMOs that analysts and other stakeholders in the food supply chain are facing, as the case studies in this chapter show.

Key words: event specific, labeling, quantification, screening, species identification.

37.1 Introduction to the cases

Genetically modified organisms (GMOs) are defined as organisms in which the genetic information (carried in the DNA) has been altered by genetic engineering techniques. The number of GMOs developed and commercialized is rapidly increasing and they are becoming more diverse. The presence of GMOs or derived products in food is regulated. Such products are also subject to specific authorization and labeling requirements in many countries around the world. Foods, feeds and seeds are routinely tested for the presence of GMOs to reveal non-compliance with authorization and labeling regulations.

Rice is one of the major cereal crops in the world and a common staple food in Asia. However, rice is also grown and exported to Europe from the USA. GM varieties of rice have been developed, just as is the case for most other important food crops. The herbicide-tolerant (Liberty Link) LL62 rice was the first and hitherto only GM rice event for which the European Union (EU) has received an application for authorization (GMO Compass 2011). The term 'event' refers to a very specific genetic modification in the very specific plant line where it is introduced. The identity of an event is therefore characterized by a specific and unique DNA sequence that can be crossed into commercial seed varieties. The event is the legal item that is subject to regulation. The application received by the EU concerned the use of LL62 rice as food and feed but not its cultivation. In 2006 the company that developed LL62 rice reported that, due to insufficient quality controls, small quantities of another similar GM rice event, LL601, was present in rice grown in and exported from the USA (Vermij 2006). This led to implementation of emergency measures calling for extensive testing of rice shipments from the USA to Europe (European Commission 2006a, 2006b, 2006c). It also had large socio-economic consequences, in particular for US rice producers and exporters (Li *et al.* 2010, Redick and Endres 2009) and later for the company Bayer CropScience (Fox 2011). The emergency measures for LL601 rice were later lifted and rice is now regularly tested in the same way as any other food and feed ingredient to enforce regulations.

The current GMO testing approach applied by the majority of laboratories in Europe is based on initial screening for particular DNA sequence motifs common to multiple GMOs. This approach, referred to as the matrix approach (Holst-Jensen 2009; Waiblinger *et al.* 2010), is considered an efficient tool to discriminate samples likely to contain GMO from samples not containing detectable GMO. Furthermore, the screening results can also be used to indicate which GMOs a sample can and cannot contain (see, e.g., Waiblinger *et al.* 2010). In some cases the application of screening can also lead to unexpected findings. Following the initial screening, it is common to apply more specific and confirmative tests. Quantitative event-specific methods in combination with quantitative species identification methods are typically used to test for compliance with labeling requirements.

The two cases discussed in this chapter both concern screening of rice products imported from outside Europe for GMOs. The first case concerns a product declared to be 'American wild rice'. The second concerns a product declared to be a processed rice derivative produced in Asia.

37.2 Significance of the cases

There are many recognized potentially harmful contaminants of food and feed: for example, *Listeria monocytogenes* and certain *E. coli* serotypes, toxigenic fungi and their mycotoxins, and allergens derived from particular ingredients such as peanuts, soybean and shrimp. In the EU, labeling of products containing or derived from GMO is required to give consumers the freedom to choose between GM and non-GM products. This labeling requirement applies only to GMOs that are authorized after a comprehensive safety assessment, and the labeled products therefore do not pose any risk of harm to health and environment. GMOs not authorized in the EU but authorized in another jurisdiction, e.g. in the USA, are also unlikely to represent a health and environmental risk if unintentionally released into the food supply chain in Europe. GMOs that have not been authorized in any jurisdiction usually have not been subjected to the same thorough risk assessments as those authorized and therefore could exceptionally represent a risk. LL601 rice is an example of a GMO that had not been authorized in any jurisdiction, and this was one of the main reasons for the establishment of emergency measures in 2006.

37.2.1 Regulatory aspects

In the EU no GMO can be released into the environment or used as a food or feed ingredient without authorization (see Further reading, European Commission 2004). Release is regulated according to Directive EC 2001/18 (European Commission 2001). Authorization for use as a food or feed and labeling are regulated according to Regulations 1829 and 1830/2003 (European Commission 2003a, 2003b), and is on a per-ingredient basis. The polymerase chain reaction (PCR) methods that are used to test for the presence of GMOs are unable to discriminate between the sources of a DNA sequence belonging to the same species but derived from different ingredients. For GMO testing the ingredient is therefore treated as equivalent to a species. The impact on analytical results of so-called botanical impurities (non-declared material from another or the same species) is often taken into account, but this is not always the case.

37.2.2 Economic and market aspects

Products not compliant with the legislation are by definition illegal. There is, however, a concern among some stakeholders that labeling can be misinterpreted as implying that the product is unsafe. At the moment there are very few labeled products on the market in the EU, probably because companies are afraid that the marketing of such products could negatively affect their activities. This could also mean that there is some degree of incitement to fraudulently put products that should have been labeled onto the market without labels. For the GMOs that have not been authorized, a major challenge is the availability of reliable detection methods and corresponding positive and negative control samples. Some GMO testing laboratories around the world apply methods with inferior specificity and/or sensitivity than the official control laboratories in the EU (see also Holst-Jensen *et al.* 2003). They may also intentionally limit the range of GMOs covered by the tests, e.g. to reduce costs. If the company putting a food or feed product on the market is not aware of this, they may be misled by incomplete information provided by other actors upstream in the supply chain. It is the responsibility of the company putting the product onto the market to be compliant with the legislation (European Commission 2004). It is therefore also very important to understand what the analytical methods detect, their limitations and the main sources of mis-communication of information.

37.3 The case histories continue

Samples of both the 'American wild rice' and the processed rice derivative produced in Asia were subjected to standard screening with PCR, targeting altogether five different sequence motifs (Waiblinger *et al.* 2010). Rice not containing any GMO should produce negative results for all five PCRs. The LL rice events are the only rice events that have (officially) been cultivated (including field trials) in the USA. These events contain two of the five screening targets, i.e. the cauliflower mosaic virus 35S promoter (P35S) and the *bar* gene coding for tolerance to the herbicide glufosinate ammonium (Liberty). Three other GM rice events developed in China – KeFeng6, KMD-1 and Bt63 – but not authorized in any jurisdiction contain a third PCR target, the nopaline synthase terminator (T-nos) from the soil bacterium *Agrobacterium tumefaciens*, and two of these GM rice events also contain P35S (Waiblinger *et al.* 2010). In both samples we found the T-nos. Only one contained P35S, while neither of the samples contained the *bar* gene. The immediate interpretation was therefore that both samples contained GMO, most likely one or more of the Chinese GM rice events. Screening, however, does not provide conclusive evidence. Verification of the results was therefore needed and had to be done using more specific methods (Reiting *et al.* 2010).

For the sample positive only for T-nos the list of candidate GMOs suspected to be present was limited to one, the Bt63 event. A method specific to Bt63 (Mäde *et al.* 2006) was therefore applied and the test result was positive. Thus, the presence of Bt63 rice in the sample was confirmed. This was an unexpected result since the product was claimed to be 'American' and contamination of rice with Bt63 was assumed to be a problem limited to China and the neighboring countries.

The presence of both P35S and T-nos indicated the possible presence of the Chinese rice GMOs KeFeng6, KMD-1 and perhaps also Bt63 (Reiting *et al.* 2010; Waiblinger *et al.* 2010) in the second sample. However, further testing for specific DNA sequence motifs from either of these GMO rice events yielded consistently negative results. Thus, the detected P35S and T-nos were assumed to originate from another GMO. The list of candidates therefore had to be modified, and the most likely source was suspected to be material from a species other than rice.

The world's most successful GMO is Monsanto's GTS 40-3-2 (Roundup Ready®) soybean (RRS). By successful, we here refer to cultivated land area and number of nations that have approved the GMO for cultivation (James 2009). Soybean is also commonly traded and transported globally, and traces of soybean can easily enter the supply chain as an impurity or contaminant. RRS contains P35S and T-nos, and the presence of RRS in the second sample could therefore explain the observed screening results (see Waiblinger *et al.* 2010). Event and species-specific quantitative testing for RRS and soybean confirmed the presence of this EU-authorized event in a quantity roughly corresponding to the observed quantities of P35S and T-nos. It was therefore concluded that the sample did not contain GM rice.

37.4 Resolution and outcomes

According to the European Food Law (European Commission 2002) it is necessary for all actors in the supply chain to be able to trace the product one step back and one step forward in the supply chain (article 18). The available traceability information revealed that the rice of 'American' origin was actually a mixture of a smaller fraction of American wild rice and a major fraction of conventional rice of Asian origin. It was concluded that the Bt63 rice detected in the sample was a contaminant of the Asian rice and not of the American wild rice. On the basis of the observed presence of the unauthorized GM rice event Bt63 in the sampled batch of the 'American wild rice' the Norwegian Food Safety Authority (NFSA) issued an immediate ban on further sales of the specific batch. A notification to the European joint Rapid Alert System for Food and Feed was also issued.

In the second rice product, the GMO detected was not a rice but a soybean GMO. Notably, however, the presence of soybean ingredients in this product was not declared. Since soybean is a known allergen, the presence of soybean in the product could be indicative of non-compliance with the requirements for labeling of products potentially causing adverse health effects (European Commission 2002, article 14). The quantity of soybean detected was very low relative to the quantity of rice in the sample. Interestingly, the GMO concentration of the soybean was significantly higher than the threshold for labeling. The NFSA concluded that the soybean present in the sample was a 'botanical impurity', and that the quantity did not have significant potential to cause adverse health effects. Thus, in this case the NFSA decided that the product was legally compliant and no further action was taken.

37.5 Commentary

These two cases exemplify that the GMO analyst and other stakeholders must be very prudent when analytical results are interpreted. However, the first example also shows that it is important not to be prejudiced when the analytical approach is chosen. All stakeholders tend to try to minimize costs. Design of simplified *ad hoc* testing schemes can be an efficient means to achieve cost reductions. An *ad hoc* scheme could, for example, be designed to test only for the GMOs authorized in the exporting jurisdiction. In this first case such a reduced testing scheme would then have been designed to test only for 'American' GM rice events, i.e. the LL rice events, and would then have included only two out of the five screening PCRs. If so, it would have been unable to detect the presence of the unauthorized GM rice event present in the sample.

The second case shows that the data obtained from analyses applied for one purpose may also have relevance to another field of food safety. In particular, the detection of DNA sequences from potentially allergenic species may be revealed as a spin-off. The range of genetic elements inserted into new GMOs is growing. Perhaps in the near future the presence of some of these genetic elements, but not of the corresponding GMO, may also lead to detection of, for example, unexpected pathogenic microorganisms or emerging and invasive weeds or other foreign species.

It is often claimed that GMO testing is costly and that, at least when testing for compliance with the labeling requirements, there is no issue of safety involved. If testing schemes are designed so that they increase the probability of detecting GMOs that are not authorized, then the holistic cost–benefit calculation is improved. Some of the potential spin-offs described above are not only hypothetical, as illustrated with the second case. So, to conclude, GMO testing is potentially about much more than just detecting GMOs.

37.6 Acknowledgments

The cases described were outcomes of the public surveillance program for GMOs in Norway financed and coordinated by the Norwegian Food Safety Authority (NFSA, Mattilsynet). The support of the NFSA is gratefully acknowledged.

37.7 Critical questions for discussion

1. What is an ingredient?
2. Is detection of GMO with methods targeting DNA more reliable than methods targeting proteins or RNA?
3. What is a botanical impurity?
4. What does labeling of a food product as 'contains GMO' indicate?
5. What does labeling of a food product as 'does not contain GMO' indicate?
6. What is the limit of detection? What determines the limit of detection?
7. Given that the resources invested in GMO testing are limited, how are they best used?

37.8 Further reading

EUROPEAN COMMISSION (2004). Recommendation 787/2004 of 4 October 2004 on technical guidance for sampling and detection of genetically modified organisms and material produced from genetically modified organisms as or in products in the context of Regulation (EC) No 1830/2003. *Official Journal of the European Union*, **L348**: 18–26.

HOLST-JENSEN A. and BERDAL K. G. (2004). The modular analytical procedure and validation approach and the units of measurement for genetically modified materials in foods and feeds. *Journal of AOAC International*, **87**: 927–36.

HOLST-JENSEN A., DE LOOSE M., and VAN DEN EEDE G. (2006). Coherence between legal requirements and approaches for detection of genetically modified organisms (GMOs) and their derived products. *Journal of Agricultural and Food Chemistry*, **54**: 2799–809.

RUTTINK T., DEMEYER R., VAN GULCK E., VAN DROOGENBROECK B., QUERCI M. *et al.* (2010). Molecular toolbox for the identification of unknown genetically modified organisms. *Analytical and Bioanalytical Chemistry*, **396**: 2073–89.

37.9 References

EUROPEAN COMMISSION (2001). Directive 2001/18/EC of the European Parliament and the Council of 12 March 2001 on the deliberate release into the environment of genetically modified organisms and repealing Council Directive 90/220/EEC. *Official Journal of the European Communities*, **L106**: 1–38.

EUROPEAN COMMISSION (2002). Regulation (EC) 178/2002 of the European Parliament and of the Council of 28 January 2002 laying down the general principles and requirements of food law, establishing the European Food Safety Authority and laying down procedures in matters of food safety. *Official Journal of the European Communities*, **L31**: 1–24.

EUROPEAN COMMISSION (2003a). Regulation (EC) No 1829/2003 of the European Parliament and of the Council of 22 September 2003 on genetically modified food and feed. *Official Journal of the European Union*, **L268**: 1–23.

EUROPEAN COMMISSION (2003b). Regulation (EC) No 1830/2003 of the European Parliament and of the Council of 22 September 2003 concerning the traceability and labelling of genetically modified organisms and the traceability of food and feed products produced from genetically modified organisms and amending Directive 2001/18/EC. *Official Journal of the European Union*, **L268**: 24–8.

EUROPEAN COMMISSION (2004). Regulation (EC) No 882/2004 of the European Parliament and of the Council of 29 April 2004 on official controls performed to ensure the verification of compliance with feed and food law, animal health and animal welfare rules. *Official Journal of the European Union*, **L191**: 1–52.

EUROPEAN COMMISSION (2006a). Decision 2006/578/EC of 23 August 2006 on emergency measures regarding the non-authorised genetically modified organism 'LL RICE 601' in rice products. *Official Journal of the European Union*, **L230**: 8–10.

EUROPEAN COMMISSION (2006b). Decision 2006/601/EC of 6 September 2006 amending Decision 2006/578/EC on emergency measures regarding the non-authorised genetically modified organism 'LL RICE 601' in rice products. *Official Journal of the European Union*, **L244**: 27–9.

EUROPEAN COMMISSION (2006c). Decision 754/2006/EC of 6 November 2006 amending Decision 2006/601/EC on emergency measures regarding the non-authorised genetically modified organism 'LL RICE 601' in rice products. *Official Journal of the European Union*, **L306**: 17–20.

FOX J.L. (2011). Bayer's GM rice defeat. *Nature Biotechnology*, **29**: 473.

GMO COMPASS (2011). *GMO Compass*. Available at http://www.gmo-compass.org.

HOLST-JENSEN A. (2009). Testing for genetically modified organisms (GMOs): Past, present and future perspectives. *Biotechnology Advances*, **27**: 1071–82.

HOLST-JENSEN A., RØNNING S.B., LØVSETH A., and BERDAL K.G. (2003). PCR technology for screening and quantification of genetically modified organisms (GMOs). *Analytical and Bioanalytical Chemistry*, **375**: 985–93.

JAMES C. (2009). *Global status of Commercialized Biotech/GM Crops: 2009*. Report: 41. ISAAA Briefs – Executive summary. Available at http://www.isaaa.org/resources/publications/briefs/41/.

LI Y., WAILES E.J., MCKENZIE A., and THOMSEN M. (2010). LL601 contamination and its impact on U.S. rice prices. *Journal of Agricultural and Applied Economics*, **42**: 31–38.

MÄDE D., DEGNER C., and GROHMANN L. (2006). Detection of genetically modified rice: A construct-specific real-time PCR method based on DNA sequences from transgenic Bt rice. *European Food Research and Technology*, **224**: 271–8.

REDICK T.P. and ENDRES A.B. (2009). Jury verdict against Bayer for Liberty Link rice breaks new ground in biotech liability. *Agricultural Law Update*, **26**: 312–13.

REITING R., GROHMANN L., and MÄDE D. (2010). A testing cascade for the detection of genetically modified rice by real-time PCR in food and its application for detection of an unauthorized rice line similar to KeFeng6. *Journal für Verbraucherschutz und Lebensmittelsicherheit*, **5**: 185–8.

VERMIJ P. (2006). Liberty Link rice raises specter of tightened regulations. *Nature Biotechnology*, **24**: 1301–2.

WAIBLINGER H.U., GROHMANN L., MANKERTZ J., ENGELBERT D., and PIETSCH K. (2010). A practical approach to screen for authorised and unauthorised genetically modified plants. *Analytical and Bioanalytical Chemistry*, **396**: 2065–72.

38

Documenting the '*terroir*' aspects of award-winning Danish conserves: a model for the development of authentic food products

L. S. Christensen and J. Hoorfar, Technical University of Denmark, Denmark, J. Sørensen, Øhavet Aps, Denmark and S. Bisgaard, Danish Veterinary and Food Administration, Denmark

Abstract: An example of the qualities embedded in the foods produced by small companies based on small Danish islands and reflections on the association between the qualities of the products and their geographical origin is presented. This case study discusses how it can be ensured that qualities in the products that can be tasted and documented truly reflect geographical origin, and ways in which authenticity can be guarded and exploited in marketing while at the same time increasing production volume.

Key words: labeling, food specialties, geographical indication.

38.1 Introduction to the case

Many nations have specialty foods that are considered exotic and fine by other nations, as well as by the nation itself. However, documenting the quality imprint established by the geographical origin and the genetic sources of the plants or animals that are ingredients in specialty foods is a challenge faced by many testing laboratories and regulators. Specialty foods, due to their nature, are also often produced, at least initially, by very small businesses. These businesses may have the desire to expand, but lack the resources needed to grow their markets. As the market for their products increases, these companies may face the challenge of scaling up operations while maintaining the '*terroir*' characteristics of their products.

The following case from Denmark illustrates the philosophy and principles of food specialties with the concept of '*terroir*', which suggests one way in which producers of such products might collaborate in order to grow their markets and discusses approaches to authenticity certification. It revolves around Øhavet Ltd, a Danish producer of award-winning specialty conserves.

38.1.1 The Danish 'small islands' and Øhavet Ltd

Denmark is a nation of more than 400 islands including 27 'small islands', each with a population of fewer than 1000 inhabitants. The small islands are characterized by great diversity in geochemical and climatic conditions, which makes each one of them unique. Yet, the small islands have similarities too that make them different from the rest of the country. These include their milder climate, the presence of sea mists, and in many cases a lack of predators such as foxes. Industry on the small islands has a downside, though: logistics costs are considerably higher than in the rest of the country. Farming on these islands in terms of conventional production of large quantities of raw materials for the feed and food manufacturing industry is therefore hardly competitive.

The founder of Øhavet Ltd was a 'sailor of the seven seas' who found a location on the island of Strynø, one of the small islands, as close to the sea as possible and established a family. Øhavet Ltd was founded in 2002 with the ambition of creating food products with unique qualities and combining production with the joy of life in a small community. The company is located on the island of Strynø, which has 200 inhabitants and is south of the larger island of Funen. The name of the company can be translated into 'the sea of islands' and refers to a location with a number of small islands characterized by a relatively mild climate as the islands are protected to the south by the European continent, from the North Sea by the peninsula of Jutland and from the Baltic Sea by a number of larger islands (see Fig. 38.1).

Øhavet Ltd is a producer of berries and prize-winning conserves based on these berries. The founder was initially distributing – in person – the products to a small group of high-end-price retail companies spread all over the country. The unique qualities of the conserves have become appreciated by consumers and the products were nominated in 2009 as 'the best conserves of Denmark'.

The philosophy and principles of Øhavet Ltd upon which the production of conserves are based can be formulated as follows:

- All berries are hand-picked to ensure that only ripe and healthy berries are used, among other reasons in order to reduce the need for sweetening by sugar.
- The majority of berries are produced on the Øhavet Ltd farm and other farms on the island of Strynø. Only occasionally have berries been imported from other islands in the 'sea of islands' and southern coastal areas of the island of Funen.
- Processing of berries, packing, labeling and storage are carried out at the production plant on Strynø (until this time this has been entirely the case).

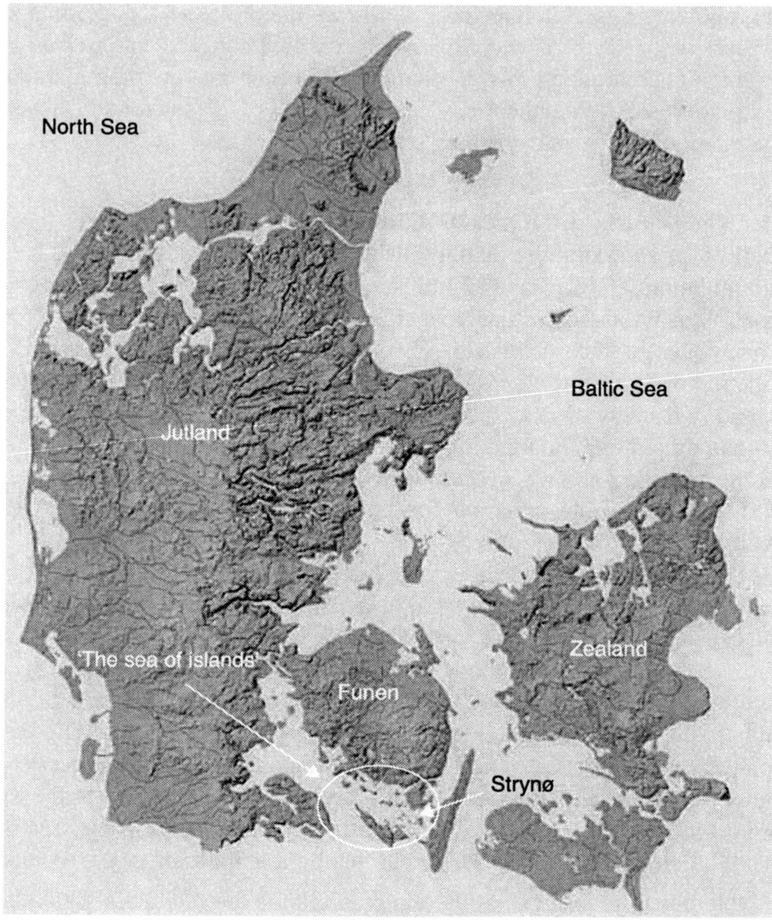

Fig. 38.1 Location of 'the sea of islands' and the island of Strynø on the map of Denmark.

- The production is certified organic.
- Varieties of berries are chosen that are optimal for their taste (in contrast to volume).
- Ancient varieties of berry plants are identified in the local area so that any advantages and adaptive characteristics of the ancestral gene resources of these plants can be exploited.

As the population of the island of Strynø has a reputation for pride in local raw materials and home-made recipes for their use, Øhavet's recipes are primarily based on ancient recipes from the island. These are then developed further and evaluated by tasting panels chosen from the population of Strynø.

38.2 Significance of the case

With the appreciation of the quality of products, and the Danish consumers' willingness to support producers on small islands of Denmark – known as 'fair trade in Danish' – there was a domestic market for Øhavet Ltd products from the first day.

There has been an increase in exports from Øhavet Ltd to Germany and Sweden; and for a number of reasons Japan is considered an interesting market, and considerable efforts have been made to initiate such exports. Obviously, volume needs to be enhanced considerably to justify the logistical expenses on these new markets. One of the lessons from this case is that as volume of production increases, a brand gains market value, and markets are expanding, there will be an increase in demand for documentation of the claimed unique qualities of the products (*terroir* and location-adaptive types of berry used in production) and the possibility to detect fraud.

38.3 The case histories continue

38.3.1 Foundation of Island Specialties®

Øhavet Ltd is part of The Small Island Association of Food Producers (www.oespecialiteter.dk) and is one of the companies that trade under the brand name Island Specialties®. The unique conditions on the small islands and their significance in the national identity of Denmark provided the basis of a strategy formulated by The Small Island Association of Food Producers to maintain sustainable communities on them. The trademark Island Specialties® was established by this Association in 2010 in order to share between the member companies the challenges of branding, marketing and distribution logistics. The member companies between them have now mastered a majority – if not all – of the operations required from primary production and processing to distribution of products, so they are nearly, if not totally, self-sufficient. The aim was to ensure that only a small volume of high-value products had to be exported from the home island of the respective companies for the businesses to be successful. Three criteria were defined to assure the quality and authenticity of each product under the brand Island Specialties® (Christensen 2011). These criteria relate to:

1. The origin of the product's essential raw materials and the qualities of the product that are specifically connected with its origin,
2. The creation of jobs and affiliation of members of the island community,
3. The place in which the product had its genesis, signaling the location's intellectual property right.

Compliance with two of the three criteria is required for the product to bear the brand.

38.3.2 Foods with Protected Geographical Status (PGS)

The European Union has created a legal framework to protect the names of regional foods, which is known as Protected Geographical Status (PGS). The framework includes three EU schemes known as PDO (protected designation of origin), PGI (protect geographical indication) and TSG (traditional speciality guaranteed) promote and protect names of quality agricultural products and foodstuffs. PDO covers agricultural products and foodstuffs which are produced, processed and prepared in given geographical areas using recognised know-how. PGI covers agricultural products and foodstuffs closely linked to the geographical area. At least one of the stages of production, processing or preparation takes place in the area. TSG highlights traditional character, either in the composition or means of production.

As origin of production and location of processing are the same, Øhavet Ltd at present complies with criteria 1 and 2 of Island Specialties® as well as the requirements for both PDO and PDI certification. Certification as PDO or PDI is therefore a possibility in the future.

Although intellectual aspects of the island of Strynø are indeed embedded in the products in compliance with criterion 3 of Island Specialties®, the time frame of their formulation makes a certification as TSG unlikely, and the difference between the products from Øhavet and other products on the market should be considered carefully to reveal whether TSG certification after the due time period is relevant.

38.3.3 Growth in volume of the business

Due to high logistics costs associated with being located on a small island in addition to other general costs of running a food production facility, Øhavet Ltd has experienced the same need for growth of volume sales that other member companies of Island Specialties® have experienced. To achieve an increase in output, the company needs to increase its supply of berries. In addition, the incidental appearance of fungal plant pathogens and subsequent death of certain varieties and species of berry plants on the island of Strynø has brought home the need not just to rely on berries produced on Strynø. Locations of berry production on the other islands in 'the sea of islands' and coastal areas of the island of Funen are to be chosen.

Various possibilities have been considered to document the authenticity of the conserves produced by Øhavet Ltd. These include collaboration between Øhavet Ltd and research institutions, firstly to measure the multi-element composition (Husted *et al.* 2004, Laursen *et al.* 2011) of the berries to document their geographical origin, and secondly to genetically fingerprint the berry-producing plants (Primrose *et al.* 2010) to determine their origin. Among the berries' unique parameters is their salt composition, which arises from the sea mist and rain with which they are in contact. Development of chemometric methods or organic compound fingerprinting to document the quality of berries harvested and processed could be other approaches.

One purpose of documenting specific fingerprints and qualities is to assure the consistent quality of the products when the locations of berry production are switched between islands or expanded from one to multiple islands and to encompass coastal areas of Funen. The identity of the location referred to by Øhavet is not restricted to one particular island but can be expanded to include islands and coastal regions of major parts of the archipelago of Denmark, provided that the general characteristics of these locations – in contrast to core areas of larger islands or the peninsula of Jutland – are reflected in the quality or fingerprints of the berries and end-products. Documentation will also provide a means by which the authenticity of the products can be guarded (i.e. detection of counterfeit products will be possible) and allow the *terroir* characteristics to be exploited in marketing.

38.4 Resolution and outcomes

At the time of writing, the company is working on the certification of the conserves as described in Section 38.3.2 and documentation of their authenticity using the methods described in Section 38.3.3. It is hoped that, as volume of production increases and the brand gains market value, these efforts will allow the company to demonstrate the qualities of the products that are considered to be unique and make the detection of fraud a possibility.

Due to high production costs and aging populations, nations such as Denmark are facing an increase in outsourcing of mass production of feed and food to low labor-cost countries; in particular to growth-economy countries. For the country of Denmark, farming and food production as an industry is still the largest generator of economic surplus, as in other countries, but the surplus is decreasing, partly due to the import of price-competitive raw materials from other countries. Denmark has a favorable position on the global market due to its reputation as having an exceptionally high zoo-sanitary status and level of food safety. However, these positions are also challenged by other feed- and food-producing countries, in part due to standardization of international regulations. To protect a favorable position of the national brand on the global market new principles of quality differentiation of food products need to be identified.

38.5 Commentary

Development of food specialties with qualities associated with *terroir* and primary production with location adaptive plant types, in combination with the highest possible level of documentation of such qualities, could be a feasible strategy of quality differentiation. If innovative food producers are to grasp this opportunity there is a need for these producers to consult with the scientific community for methods to identify and document these qualities. Producer- and consumer-driven research into the identification of suitable methodological approaches and fingerprints unique for the products is warranted.

38.6 Acknowledgments

The Small Island Network of Food Producers (www.ospecialiteter.dk) is thanked for financial support and intellectual sparring during development of the principles of Island Specialties®.

38.7 Critical questions for discussion

1. What are the main parameters of Øhavet Ltd's products that may potentially result in their unique qualities and fingerprints?
2. What are the candidate methodological approaches to generate fingerprints that are unique to Øhavet Ltd?
3. Why is PGI, but not PDO, considered a relevant certification for Øhavet Ltd's products?
4. Is the intellectual property of the location of Øhavet Ltd sufficient for GTS certification?
5. What is the difference between criterion 3 of Island Specialties® and the principles of GTS certification? What could be the reason for this difference?

38.8 References

CHRISTENSEN, L.S. (2011). De danske småøer – en erhvervsudviklingsstrategi. Småøernes Erhvervsudviklingscenter.

EUROPEAN COUNCIL (2006a). Council Regulation (EC) No 510/2006 of 20 March 2006 on the protection of geographical indications and designations of origin for agricultural products and foodstuffs (Oj L 93, 23.12.2006 p. 12).

EUROPEAN COUNCIL (2006b). Commission Regulation (EC) No 1898/2006 of 14 December 2006 laying down detailed rules of implementation of Council Regulation (EC) No 510/2006 (Oj L 369, 23.12.2006 p. 1).

HUSTED, S., MIKKELSEN, B.F., JENSEN, J. and NIELSEN, N.E. (2004). Determining the geographical origin of barley genotypes using inductively coupled plasma mass spectrometry, isotope ratio mass spectrometry and multivariate statistics. *Analytical and Bioanalytical Chemistry*, **378**: 171–82.

LAURSEN, K.H., SCHJOERRING, J.K., OLESEN, J.E., ASKEGAARD, M., HALEKOH, U. *et al.* (2011). Multi-elemental fingerprinting as a tool for authentication of organic wheat, barley, faba bean and potato. *J Agric Food Chem*, **59**: 4385–96.

PRIMROSE, S., WOOLFE, M. and ROLLINSON, S. (2010). Food forensics: Methods for determining the authenticity of foodstuffs. *Trends in Food Science and Technology*, **21**: 582–90.

Index